T0206232

Change-Point Analysis in Nonstationary Stochastic Models

Change-Point Analysis in Nonstationary Stochastic Models

Boris Brodsky

CRC Press is an imprint of the
Taylor & Francis Group, an **informa** business
A CHAPMAN & HALL BOOK

CRC Press
Taylor & Francis Group
6000 Broken Sound Parkway NW, Suite 300
Boca Raton, FL 33487-2742

First issued in hardback 2019

First issued in paperback 2022

ISBN 13: 978-1-03-240220-8 (pbk)
ISBN 13: 978-1-4987-5596-2 (hbk)
ISBN 13: 978-1-315-36798-9 (ebk)

DOI: 10.1201/9781315367989

Publisher's Note
The publisher has gone to great lengths to ensure the quality of this reprint but points out that some imperfections in the original copies may be apparent.

Visit the Taylor & Francis Web site at
http://www.taylorandfrancis.com

and the CRC Press Web site at
http://www.crcpress.com

Contents

List of Figures

List of Tables

Preface

This book, *Change-Point Analysis in Nonstationary Stochastic Models* is aimed at development of methods for detection and estimation of changes in complex systems. These systems are generally described by nonstationary stochastic models, which comprise both static and dynamic regimes, linear and nonlinear dynamics, and constant and time-variant structures of such systems.

I proceed from the method of collection of information about these systems. For retrospective problems, I deal with the whole array of obtained information and are supposed to test hypotheses about existence of change-points in this array. In case we make a decision about such change-points, we need to estimate them correctly. For sequential problems, we make decisions about change-points online, i.e., simultaneously with the sequential process of data collection. Here we need to detect real changes and minimize the number of "false alarms".

In order to detect change-points properly, I need to consider theoretical methods of optimal (asymptotically optimal) detection of changes in retrospective and sequential formulation of this problem. In this book, such methods are constructed and their characteristics are analyzed both theoretically and experimentally. Results of this book include theorems about characteristics of the proposed methods, computer simulations of resulting algorithms, and practical applications of the proposed methods to detect of changes in real-world systems. In this book, I consider in detail financial applications and methods for detection of structural changes in various financial models.

The following is the author's statement of aims for this book:

1. To consider retrospective and sequential problems of detection of nonstationarities.

2. To prove optimality and asymptotic optimality of proposed methods.

3. To consider parametric and nonparametric methods of detection of nonstationarities.

4. To consider different types of changes: abrupt changes in statistical characteristics ("change-points" or instants of "structural changes"); gradual changes in statistical characteristics (deterministic and stochastic trends or "unit roots"); purely random and disappearing changes in statistical characteristics ("outliers", switches in coefficients of stochastic models).

5. To consider changes in parameters of univariate and multivariate nonstationary stochastic models.

6. To consider hypothesis testing and change-point detection methods.

In Part I of this book, I consider different statements of the retrospective change-point problem with the special sections devoted to the asymptotically optimal choice of parameters of decision statistics and a priori lower bounds for performance efficiency in different change-points problems.

I propose methods for detection of changes in parameters of multivariate stochastic models (including multifactor regressions, systems of simultaneous equations, etc.) and prove that probabilities of type 1 and type 2 errors for these methods converge to zero as the sample size tends to infinity. Then, I prove the asymptotic lower bounds for the probability of error of estimation for the case of dependent observations in situations of one change-point and multiple change-points.

Applied problems of the retrospective change-point detection for different financial models including GARCH and SV models, as well as Copula models, are considered. I propose methods for the retrospective detection of structural changes and give experimental results demonstrating their efficiency.

In the second part of this book, I study sequential problems of changepoint detection. In sequential problems, we collect observations online, i.e., in a sequential way, and are obliged to test hypotheses and to make decisions about the presence of a change-point at every step of data collection. Therefore, in sequential problems, the notions of a false decision about a change-point (false alarm, type 1 error) and a false tranquility (type 2 error), as well as the delay time in change-point detection, naturally appear.

Prior lower bounds for performance efficiency of sequential tests are proved, including the a priori lower estimate for the delay time, the Rao-Cramer type inequality for sequential change-point detection methods, and the lower bound for the probability of the error in the normalized delay time estimation. Then I formulate results of the asymptotic comparative analysis and asymptotic optimality of sequential change-point detection methods.

In this book, a new problem of early change-point detection is examined. The main difference between early and sequential change-point detection is that in early detection problems we consider gradual changes in statistical characteristics of observations, while in classic sequential detection problems only abrupt changes are analyzed.

In a separate chapter sequential detection of switches in models with changing structures is considered. These problems include decomposition of mixtures of probabilistic distributions (sequential context), classification of multivariate observations, and classifications of observations in regression models with randomly changing coefficients of regression equations.

Looking at the rich history of the change-point analysis, which has now become a self-sufficient area of mathematical statistics, I can mark the following main stages of its development:

1. 1960–1980s: The initial stage which is characterized by appearance of ideas and papers by Kolmogorov, Shiryaev, Page, Girshick and Rubin, Lorden, Siegmund, Lai, et al. At that time, the sphere of change-point analysis seemed to be a terra incognita with rare and courageous research projects of pioneers of science.

2. 1980–2000s: The classic stage of the universal development of the stage of change-point analysis; the main bulk of theoretical results is created.

3. 2000–present: The modern stage, history of which is not yet completed.

Large-scale change-point problems and real-world applications, including the analysis of multivariate nonstationary models and changes of different types (disorders, stochastic trends, random switches), are considered.

Nonstationarity modeling has received great attention in time series analysis, and the interest increases in nonstationarity modeling for multivariate, possibly high-dimensional data. Change-point detection allows flexible modeling with regard to the detected change-points. In some papers, it is noted that stochastic regime switching may be confused with the long memory property of a model, and the presence of structural breaks may easily be regarded as fractional integration (see, e.g., Diebold and Inoue [2001], "Long range dependence and regime switching," *Journal of Econometrics*, v. 101); Mikosch and Starica [2004], "Non-stationarities in financial time series: Long-range dependence and IGARCH effects," *Review of Economics and Statistics*, v.86). Moreover, the effect of high dimensionality on change-point detection was reported in certain new papers (see, e.g., Kirch and Aston [2014], "Change-points in high-dimensional settings"), where the role of increasing dimensionality on the asymptotics of change-point detection consistency is emphasized.

In this book, a new method is proposed for discrimination between the structural change and the fractional integration hypothesis (see Chapter 7). I consider multivariate nonstationary models and various financial applications, retrospective, and sequential methods of change-point detection, thus, searching for ways and methods of theoretical and practical implementation of these new ideas.

This book was written during the last several years. Separate chapters of it first appeared in Taylor and Francis journals *Sequential Analysis* and *Stochastics*. I am grateful to Dr. Boris Darkhovsky for the immense volume of collaboration and the privilege to benefit from scientific contact with him. I am also indebted to editors and referees of this book for their valuable comments. Only after these initial steps the whole text of this book was collected and prepared for publication. In formatting of this book, the help of CRC Press is highly appreciated. I am grateful to Acquisitions Editor, Rob Calver and Editorial Assistant, Alex Edwards for their professional assistance and eagerness to improve the quality of the manuscript. The help of Production Editor, Robin Lloyd-Starkes, is also highly appreciated.

Part I

Retrospective Problems

The first part of this book is devoted to retrospective change-point problems. Here, I consider preliminary results from mathematical statistics and the theory of random processes, including Cramer's condition, different mixing conditions, Markov processes, etc. I consider here invariance principles for dependent random variables, the exponential estimate for the maximum of sums of dependent random variables, and the lower bound for performance efficiency in change-point problems.

In Chapter 2, I consider different statements of the retrospective change-point problem, with special sections devoted to the asymptotically optimal choice of parameters of decision statistics and a priori lower bounds for performance efficiency in multiple change-points problems.

In Chapter 3, retrospective detection and estimation problems for stochastic trends are considered. Here, I consider a new method for retrospective detection of a stochastic trend and the problem of discrimination of a stochastic trend and a structural change hypothesis.

In Chapter 4, I consider retrospective detection and estimation of switches in models with changing structures. Here, the case of outliers in univariate models is analyzed, and the asymptotically optimal method for detection of these outliers is proposed. I analyze here methods for splitting mixtures of probabilistic distributions (nonparametric problem statement), methods for classification of multivariate observations, and classification problems for non-stationary multi-factor regression models.

In Chapter 5, problems of retrospective detection and estimation of change-points in multivariate models are considered. Here, I propose methods for detection of changes in parameters of multivariate stochastic models (including multifactor regressions, systems of simultaneous equations, etc.) and prove that probabilities of type 1 and type 2 errors for these methods converge to zero as the sample size tends to infinity. Then I prove the asymptotic lower bounds for the probability of the error of estimation for the case of dependent multivariate observations in situations of one change-point and multiple change-points.

In Chapter 6, problems of the retrospective detection of change-points in state-space models are considered. This is the important particular case of the previous chapter that has many interesting applications. Again, I prove theorems about convergence to zero of the probabilities of type 1 and type 2 errors as the sample size tends to infinity and experimentally test characteristics of the proposed method in different state-space models.

In Chapter 7, applied problems of the retrospective change-point detection for different financial models, including GARCH and SV models, as well as Copula models, are considered. I propose methods for the retrospective detection of structural changes and give experimental results demonstrating their efficiency.

1

Preliminary Considerations

In this chapter, for the sake of the reader's convenience, we present necessary results from the theory of random processes and mathematical statistics. These results will be used in subsequent chapters of this book.

We also consider here main ideas that are used for solving different problems of statistical diagnosis and formulate basic assumptions of our approach to these problems.

1.1 Cramer's Condition

In this section, we give the definition and certain results about Cramer's condition (see Brodsky, and Darkhovsky, 2000).

We say that the r.v. X satisfies *Cramer's condition*, if for a certain $t > 0$ the following inequality holds:

$$\mathbf{E}\exp(t|X|) < \infty. \tag{1.1}$$

The value

$$T(X) \stackrel{\text{def}}{=} \sup\{t : \mathbf{E}\exp(t|X|) < \infty\} \tag{1.2}$$

is called *the Cramer parameter.*

If $\mathbf{E}X = 0$, then Cramer's condition is equivalent to

$$\mathbf{E}\exp(tX) \le \exp(gt^2) \tag{1.3}$$

for a certain constants $g > 0$ and $H > 0$: $t < H$.

Let $\{\xi_n\}$ be a sequence of r.v.'s. We say that the uniform Cramer's condition is satisfied for this sequence if

$$\sup_n \mathbf{E}(t|\xi_n|) < \infty \tag{1.4}$$

for some $t > 0$.

The value

$$T \stackrel{\text{def}}{=} \sup\{t : \sup_n \mathbf{E}\exp(t|\xi_n|) < \infty\} \tag{1.5}$$

is called *the uniform Cramer's parameter.*

Let $X = (x_1, \dots, x_d)^*$ be a random vector. We say that X satisfies *Cramer's condition* if

$$\mathbf{E}\exp(t\|X\|) < \infty \tag{1.6}$$

for some $t > 0$.

Definitions of Cramer's parameter and the uniform Cramer condition are made in analogy with the scalar case.

From these definitions, it follows that if the random vector $X = (x_1, \dots, x_d)$ satisfies Cramer's condition, then each of its components also satisfies Cramer's condition, and in case these components are independent r.v.'s then the r.v.'s, $x_i x_j$, $i \neq j, i = 1, \dots, d$, $j = 1, \dots, d$ also satisfy Cramer's condition.

In some cases, we use a stronger condition for vector-valued r.v.'s. Suppose $X = (x_1, \dots, x_d)$ is a random vector. We say that vector X satisfies *the augmented Cramer's condition* if all scalar variables $x_i x_j$, $i = 1, \dots, d$, $j = 1, \dots, d$ satisfy *Cramer's condition.*

The definition of the *uniform augmented Cramer's condition* for a sequence of random vectors is made in analogy.

1.2 Mixing Conditions (Bradley, 2005)

Let $(\Omega, \mathfrak{F}, \mathbf{P})$ be a probability space. Let $\mathbf{H}_1$ and $\mathbf{H}_2$ be two σ-algebras contained in $\mathfrak{F}$. Let $L_p(\mathbf{H})$ be a collection of L_p-integrated random variables measurable with respect to some σ-algebra $\mathbf{H} \subseteq \mathfrak{F}$. Define the following measures of dependence between $\mathbf{H}_1$ and $\mathbf{H}_2$:

$$\begin{aligned}
\rho(\mathbf{H}_1, \mathbf{H}_2) &= \sup_{X \in L_2(\mathbf{H}_1), Y \in L_2(\mathbf{H}_2)} \frac{|\mathbf{E}(X - \mathbf{E}X)(Y - \mathbf{E}Y)|}{\sqrt{\mathbf{E}(X - \mathbf{E}X)^2 \mathbf{E}(Y - \mathbf{E}Y)^2}} \\
\alpha(\mathbf{H}_1, \mathbf{H}_2) &= \sup_{A \in \mathbf{H}_1, B \in \mathbf{H}_2} |\mathbf{P}(AB) - \mathbf{P}(A)\mathbf{P}(B)| \\
\varphi(\mathbf{H}_1, \mathbf{H}_2) &= \sup_{A \in \mathbf{H}_1, B \in \mathbf{H}_2, \mathbf{P}(B) \neq 0} \left| \frac{\mathbf{P}(AB) - \mathbf{P}(A)\mathbf{P}(B)}{\mathbf{P}(B)} \right| \\
\psi(\mathbf{H}_1, \mathbf{H}_2) &= \sup_{A \in \mathbf{H}_1, B \in \mathbf{H}_2, \mathbf{P}(A)\mathbf{P}(B) \neq 0} \left| \frac{\mathbf{P}(AB)}{\mathbf{P}(A)\mathbf{P}(B)} - 1 \right|.
\end{aligned}$$

Let $(X \stackrel{\text{def}}{=} \{x_i\}_{i=1}^{\infty})$ be a sequence of real random variables on $(\Omega, \mathfrak{F}, \mathbf{P})$. Let $\mathfrak{F}_s^t = \sigma\{x_i : s \leq i \leq t\}, 1 \leq s \leq t < \infty$, be the minimal σ-algebra generated by random variables $x_i, s \le i \le t$. Put

$$\begin{aligned}\rho(n) &= \sup_{t\ge1} \rho(\mathfrak{F}_1^t, \mathfrak{F}_{t+n}^\infty), \quad \alpha(n) = \sup_{t\ge1} \alpha(\mathfrak{F}_1^t, \mathfrak{F}_{t+n}^\infty) \\ \varphi(n) &= \sup_{t\ge1} \varphi(\mathfrak{F}_1^t, \mathfrak{F}_{t+n}^\infty), \quad \psi(n) = \sup_{t\ge1} \psi(\mathfrak{F}_1^t, \mathfrak{F}_{t+n}^\infty).\end{aligned} \tag{1.5}$$

A sequence X is said to be a sequence with *ρ-mixing* (respectively, *α-mixing, φ-mixing, ψ-mixing*) if the function $\rho(n)$ (respectively, $\alpha(n)$, $\varphi(n)$, $\psi(n)$), which is also called the *coefficient of ρ-mixing* (respectively, α-mixing, φ-mixing, ψ-mixing), tends to zero as n tends to infinity.

It is easy to see that $\alpha(n) \le 0.25\rho(n)$, $\varphi(n) \le \psi(n)$. Moreover, it is known that $\rho(n) \le 2\sqrt{\varphi(n)}$.

Therefore, the ψ-mixing sequence is the φ-mixing sequence, the φ-mixing sequence is the ρ-mixing sequence, and the ρ-mixing sequence is the α-mixing sequence.

Suppose now that $\mathcal{X} \stackrel{\text{def}}{=} \{X_i\}_{i=1}^\infty$ is a sequence of real-valued random vectors $X_i = \left(x_i^{(1)}, \dots, x_i^{(d)}\right)$. Let $\mathfrak{F}_s^t(i) \stackrel{\text{def}}{=} \sigma\{x_k^{(i)} : s \le k \le t\}, 1 \le s \le t < \infty$ be the minimal σ-algebra generated by the trajectory of ith component of the random sequence $\mathcal{X}$ in the interval $[s,t]$. Define

$$\mathbb{F}_s^t(i,j) \stackrel{\text{def}}{=} \mathfrak{F}_s^t(i) \otimes \mathfrak{F}_s^t(j).$$

The matrix $A(n) = \{a_{ij}(n)\}$ of coefficients of (i,j)-mixing for a vector sequence (the symbol a_{ij} denotes here an arbitrary mixing coefficient) is defined according to (1.5) if we substitute σ-algebra $\mathfrak{F}_s^t$ for $\mathbb{F}_s^t$. For example, the coefficient of ρ_{ij}-mixing is equal to

$$\rho_{ij}(n) = \sup_{t\ge1} \rho\left(\mathbb{F}_1^t(i,j), \mathbb{F}_{t+n}^\infty(i,j)\right). \tag{1.6}$$

From this definition, it follows that $a_{ii}(\cdot)$ is a mixing coefficient for the ith component of the vector-valued sequence $\mathcal{X}$.

A-mixing of the vector-valued random sequence (here, A denotes α, ρ, ϕ, ψ) is equivalent to the condition $\max\limits_{i,j} a_{ij}(n) \to 0$ for $n \to \infty$.

Let *ρ_j-mixing* (respectively, *α_j-mixing, φ_j-mixing, ψ_j-mixing*), $j = 1, \dots, d$ be the coefficient of ρ-mixing (respectively, α, φ, ψ-mixing) in j-th component of the vector sequence $\mathcal{X}$.

From the definition, it follows that if a vector sequence satisfies ρ-mixing condition (respectively, α-mixing, φ-mixing, ψ-mixing), then $\max\limits_{1\le i\le d} \rho_i(n)$ (respectively, $\max\limits_{1\le i\le d} \alpha_i(n)$, $\max\limits_{1\le i\le d} \varphi_i(n)$, $\max\limits_{1\le i\le d} \psi_i(n)$) converge to zero as $n \to \infty$.

Thus, if a vector random sequence satisfies a certain mixing condition, then each component of this sequence satisfies the same mixing condition.

The reversed statement is true if all components of a vector random sequence are independent random sequences.

Assumptions that guarantee validity of mixing conditions of a definite type can be found in Bradley (2005).

In some cases, we use some stronger mixing conditions for vector-valued random sequences. Suppose $\mathcal{X} \stackrel{\text{def}}{=} \{X_i\}_{i=1}^{\infty}$ is a sequence of real-valued random vectors $X_i = \left(x_i^{(1)}, \ldots, x_i^{(d)}\right)$. We say that $\mathcal{X}$ satisfies *the strengthened u-mixing condition* ($u = \rho, \alpha, \varphi, \psi$) in all scalar sequences $\{x_i^{(k)} x_i^{(s)}\}_{i=1}^{\infty}$, $k = 1, \ldots, d$, $s = 1, \ldots, d$ satisfy the same mixing condition.

1.3 Invariance Principles for Dependent Random Sequences (Peligrad, 1982)

Suppose $X = \{x_i\}_{i=1}^{\infty}$ is a random sequence, such that $\mathbf{E}x_i \equiv 0$. Denote

$$S(n) = \sum_{i=1}^{n} x_i$$

$$\sigma^2 = \lim_{n\to\infty} n^{-1}\mathbf{E}S^2(n) \tag{1.7}$$

$$W_n(t) = (\sigma\sqrt{n})^{-1} \sum_{i=1}^{[nt]} x_i.$$

The function $\omega \longrightarrow W_n(t, \omega)$ is a measurable map from $(\Omega, \mathfrak{F})$ into Skorohod space $(D, \mathfrak{D})$. Below, we will denote by W the standard Wiener process, and the weak convergence will be considered in $(D, \mathfrak{D})$.

Remember that a family of random variables $\{x_n\}_{n=1}^{\infty}$ is called *uniformly integrable* if

$$\sup_n \mathbf{E}[|x_n|\mathbb{I}(|x_n| > c)] \to 0$$

as $c \to \infty$.

A sufficient condition of uniform integrability.

Lemma 1.3.1. *Let $\{x_n\}_{n=1}^{\infty}$ be a sequence of integrable r.v.'s, and $G = G(t)$ is a nonnegative increasing function defined for all $t \geq 0$ and such that*

$$\lim_{t\to\infty} [G(t)/t] = \infty$$

$$\sup_n \mathbf{E}[G(|x_n|)] < \infty.$$

Then the family of r.v.'s $\{x_n\}$ is uniformly integrable.

Evidently, we can take $G(t) = t^{1+a}$, $a > 0$.

Theorem 1.3.1. *Suppose for the sequence X there exists Limit (1.7) and, moreover,*

a) the sequence $X^2 \stackrel{def}{=} \{x_i^2\}_{i=1}^\infty$ is uniformly integrable; and
b) $\sum_i \sqrt{\rho(2^i)} < \infty$, where $\rho(\cdot)$ is the mixing coefficient for X.
Then the process W_n weakly converges to W.

From Lemma 1.3.1 it follows that for the uniform integrability of the sequence $\{X^2\}$ that

$$\sup_n \mathbf{E}x_n^{2+a} < \infty,\ a > 0$$

Now, we give a sufficient condition for existence of Limit (1.7). Suppose $X = \{x(n)_{n=1}^\infty\}$ is a stationary random sequence with zero mathematical expectation and the correlation function $R(k) = \mathbf{E}x(n)x(n+k)$.

Lemma 1.3.2. *If*

$$\sum_k | R(k) | < \infty,$$

then limit (1.7) exists and is equal to

$$\sigma^2 = \mathbf{E}x^2(1) + 2\sum_{k=1}^\infty R(k).$$

The proof easily follows from the following considerations:

$$\mathbf{E}S^2(n) = nR(0) + 2\sum_{k=1}^{n-1}(n-k)R(k)$$

$$| \sigma^2 - n^{-1}\mathbf{E}S^2(n) | \le 2\left(\sum_{k=n}^\infty | R(k) | + (n_0/n)\sum_{k=1}^{n_0} | R(k) | + \epsilon\right),$$

where $\epsilon > 0$ is an arbitrary fixed number and n_0 is chosen, such that $\sum_{k=n_0+1}^\infty |R(k)| \le \epsilon$.

Below, we use the scheme of gluing of several random sequences. Now, we give conditions for the existence of Limit (1.7) for such scheme.

Let

$$0 \equiv \vartheta_0 < \vartheta_1 < \vartheta_2 < \cdots < \vartheta_k < \vartheta_{k+1} \equiv 1, \quad \vartheta \stackrel{\text{def}}{=} (\vartheta_1, \ldots, \vartheta_k).$$

Consider a collection of random sequences

$$\mathcal{X} = \{X^{(1)}, X^{(2)}, \ldots, X^{(k+1)}\}, X^{(i)} = \{x^{(i)}(n)\}_{n=1}^\infty,\ i = 1, \ldots, (k+1).$$

Now, define a family of random sequences

$$\mathfrak{X} = \{X^N\}, N = N_0, N_0+1, N_0+2, \ldots, N_0 > 1, X^N = \{x^N(n)\}_{n=1}^N$$

as follows:

$$x^N(n) = x^{(i)}(n),$$

if $[\vartheta_{i-1}N] \le n < [\vartheta_i N], i = 1, \dots, k+1$.

We say that the family $\mathfrak{X} = \{X^N\}$ is generated by the process of *"gluing" (concatenation)*.

So, the family $\mathfrak{X}$ is the *"glued"* random sequence generated by the collection $\mathcal{X}$, and the collection $\{\vartheta\}$ are the *points of "gluing"* or *change-points.*

Evidently, the process of "gluing" is the variant of the *triangular array scheme*, which is often used in the probability theory.

Suppose $\mathfrak{X} = \{X^N\}$ is a "glued" random sequence generated by the collection $\mathcal{X} = \{X^{(1)}, X^{(2)}, \dots, X^{(k+1)}\}$ of random sequences with zero expectations. Suppose there exists Limit (1.7) for each of these sequences. Denote

$$\sigma_i^2 = \lim_{n\to\infty} n^{-1}\mathbf{E}S_i^2(n),$$

where

$$S_i(n) = \sum_{s=1}^{n} x^{(i)}(s), \quad i = 1, 2, \dots, k+1,$$

and let $\rho_{ij}(\cdot)$ be a ρ-mixing coefficient between i-th and j-th components of the vector sequence $\mathcal{X}$ defined above, and $\rho_{ii}(\cdot) \stackrel{\text{def}}{=} \rho_i(\cdot)$ is the coefficient of ρ-mixing for the sequence $\{x^{(i)}(n)\}$).

Lemma 1.3.3. *Suppose the following assumptions are satisfied:*

i)

$$\sup_k \max_i \mathbf{E}\big(x^i(k)\big)^2 < \infty \tag{1.8}$$

ii)

$$\sum_{k=1}^{\infty} \max_{i,j} \rho_{ij}(k) < \infty. \tag{1.9}$$

Then for the "glued" random sequence $\mathfrak{X}$, there exists Limit (1.7) and

$$\lim_{N\to\infty} N^{-1}\mathbf{E}(\sum_{k=1}^{N} x^N(k))^2 = \sigma^2,$$

and

$$\sigma^2 = \sum_{s=1}^{k+1} \alpha_s \sigma_s^2,$$

where $\alpha_1 = \vartheta_1, \alpha_2 = \vartheta_2 - \vartheta_1, \dots, \alpha_k = \vartheta_k - \vartheta_{k-1}, \alpha_{k+1} = 1 - \vartheta_k$, and points ϑ_i determine the "gluing" process.

Proof. We will give proof for the case of "gluing" of two random sequences $X^{(1)}$ and $X^{(2)}$.

Without loss of generality, we will assume that the left-hand side in (1.3.2) is no more than 1. Then by definition of ρ_{12}-mixing coefficient,

$$|\mathbf{E}x^{(1)}(k)x^{(2)}(k+s)| \leq \rho_{12}(s). \tag{1.10}$$

In our case, we have

$$N^{-1}(\sum_{k=1}^{N} x^N(k))^2 = N^{-1}\Big(\sum_{k=1}^{[\vartheta N]} x^{(1)}(k) + \sum_{[\vartheta N]+1}^{N} x^{(2)}(k)\Big)^2. \tag{1.11}$$

Let us estimate the term $\mathfrak{J} \equiv |\mathbf{E}\Big(\sum_{k=1}^{[\vartheta N]} x^{(1)}(k)\Big)\Big(\sum_{k=[\vartheta N]+1}^{N} x^{(2)}(k)\Big)|$.

Fix some positive number p and, taking into account (1.10) and positiveness of the sequences $\rho_{ij}(k)$, we have

$$\begin{aligned}\mathfrak{J} \leq & \sum_{k=1}^{N-[\vartheta N]} \rho_{12}(k) + \sum_{k=2}^{N-[\vartheta N]+1} \rho_{12}(k) + \ldots + \sum_{k=[\vartheta N/p]}^{N-[\vartheta N]+[\vartheta N/p]-1} \rho_{12}(k) + \\ + & \sum_{[\vartheta N/p]+1}^{N-[\vartheta N]+[\vartheta N/p]} \rho_{12}(k) + \ldots \sum_{k=[\vartheta N]-1}^{N-2} \rho_{12}(k) + \sum_{k=[\vartheta N]}^{N-1} \rho_{12}(k) \leq \\ \leq & [(\vartheta N/p)] \sum_{k=1}^{N-[\vartheta N]+[\vartheta N/p]} \rho_{12}(k) + ([\vartheta N] - [\vartheta N/p]) \sum_{k=[\vartheta N/p]+1}^{N} \rho_{12}(k).\end{aligned} \tag{1.12}$$

Due to (1.9), we can choose p sufficiently large, such that

$$N^{-1}\mathfrak{J} \leq \epsilon + \sum_{k=[(\vartheta N/p)]+1}^{N} \rho_{12}(k). \tag{1.13}$$

where ϵ is given a positive number.

Now, again using (1.9), we see that

$$\lim_{N\to\infty} N^{-1}\mathfrak{J} = 0 \tag{1.14}$$

and, therefore, taking into account (1.11), (1.14) we have

$$\lim_{N\to\infty} N^{-1}\mathbf{E}\Big(\sum_{k=1}^{N} x^N(k)\Big)^2 = \vartheta\sigma_1^2 + \lim_{N\to\infty} N^{-1}\mathbf{E}\Big(\sum_{k=[\vartheta N]+1}^{N} x^{(2)}(k)\Big)^2.$$

Taking into account that

$$\sum_{k=[\vartheta N]+1}^{N} x^{(2)}(k) = \sum_{k=1}^{N} x^{(2)}(k) - \sum_{k=1}^{[\vartheta N]} x^{(2)}(k),$$

and using condition $\sum_k \rho_2(k) < \infty$, we obtain the lemma's result.

Remark. If components of a vector-valued random sequence $\mathcal{X}$ are independent, then $\rho_{ij}(n) = 0$ for $i \neq j$. Assumption ii) of Lemma 1.3.3 is substituted by $\max_i \sum_k \rho_i(k) < \infty$.

Taking into account results of Lemma 1.3.3 and Theorem 1.3.1, we obtain the following theorem for the above introduced sequence $\mathfrak{X}$:

Theorem 1.3.2. *Suppose for the vector-valued random sequence $\mathcal{X}$ the following assumptions are satisfied:*

i) there exists limit (1.7) for each component;

i) $\max_i \sup_k \mathbf{E}\left(x^{(i)}(k)\right)^{2+a} < \infty$, $a > 0$ *and*

ii) $\sum_{k=1}^{\infty} \max_{(i,j)} \sqrt{\rho_{ij}(k)} < \infty$.

Then the process $W_N(t) \stackrel{def}{=} N^{-1/2} \sum_{k=1}^{[Nt]} x^N(k)$, $t \in [0,1]$ *weakly converges (in Skorokhod space* $(D, \mathfrak{D})$*) to the process* $\tilde{\sigma}(t)W(t)$*, where*

$$\tilde{\sigma}(t) = t^{-1} \int_0^t \sigma^2(s)ds, \ \sigma^2(s) = \sum_{i=1}^{k+1} \sigma_i^2 \mathbb{I}(\vartheta_{i-1} \leq s < \vartheta_i).$$

1.4 Exponential Estimate for the Maximum of Sums of Dependent r.v.'s (Brodsky, Darkhovsky (2000))

Suppose $S_N = \sum_{k=1}^{N} \xi^N(k)$, where $\xi^N = \{\xi^N(n)\}_{n=1}^{\infty}$ is a random sequence "glued" according to scheme from subsection 1.3 from the vector-valued sequence Ξ, i.e., $\Xi \stackrel{\text{def}}{=} \{\Xi^{(1)}, \ldots, \Xi^{(k+1)}\}$, $\Xi^{(i)} = \{\xi^{(i)}(n)\}_{n=1}^{\infty}$ and $\xi^N(n) = \xi^{(i)}(n)$ if $[\vartheta_{i-1}N] \leq n < [\vartheta_i N]$, $i = 1, \ldots, k+1$, and $\mathbf{E}_\theta \xi^N(n) \equiv 0$.

Suppose for the vector-valued random sequence Ξ the uniform Cramer condition is satisfied, as well as ψ-mixing condition.

Then for any $0 < \beta < 1/2$, $x > 0$ and large enough N the following inequalities hold:

$$\begin{aligned} &\mathbf{P}\left\{\max_{[\beta N] \leq n \leq N} |S_n|/N > x\right\} \leq \mathbf{P}\left\{\max_{[\beta N] \leq n \leq N} (|S_n|/n) > x\right\} \leq \\ &\leq A(x) \exp\left(-\tilde{B}(x)\beta N\right) \stackrel{\text{def}}{=} A(x) \exp\left(-B(x)N\right), \end{aligned} \tag{1.15}$$

where the functions $A(\cdot)$, $B(\cdot)$ are positive for positive values of their arguments and can be written explicitly. In particular, $B(x) = \min\left(ax, bx^2\right)$ for some $a > 0, b > 0$, but, at the same time, note that $A(x) \to \infty$ as $x \to 0$ (see details in Brodsky and Darkhovsky, 2000, pp. 33–35).

In particular, it follows from here that for each $x > 0$

$$\mathbf{P}\{|S_N|/N > x\} \leq A(x) \exp\left(-B(x)N\right).$$

Let us explain the main moments of the proof of this inequality. By a fixed $x > 0$ п»„hoose, the number $\epsilon(x) > 0$ from the following condition

$$\ln(1 + \epsilon(x)) = \begin{cases} \frac{x^2}{4g} & \text{if } x \leq gT, \\ \frac{xT}{4} & \text{if } x > gT, \end{cases}$$

where the constants g and T are taken from the uniform Cramer condition.

Then, for a chosen $\epsilon(x) \triangleq \epsilon$, find from the ψ-mixing condition such $m_0(x) \geq 1$ that $\psi(m) \leq \epsilon(x)$ for $m \geq m_0(x)$. Then, we split the whole sum into groups of weakly dependent summands, i.e., ψ-mixing coefficient between them of no more than ϵ.

Then for each group of weakly dependent terms, we estimate the probability of the event that the absolute value of the sum of these terms exceeds a certain limit. This is done using the following inequality for r.v.'s satisfying ψ-mixing condition:

$$|\mathbf{E}\xi\eta - \mathbf{E}\xi\mathbf{E}\eta| \leq \psi(s)\mathbf{E} \mid \xi \mid \mathbf{E}|\eta|.$$

Here, the random variables ξ and η are measurable with respect to the σ-algebras $\mathfrak{J}_1^t$ and $\mathfrak{J}_{t+s}^{\infty}$, respectively, where σ-algebras $\mathfrak{J}_s^t$ are generated by some mixing random sequence.

After that, the required inequality is readily obtained.

1.5 Lower Bound for Performance Efficiency of Parameter Estimation

1.5.1 Problem Statement

Let $\{x_n\}$ be a random sequence. Suppose for each $N \geq 1$ there exists (w.r.t. a certain σ-finite measure) the density function $f(X^N, \theta)$ of $X^N \stackrel{\text{def}}{=} (x_1, \ldots, x_N)$, which depends on the parameter $\theta \in \Theta \subset \mathbb{R}^k$ (below, for simplicity of notation, we take densities w.r.t. the corresponding Lebesgue measure).

Suppose $\theta \in \Theta$ is a certain fixed point. We consider the problem of the asymptotic (as $N \to \infty$) a priori lower estimate for the probability $\mathbf{P}_\theta\{\|\theta - \hat{\theta}_N\| > \epsilon\}$, where $\hat{\theta}_N$ is an arbitrary estimate of the parameter θ for the sample X^N.

Such a problem is traditional for mathematical statistics, but (as far as the authors are informed) in most papers, only the case of independent r.v.'s is considered.

1.5.2 Assumptions and Formulation of a Result

Below, we suppose that for all θ the sequence $\{x_n\}$ is a stationary and strongly ergodic (see, e.g., Borovkov (1999)) k-step Markov sequence. In particular, such is the sequence defined by the following equation,

$$x_{n+1} = \sum_{i=1}^{s} a_i(\theta)x_{n-i} + \epsilon_{n+1},$$

where $\{\epsilon_n\}$ is a sequence of i.r.v.'s with the finite second moment. We suppose that the characteristic polynom here is stable for all θ.

From this assumption, it follows that if $f : \mathbb{R}^s \to \mathbb{R}^1$, $s \geq 1$ is a measurable function and $\mathbf{E}_\theta|f(x_1, \dots, x_s)| < \infty$, then there exists $\mathbf{P}_\theta$-a.s. the limit

$$\lim_{N\to\infty} N^{-1}\sum_{i=1}^{N} f(x_i, \dots, x_{i+s-1}) = \mathbf{E}_\theta f(x_i, \dots, x_{i+s-1}). \tag{1.16}$$

Put $x_1^s \stackrel{\text{def}}{=} (x_1, \dots, x_s)$ and denote by $\varphi(\cdot, \theta|x_1^{k-1})$ the corresponding conditional densities, $\varphi(\cdot, \theta|x_1^0) \stackrel{\text{def}}{=} f(\cdot, \theta)$.

For all $(\theta_1, \theta_2) \in \Theta \times \Theta$, $\theta_1 \neq \theta_2$, define

$$\zeta_n \stackrel{\text{def}}{=} \ln \frac{\varphi(x_n, \theta_1|x_{n-k}^{n-1})}{\varphi(x_n, \theta_2|x_{n-k}^{n-1})}, n = k+1, \dots,$$

and

$$\zeta_1 \stackrel{\text{def}}{=} \ln \frac{f(x_1, \theta_1)}{f(x_1, \theta_2)}, \zeta_2 \stackrel{\text{def}}{=} \ln \frac{f(x_1, x_2, \theta_1)}{f(x_1, x_2, \theta_2)}, \dots, \zeta_k \stackrel{\text{def}}{=} \ln \frac{f(x_1, \dots, x_k, \theta_1)}{f(x_1, \dots, x_k, \theta_2)},$$

where $\varphi(\cdot, \theta|\cdot)$ is the corresponding conditional density function, and $f(x_1^s, \theta)$ is the corresponding stationary density function.

The r.v. ζ_n is the measurable function of the set $(x_{n-k}, \dots, x_{n-1})$. Therefore, if $\mathbf{E}_{\theta_1}|\zeta_n| < \infty$, then in virtue of (1.16)

$$\lim_{N\to\infty} N^{-1}\sum_{s=1}^{N} \zeta_s = \mathbf{E}_{\theta_1}\zeta_n \stackrel{\text{def}}{=} \mathbf{I}(\theta_1, \theta_2), \quad \mathbf{P}_{\theta_1} \text{ a.s.} \tag{1.17}$$

In particular, if we consider a one-step stationary Markov sequence, then

$$\mathbf{I}(\theta_1, \theta_2) = \int\int \left(\ln \frac{\varphi(x, \theta_1|y)}{\varphi(x, \theta_2|y)}\right) \varphi(x, \theta_1|y) f(y, \theta_1) dx dy,$$

where $f(\cdot, \theta)$ is one-dimensional density function.

Now, let us formulate the main result.

Theorem 1.5.1. *Let θ be an inner point of the set Θ. Suppose for all $z \in \Theta$ with a small enough norm there exists $\mathbf{I}(\theta_1, \theta_2)$ for $\theta_1 = \theta$, $\theta_2 = \theta + z$ and the integral $\mathbf{E}_{\theta+z} \left| \ln \frac{\varphi(x_n, \theta + z | x_{n-k}^{n-1})}{\varphi(x_n, \theta | x_{n-k}^{n-1})} \right|$. Then, for all small enough, ϵ the following inequality holds:*

$$\liminf_{N} N^{-1} \ln \inf_{\hat{\theta}_N} \mathbf{P}_\theta\{\|\hat{\theta}_N - \theta\| > \epsilon\} \geq - \liminf_{\eta \downarrow \epsilon} \inf_{\{z: \|z\| = \eta\}} \mathbf{I}(\theta + z, \theta), \quad (1.18)$$

where the infimum in the left hand of (1.18) is taken by all estimators $\hat{\theta}_N$.

Before the proof of this theorem, let us give one example of an explicit computation of the right hand of (1.18). Let $\{x_n\}$ be a sequence of i.i.d.r.v.'s with the density function $f(x, \theta)$. Suppose the function $\mathbf{J}(u) \stackrel{\text{def}}{=} \int \left(\ln \frac{f(x, \theta + u)}{f(x, \theta)} \right) f(x, \theta + u) dx$ is continuous, and for all small enough u there exists the integral $\int \left| \ln \frac{f(x, \theta + u)}{f(x, \theta)} \right| f(x, \theta + u) dx$.

Then the right hand of (1.18) becomes $(-\min\{\mathbf{J}(\epsilon), \mathbf{J}(-\epsilon)\})$.

1.5.3 Proof

Without loss of generality, we can consider only consistent estimators of the parameter θ, because for nonconsistent estimators, Iinequality (1.18) is trivially fulfilled.

Suppose $\hat{\theta}_N$ is a certain consistent estimate of the parameter θ constructed by the sample $X^N = \{x_1, \ldots, x_N\}$. Define the r.v. $\lambda_N = \lambda_N(x_1, \ldots, x_N) = \mathbb{I}\{\|\hat{\theta}_N - \theta\| > \epsilon\}$.

Let us fix a small enough $\epsilon > 0$. Suppose $d > 0$ and $z \in \Theta$ is a fixed vector, such that $\|z\| = \tilde{\epsilon} > \epsilon$. Then,

$$\begin{aligned} &\mathbf{P}_\theta\{\|\hat{\theta}_N - \theta\| > \epsilon\} = \mathbf{E}_\theta \lambda_N \geq \mathbf{E}_\theta \left(\lambda_N \mathbb{I}(f(X^N, \theta + z)/f(X^N, \theta) < e^d) \right) \geq \\ &\geq e^{-d} \left(\mathbf{E}_{\theta+z} \{ \lambda_N \mathbb{I}(f(X^N, \theta + z)/f(X^N, \theta) < e^d \} \right) \geq \\ &\geq e^{-d} \left(\mathbf{P}_{\theta+z}\{\|\hat{\theta}_N - \theta\| > \epsilon\} - \mathbf{P}_{\theta+z}\{f(X^N, \theta + z)/f(X^N, \theta) \geq e^d\} \right). \end{aligned} \quad (1.19)$$

Here we used the elementary inequality

$$\mathbf{P}(AB) \geq \mathbf{P}(A) - \mathbf{P}(\Omega \backslash B).$$

Since $\hat{\theta}_N$ is a consistent estimate of the parameter θ, we obtain

$$\mathbf{P}_{\theta+z}\{\|\hat{\theta}_N - \theta\| > \epsilon\} \to 1 \quad \text{п»„п»„п»„} \quad N \to \infty. \quad (1.20)$$

For estimation of the second inequality in brackets of the right hand, remark that

$$\mathbf{P}_{\theta+z}\{f(X^N,\theta+z)/f(X^N,\theta) \geq e^d\} = \mathbf{P}_{\theta+z}\left\{\ln\frac{f(X^N,\theta+z)}{f(X^N,\theta)} \geq d\right\} =$$

$$= \mathbf{P}_{\theta+z}\left\{N^{-1}\sum_{s=1}^{N}\ln\frac{\varphi(x_s,\theta+z|x_{s-k}^{s-1})}{\varphi(x_s,\theta|x_{s-k}^{s-1})} - \mathbf{I}(\theta+z,\theta) \geq \right.$$

$$\left. \geq \frac{d}{N} - \mathbf{I}(\theta+z,\theta)\right\}. \tag{1.21}$$

In virtue of our assumptions, $\mathbf{P}_{\theta+z}$-a.s.

$$\lim_{N\to\infty} N^{-1}\sum_{s=1}^{N}\ln\frac{\varphi(x_s,\theta+z|x_{s-k}^{s-1})}{\varphi(x_s,\theta|x_{s-k}^{s-1})} = \mathbf{I}(\theta+z,\theta). \tag{1.22}$$

Therefore, we choose $d = d_1(N) = N\left(\mathbf{I}(\theta+z,\theta)+\delta\right)$ for any $\delta > 0$ and obtain from (1.21) that

$$\mathbf{P}_{\theta+z}\{f(X^N,\theta+z)/f(X^N,\theta) \geq \exp(d_1(N))\} \to 0 \text{ п»„п»„п»„ } N \to \infty. \tag{1.23}$$

Taking into account that (1.19) is valid for an arbitrary estimator, taking the logarithms from both parts and remembering (1.20) and (1.23), we obtain

$$\liminf_{N} N^{-1}\ln\inf_{\hat{\theta}_N}\mathbf{P}_\theta\{\|\hat{\theta}_N-\theta\| > \epsilon\} \geq -\mathbf{I}(\theta+z,\theta). \tag{1.24}$$

The vector z in the right hand of (1.24) was fixed on condition that $\|z\| = \tilde{\epsilon} > \epsilon$. However, since the left hand of (1.24) does not depend on this vector, we can first take the supremum by the set $\{z \in \Theta : \|z\| = \tilde{\epsilon}\}$ and then – the inf lim by the parameter $\tilde{\epsilon} \downarrow \epsilon$. The result of this theorem follows immediately.

1.6 Some Results for the Maximum Type Functionals

In this section, we remember certain results (see, Brodsky, and Darkhovsky (2000)), which will be used below. All proofs can be found in Brodsky, Darkhovsky (2000).

Let T be any compact set in a space with the norm $\|\cdot\|$, and let $g(t)$ be a real function on T. For every $\varkappa \geq 0$, define sets

$$A_\varkappa(g) = \{\tilde{t} \in T : \sup_{t\in T} g(t) - \varkappa \leq g(\tilde{t})\}$$

$$B_\varkappa(g) = \{\tilde{t} \in T : g(\tilde{t}) \leq \inf_{t\in T} g(t) + \varkappa\}$$

$$(A_{\varkappa}(g) \triangleq \emptyset,\ B_{\varkappa}(g) \triangleq \emptyset, \text{if} \varkappa < 0).$$

Below, the sets $A_{\varkappa}(g)(B_{\varkappa}(g))$ will be called the sets of $\varkappa$-*maximum* (respectively, $\varkappa$-*minimum*) of the function g on T.

For every real function $h(t)$ on T, we define its sup-norm

$$\|h\|_c \triangleq \sup_{t\in T} |h(t)|.$$

Lemma 1.6.1. *If $g(t), h(t)$ are real functions on T, then for every $\varkappa \geq 0$*

$$A_{\varkappa-2\|h\|_c}(g) \subset A_{\varkappa}(g+h) \subset A_{\varkappa+2\|h\|_c}(g), \tag{1.25}$$

$$B_{\varkappa-2\|h\|_c}(g) \subset B_{\varkappa}(g+h) \subset B_{\varkappa+2\|h\|_c}(g). \tag{1.26}$$

For every two sets A, B in a normed space, the Hausdorf distance *Dist*(A, B) is defined as follows:

$$Dist(A, B) = \max\{\sup_{x\in A} dist(x, B), \sup_{x\in B} dist(x, A)\}.$$

In particular, it follows from here that for a point x and a set M,

$$Dist(x, M) = \sup_{y\in M} \|x - y\|$$

Corollary 1.6.1 from Lemma 1.6.1

Let $\|h\|_c \to 0$. Then, for every $\varkappa > 0$, the following relationships hold:

$$\sup_{x\in A_{\varkappa}(g+h)} dist(x, A_{\varkappa}(g)) \to 0,$$

$$\sup_{x\in B_{\varkappa}(g+h)} dist(x, B_{\varkappa}(g)) \to 0.$$

If, however, $A_0(g) \neq \emptyset$, $A_0(g+h) \neq \emptyset$, $B_0(g) \neq \emptyset$, $B_0(g+h) \neq \emptyset$, then

$$\sup_{x\in A_0(g+h)} dist(x, A_0(g)) \to 0,$$

$$\sup_{x\in A_0(g+h)} dist(x, A_0(g)) \to 0,$$

i.e., any point of the maximum (minimum) of the function $(g+h)$ converges as $\|h\|_c \to 0$ to a certain point of the maximum(minimum) of the function g.

This corollary is often used in situations when g is a continuous function and h is a piecewise-constant and right continuous function, and the maximum(minimum) of (g_h) is attained.

Corollary 1.6.2 from Lemma 1.6.1

For all small enough $\varkappa$, $\|h\|_c$, the following inclusion holds:

$$A_{\varkappa}(|g+h|) \subset A_{\varkappa+2\|h\|_c}(|g|), \tag{1.27}$$

and, therefore, for $\|h\|_c \to 0$ for all small enough $\varkappa > 0$, the following relationship holds:

$$\sup_{x \in A_\varkappa(|g+h|)} dist(x, A_\varkappa(|g|)) \to 0,$$

which is valid also for $\varkappa = 0$ in case of nonemptyness of corresponding sets.

Lemma 1.6.2. *Suppose that there exists a function $\rho(\cdot)$, such that*

$$\mathrm{Dist}(A_{\varkappa_1}(f), A_{\varkappa_2}(f) \leq \rho(|\varkappa_2 - \varkappa_1|)$$

for all $\varkappa_1, \varkappa_2 \in [0, \varkappa_0]$. Then, for all $\varkappa \leq \varkappa_0$, $|h\|_c < \varkappa/2$, the following inequality holds true:

$$\mathrm{Dist}(A_\varkappa(f+h), A_\varkappa(f)) \leq \rho(4\|h\|_c).$$

A similar property holds true for the sets $B_\varkappa(f)$.

Lemma 1.6.3. *Let $x(t) \in C(T)$ and t_0 be a unique point of maximum of $x(t)$ in T. Suppose there exists a continuous and monotonously increasing function $F : \mathbb{R}^+ \to \mathbb{R}^+, F(0) = 0$, such that*

$$x(t_0) - x(t) \geq F(\|t - t_0\|) \text{ for all } t \in T. \tag{1.28}$$

Then, for any function $h(t)$ and every $\varkappa > 0$, the following inequality holds:

$$\mathrm{Dist}(t_0, A_\varkappa(x+h)) \leq F^{-1}(2\|h\|_c + \varkappa).$$

Remark 1.6.1. An analogous inequality holds for all points of the almost-minimum of the function $(x + h)$:

if t_0 is a unique point of the minimum of $x(t)$, and

$$x(t) - x(t_0) \geq F(\|t - t_0\|) \text{ for all} t \in T.$$

then for any $\varkappa > 0$,

$$Dist(t_0, B_\varkappa(x+h)) \leq F^{-1}(2\|h\|_c + \kappa).$$

Remark 1.6.2. If $A_0(x + h) \neq \emptyset$, then

$$Dist(t_0, A_0(x+h)) \leq F^{-1}(2\|h\|_c)$$

Remark 1.6.3. Let t_0 be a unique point of the maximum of $|x(t)|$ in T and

$$|x(t_0)| - |x(t)| \geq F(\|t - t_0\|) \text{ for all } t \in T.$$

Then, for all small enough $\varkappa > 0$ and $\|h\|_c$, the following inequality holds:

$$Dist(t_0, A_\varkappa(|x+h|) \leq F^{-1}(2\|h\|_c + \varkappa),$$

and in case $A_0(|x + h|) \neq \emptyset$,

$$Dist(t_0, A_0(|x+h|) \leq F^{-1}(2\|h\|_c).$$

The proof follows from the fact that for all small enough $\varkappa$ and $\|h\|_c$, the set of almost-maximums of $|x(t) + h(t)|$ coincides with the set of almost-maximums or with the set of almost-minimums of $(x(t) + h(t))$.

2

General Retrospective Disorder Problem

2.1 Introduction

Retrospective problems of detection of changes in characteristics of random processes occur in econometrics, financial mathematics, biomedicine, and many other applications. These problems constitute an indispensable and important part of the general body of change-point analysis.

The retrospective change-point problem was first formulated by Page (1954,1955). A sample composed of independent random variables $x_1, x_2, \ldots, x_n$ was considered, and an assumption was made that the density function of these observations are different before and after an unknown change-point t_0, $1 \leq t_0 \leq N$. Page proposed the CUSUM rule for detection the change-point, which is closely connected with the maximum likelihood method.

The idea of the Bayesian approach to the retrospective change-point problem was proposed by Chernoff and Zacks (1964) for a sequence of i.r.v.'s with the Gaussian d.f. $\mathcal{N}(\theta_i, 1)$, $i = 1, \ldots, n$.

The maximum likelihood approach constitutes another main research trend. The maximum likelihood statistic was first proposed by Hinkley (1969, 1970) in the context of i.r.v.'s First, suppose that parameters θ_1, θ_2 are known. Then the maximum likelihood estimate of the parameter τ is

Hinkley proposed the idea of computing the asymptotical distribution of this MLE via the asymptotics (as $n \to \infty$) of a two-side random walk. Bhattacharya and Brockwell (1976) studied this asymptotic distribution of MLE for the problem of changing mean value of an observed sequence. Generalizations of these results to multiparametric families was obtained by Bhattacharya (1987).

The problem of computing confidence intervals for MLE of the change-point τ was studied by Siegmund (1988) in the Gaussian case. James, James, and Siegmund (1987) obtained approximation of the large deviations probability for the null hypothesis and MLE of an unknown change-point.

The above papers considered independent observations. Generalizations to the case of dependent observations constitute a separate chapter in the retrospective change-point analysis. One of the first papers in this direction Box, Tiao (1965) considered ARIMA model for an observed random sequence with a change in mean. For detection of this change-point, Student t-statistic was used.

Bhattacharya and Frierson (1981) proposed the first nonparametric *retrospective* test for change-point detection. They considered a sequence of independent random variables $x_1, \ldots, x_n$ with one-dimensional distribution functions $F_0(x)$ and $F_1 = F_0(x - \Delta)$ before and after the change-point correspondingly, where Δ is an unknown shift. The test statistic proposed in Bhattacharya and Frierson (1981) for testing the hypothesis about the presence of a change-point, is the function of ranks of observations R_k, $k = 1, \ldots, n$ (remind that $R_k = \sum_{i \le n} \mathbb{I}(x_i \le x_k)$, n is the number of observations).

Darkhovsky (1976) considered the change-point problem for the sequence of independent continuously distributed random variables $x_1, \ldots, x_N$. The method of detection was based on the Mann-Whitney statistic. Suppose that the one-dimensional distribution function (d.f.) of observations $x_1, \ldots, x_{n_0}$ is equal to F_1, and the d.f. of $x_{n_0+1}, \ldots, x_N$ is equal to F_2, where

$$a = \int_{-\infty}^{\infty} F_2(y)\, dF_1(y) \neq 1/2.$$

The problem of *a posteriori* detection of a change-point $n_0(N) = [\theta N]$, where $0 < \theta < 1$ is an unknown parameter, was considered. The Mann-Whitney statistic $V_N(n)$ was constructed by the sample $(x_1, \ldots, x_N)$:

$$\begin{aligned} V_N(n) &= u_N(n)/[n(N-n)], \\ u_N(n) &= \sum_{k=n+1}^{N} \sum_{i=1}^{n} z_{ik}, \ z_{ik} = \mathbb{I}(x_i > x_k), \end{aligned}$$

and its asymptotics investigated as $N \to \infty$.

For any $N > 1/\alpha$, a continuous random process $\xi_N(t)$ was constructed on the interval $[\alpha, 1-\alpha]$, $\alpha > 0$:

$$\xi_N(t) = (1 - \{Nt\})V_N([Nt]) + \{Nt\}V_N(1 + [Nt]), \ k/N \le t < (k+1)/N,$$

by means of the linear interpolation with the points $(n/N, V_N(n))$ ($[a]$ and $\{a\}$ are, correspondingly, an integral and a fractional part of a). Then the set of minimums $\mathcal{M}_N$ and maximums $\widetilde{\mathcal{M}_N}$ of the random process $\xi_N(t)$ on $[\alpha, 1-\alpha]$ was considered. For $0 \le a < 0.5$, an arbitrary point of $\mathcal{M}_N$ was taken to be the estimate of the change-point θ, and, for $0.5 < a \le 1$, an arbitrary point of $\widetilde{\mathcal{M}_N}$ was taken to be the estimate of θ. Weak convergence of this estimate $\hat{\theta}$ to the change-point θ was proved as $N \to \infty$, and the rate of convergence was estimated for any $\epsilon > 0$

$$\mathbf{P}_\theta[|\hat{\theta} - \theta| > \epsilon] = O(N^{-1}).$$

Darkhovsky and Brodsky (1979, 1980) considered the problem of *a posteriori* change-point detection for a sequence of dependent random variables with a finite number of values. At the change-point n_0, the probabilities of states a_j, $j = 1, \ldots, k$ changed, i.e.,

$$\mathbf{P}\{x^N(n) = a_j\} = \begin{cases} p_j, & 1 \leq n \leq n_0(N), \\ q_j, & n_0(N) < n \leq N, \end{cases} \quad j = 1, \ldots, k,$$

and

$$\sum_{j=1}^{k} (p_j - q_j)^2 = V \geq \delta > 0,$$

where the change-point $n_0(N) = [\theta N]$, and $0 < \theta < 1$ was assumed to be an unknown estimated parameter. It was supposed that the family of processes $X_N = \{x^N(n)\}_{n=1}^N$ fulfils α-mixing condition (see definition below).

For estimation of the change-point, the following statistic was used:

$$Y_N(n) = \sum_{j=1}^{k} \left(\frac{1}{n} \sum_{i=1}^{n} y_{ij} - \frac{1}{N-n} \sum_{i=n+1}^{N} y_{ij} \right)^2, \tag{2.1}$$

where

$$y_{ij} = \mathbb{I}(x^N(i) = a_j), \; i = 1, \ldots, N, \qquad j = 1, \ldots, k.$$

The estimate $\hat{n}(N)$ of the change-point was defined as an arbitrary point of the set

$$\arg \max_{[\alpha N] \leq n \leq [\beta N]} Y_N(n), \qquad (0 < \alpha \leq \theta \leq \beta < 1). \tag{2.2}$$

Weak convergence of the normalized estimate $\hat{\theta}_N = \hat{n}(N)/N$ to the parameter θ was proved as $N \to \infty$. For any $\epsilon > 0$, the following relationship was proved:

$$\mathbf{P}_\theta\{|\hat{\theta}_N - \theta| > \epsilon|\} = O\left(k^3 \tau_N / \epsilon \delta^3 N\right), \quad N \to \infty, \tag{2.3}$$

where

$$\tau_N = 1 + \sum_{i=1}^{N} \alpha(i)$$

and $\alpha(i)$ is α-mixing coefficient.

Different modifications of statistic (2.1) and estimate (2.2) were proposed in works of Brodsky and Darkhovsky (1990, 1993, 1995). In particular, in Brodsky and Darkhovsky (1990), the exponential analogs of (2.3) and strong consistency of change-point estimates were established if an observed random sequence fulfills Cramer's and ψ-mixing conditions.

This approach to change-point estimation was generalized by Carlstein (1988) and D́umbgen (1991).

Different modifications of Kolmogorov's test were proposed by Vostrikova (1983) and Deshayes and Picard (1986). The idea of Kolmogorov's test for detecting change-points was rediscovered many times in subsequent works on the retrospective change-point detection, and nowadays it is (we do not know why) called the CUSUM test (see papers by Inclan and Tiao (1994), Lee et al. (2003), etc.,

So, originally the problem began with i.i.d. samples (see Page (1954, 1955), Chernoff and Zacks (1964), Hinkley (1971) et al.). It then moved naturally to dependent context (see Brodsky and Darkhovsky (1980, 1983)), and time series context (see, e.g., Basseville and Nikiforov (1993), Picard (1985), Kim et al. (2000), Lee and Park (2001) and the papers cited therein). Much attention was paid to detection of change-points in linear models (like AR(p) model): see, e.g., Lee et al. (2003), Davis et al. (2006). Nowadays, we see attempts to consider the multivariate time series case (see, e.g., Aue et al. (2008), Cho and Fryzlewicz (2013)) and large-scale problems (see, e.g., Horvath (2012)). Many of these results are mentioned in subsequent chapters.

2.2 Problem Statement

We use the following construction. Let

$$0 \equiv \vartheta_0 < \vartheta_1 < \vartheta_2 < \cdots < \vartheta_k < \vartheta_{k+1} \equiv 1, \quad \vartheta \stackrel{\text{def}}{=} (\vartheta_1, \ldots, \vartheta_k).$$

We call ϑ *unknown parameter.*

Consider a collection of random sequences (in other words, a vector-valued random sequence)

$$\mathcal{X} = \{X^{(1)}, X^{(2)}, \ldots, X^{(k+1)}\}, X^{(i)} = \{x^{(i)}(n)\}_{n=1}^{\infty}.$$

Now, define a family of random sequences

$$\mathfrak{X} = \{X^N\}, N = N_0, N_0 + 1, N_0 + 2, \ldots, N_0 > 1, X^N = \{x^N(n)\}_{n=1}^{N}$$

as follows:

$$x^N(n) = x^{(i)}(n),$$

if $[\vartheta_{i-1}N] \leq n < [\vartheta_i N], i = 1, \ldots, k+1$.

We say that the family $\mathfrak{X} = \{X^N\}$ is generated by the process of *'gluing' (concatenation).*

In other words, the family $\mathfrak{X}$ is the *"glued"* random sequence generated by the collection $\mathcal{X}$, and the collection $\{\vartheta\}$ are the *points of 'gluing'* or *change-points.*

Evidently, the process of 'gluing' is the variant of the *triangular array scheme,* which is often used in the probability theory.

Therefore, our statement of the offline change-point problem for random sequences is as follows:

find an estimate of unknown parameter ϑ in described scheme by given sample X^N.

Evidently, as soon as we have an estimate $\hat{\vartheta}_N$ of parameter ϑ, we get the estimate of the change-points in the form $\hat{n}_i = [\hat{\vartheta}_i N]$, $i = 1, \ldots, k$.

Remark 2.2.1. Everywhere below the symbol $\vartheta \equiv 0$ denotes the situation of no change-points in an observed random sequence. We also use the symbols $\mathbf{P}_0$ ($\mathbf{E}_0$) and $\mathbf{P}_\theta$ ($\mathbf{E}_\theta$) for denoting probability measures (expectations) corresponding to the case of absence and presence of change-points, respectively.

2.3 Main Ideas

There are two main ideas in our approach to the problem. These ideas were first proposed in the works of Brodsky and Darkhovsky in 1979–1980s.

The first idea

The *first idea* is as follows: detection of changes in any d.f. or some probabilistic characteristic *can be (with an arbitrary degree of accuracy) reduced to detection of changes in the mean value of some new sequence constructed from an initial one.*

Let us explain this idea by the following example. Suppose we observe a random sequence $X = \{x(t)\}_{t=1}^N$ "glued" from two strictly stationary sequences

$$X_1 = \{x_1(t)\}_{t=1}^{n^\star}, X_2 = \{x_2(t)\}_{t=n^\star+1}^{N}, n^\star = [\vartheta N],\ 0 < \vartheta < 1,$$

and it is required to estimate the change-point $n^\star$.

Suppose we know that X_1 and X_2 differ from each other by some two-dimensional d.f., namely,

$$\mathbf{P}\{x(t) \leq u_0, x(t+2) \leq u_1\} = F_1(u_0, u_1),$$

before the instant $t_1 = n^\star - 2$, and for $t \geq t_2 = n^\star + 1$ is equal to $F_2(\cdot)$, and $\|F_1(\cdot) - F_2(\cdot)\| \geq \epsilon > 0$, where $\|\cdot\|$ is the usual sup-norm.

It is well known that the d.f. of a finite-dimensional random vector can be uniformly approximated to an arbitrary accuracy by the d.f. of some random vector with a finite number of values.

Hence, after a partition of the plane $\mathbb{R}^2$ into some number of nonintersecting areas A_i, $i = 1, \ldots, r$, the vector (x_t, x_{t+2}) can be approximated by some vector with a finite number of values.

Therefore, if we introduce new sequences $V_t^i = \mathbb{I}((x_t, x_{t+2}) \in A_i)$, $1 \leq i \leq r$, then at least in one of them the mathematical expectation changes.

Therefore, if there exists a method detecting changes in the mean value, then the same method will detect changes in the d.f.

In the same way, we can detect changes in any probabilistic characteristic.

For example, if the correlation function of an observed sequence changes, then considering for every fixed $\tau = 0, 1, 2, \ldots$ new sequences $V_t(\tau) = x_t x_{t+\tau}$, we reduce the problem to detection of changes in the mathematical expectation of one of the sequences $V_t(\tau)$.

This argument enables us to develop only one **basic method** of diagnosis that can detect changes in the mathematical expectation instead of creating an infinite number of algorithms for detection of changes in arbitrary probabilistic characteristics.

The problem of detection of changes in the mathematical expectation will be called the *basic problem*, and the method of its solution will be called the *basic method*. The random sequence formed from an initial one for detection of changes in the mean value will be called the *diagnostic sequence*.

According to the first idea, we suppose that the diagnostic sequence $X^N = \{x^N(n)\}$ can be written as follows:

$$x^N(n) = \varphi(\vartheta, n/N) + \xi^N(n), \; n = 1, \ldots, N, \tag{2.4}$$

where $\xi^N = \{\xi^N(n)\}_{n=1}^{\infty}$ is the "glued" (according to the scheme of 1.1, i.e., $\Xi \stackrel{\text{def}}{=} \{\Xi^{(1)}, \ldots, \Xi^{(k+1)}\}$, $\Xi^{(i)} = \{\xi^{(i)}(n)\}_{n=1}^{\infty}$) and $\xi^N(n) = \xi^{(i)}(n)$ if $[\vartheta_{i-1}N] \le n < [\vartheta_i N]$, $i = 1, \ldots, k+1$) is the random sequence, $\mathbf{E}_\theta \xi^N(n) \equiv 0$, and the function $\varphi(\vartheta, t)$, $t \in [0, 1]$ has the following form,

$$\varphi(\vartheta, t) = \sum_{i=1}^{k} a_i \mathbb{I}(\vartheta_{i-1} \le t < \vartheta_i) + a_{k+1} \mathbb{I}(\vartheta_k \le t \le 1), \tag{2.5}$$

with $a_i \ne a_{i+1}$, $i = 1, 2, \ldots, k$.

So, Scheme (2.4) in the general case allows for a rather sophisticated noise component of the diagnostic sequence.

The **second idea** of our (nonparametric) approach consists in the use of the following family of statistics for detection of change-points:

$$Y_N(n, \delta) = \left[\frac{(N-n)n}{N^2}\right]^{\delta} \left(n^{-1} \sum_{k=1}^{n} x^N(k) - (N-n)^{-1} \sum_{k=n+1}^{N} x^N(k)\right), \tag{2.6}$$

where $0 \le \delta \le 1$, $1 \le n \le N-1$, $X^N = \{x^N(k)\}_{k=1}^{N}$ is an observed realization (or a diagnostic sequence).

Let us explain the origin of family (2.4). Suppose an observed sample is "glued" from two random sequences with different one-dimensional d.f.'s and we want to test the hypothesis that some point $n^\star$ of this sample is the "gluing" point.

This is the well-known statistical problem of testing the hypothesis about equivalence of the d.f.'s of two subsamples: the first subsample with numbers of elements from 1 to $n^\star$, and the second subsample with numbers of elements from $(n^\star + 1)$ to N.

For solving such problems when there is no *a priori* information the well-known Kolmogorov–Smirnov statistic is used:

$$Z_N(n^\star) = \max_u \Big| (n^\star)^{-1} \sum_{k=1}^{n^\star} \mathbb{I}(x^N(k) \le u) - \\ -(N - n^\star)^{-1} \sum_{k=n^\star+1}^{N} \mathbb{I}(x^N(k) \le u) \Big|. \tag{2.7}$$

The sums under the max symbol are the empirical d.f.'s of the 1st and the 2nd sample, and the statistic is the norm of the difference of the empirical d.f.'s.

Suppose that the random variables take only two values. Then we obtain the statistic of type (2.6) for $n = n^\star$.

The change-point problem differs from the problem of testing equivalence of distributions by one substantial detail: the point n^* of "gluing" is *unknown* and it is required to find it in the sample X^N.

Therefore, it is quite natural to generalize (2.7) and to try all numbers $1 \le n \le N$ in the sample X^N as candidates for a change-point.

The use of indicators corresponds to our first idea and the additional multiplier enables us to optimize characteristics of estimates (this will be explained in the following section).

Thus, the family of statistics (2.6) is the generalization of the Kolmogorov–Smirnov test.

Remark 2.3.1. The above-considered scheme of "gluing" assumes, in essence, that all changes in parameters of a stochastic system occur instantaneously. In practical applications, we can obtain other situations. For example, a transition from certain values of parameters to other values of these parameters occurs gradually (we can call this case a gradual disorder) or is not finished at all (e.g., the case of a deterministic or stochastic trend). Some of these cases will be considered in special chapters (see, e.g., Chapter 3). Remark, however, that this fact does not violate consistency of change-point estimates if stability of a dynamic system preserves after a transition period. The typical example here is an abrupt change of autoregressive parameters in a stable dynamical system.

In this example, any abrupt change of parameters leads to a gradual change in expectation of a diagnostic sequence (which here can be chosen as $x(t)x(t+\tau), \tau = 0, 1, 2.$, where $x(t)$ is an observed sequence). However, for every $\epsilon > 0$, we can determine the number $n(\epsilon) > 0$ (which does not depend on the sample size N), such that for $n > n^* + n(\epsilon)$ (n^* is the change-point), the mathematical expectation of a diagnostic sequence differs from the expectation under an ideal change no more than for ϵ.

It follows from here that the share of time when the real mathemativ=cal expectation of a diagnostic sequence differs from the corresponding expectation (under an ideal change) is of order $n(\epsilon)/N$ and converges to zero if the sample size tends to infinity. This fact helps us to obtain consistent estimators of the parameter ϑ on the basis of ideas described in this section.

2.4 Asymptotically Optimal Choice of Parameters of Decision Statistic

2.4.1 Weak Convergence of a Process Generated by Decision Statistic

Consider the case when $\varphi(\cdot) \equiv 0$ in Scheme (2.4), i.e., changes in the mathematical expectation, are absent. Statistic (2.6) in this case takes the form

$$Y_N(n,\delta) = \left[\frac{(N-n)n}{N^2}\right]^\delta \left(n^{-1}\sum_{k=1}^{n}\xi^N(k) - (N-n)^{-1}\sum_{k=n+1}^{N}\xi^N(k)\right). \quad (2.8)$$

We associate the random process $y_N(t,\delta) \stackrel{\text{def}}{=} Y_N([Nt],\delta)$ with Statistic (2.8), which will be considered in the Skorokhod space $D[a,b], 0 < a < b < 1$.

From Theorem 1.3.2 from Chapter 1, we immediately obtain

Theorem 2.4.1. *Suppose for a vector-valued sequence Ξ the following conditions are satisfied:*

i) for each component there exists

$$\sigma_i^2 \stackrel{def}{=} \lim_{n\to\infty} n^{-1}\left(\sum_{k=1}^{n}\xi^{(i)}(k)\right)^2;$$

ii) $\max_i \sup_k \mathbf{E}_0\left(\xi^{(i)}(k)\right)^{2+a} < \infty$, *$a > 0$; and*

iii) $\sum_{k=1}^{\infty}\max_{(i,j)}\sqrt{\rho_{ij}(k)} < \infty$, *where $\rho_{ij}(\cdot)$ is the coefficient of ρ-mixing between ith and jth components of the vector-valued sequence Ξ.*

Then the process $N^{1/2}y_N(t,\delta)$ weakly converges in $D[a,b]$ to the process $\sigma(t)\,(t(1-t))^{\delta-1}\,W^0(t)$, where $W^0(t)$ is the standard Brownian bridge, and

$$\sigma(t) = t^{-1}\int_0^t \sigma^2(s)ds, \;\; \sigma^2(s) = \sum_{i=1}^{k+1}\sigma_i^2\mathbb{I}(\vartheta_{i-1} \le s < \vartheta_i).$$

In a particular case when all components of the vector-valued random sequence Ξ are equal (i.e., in the sequence $\xi^N(n) \equiv \xi(n)$, there are no points of "gluing"), the limit process takes the form:

$$\sigma\,(t(1-t))^{\delta-1}\,W^0(t), \quad (2.9)$$

where $\sigma^2 = \lim_{n\to\infty} n^{-1}\left(\sum_{k=1}^{n}\xi(k)\right)^2$.

Here, assumption iii) is substituted by the following assumption $\sum_{k=1}^{\infty}\sqrt{\rho(k)} < \infty$, *where $\rho(\cdot)$ is the coefficient of ρ-mixing for the sequence* $\{\xi(n)\}_{n=1}^{\infty}$.

2.4.2 Asymptotical Analysis of Decision Statistic and Choice of Parameter δ

In this subsection, we give recommendations for the choice of the parameter δ in the main Statistic (2.6). These recommendations are based upon the asymptotical analysis of this statistic.

We consider the following model of a *diagnostic sequence* $X^N = \{x^N(n)\}_{n=1}^N$:

$$x^N(n) = a_1\mathbb{I}(0 \le n/N < \vartheta) + a_2\mathbb{I}(\vartheta \le n/N \le 1) + \xi(n),\ n = 1, \dots, N, \quad (2.10)$$

where $a_1 \neq a_2$, and the parameter ϑ determines a unique change-point. It is supposed that $0 < a \le \vartheta \le b < 1$, and for the random sequence $\Xi \stackrel{\text{def}}{=} \{\xi(n)\}_{n=1}^\infty$, the following conditions are satisfied:

$$\begin{aligned}
&\mathbf{E}_\vartheta \xi(n) \equiv 0.\\
&\text{There exists the limit } \sigma^2 \stackrel{\text{def}}{=} \lim_{n\to\infty} n^{-1}\left(\sum_{k=1}^{n} \xi(k)\right)^2\\
&\sup_k \mathbf{E}_0 \left(\xi(k)\right)^{2+\epsilon} < \infty,\ \epsilon > 0\\
&\sum_{k=1}^{\infty} \sqrt{\rho(k)} < \infty,
\end{aligned} \quad (2.11)$$

where $\rho(\cdot)$ is the coefficient of ρ-mixing for the sequence Ξ.

Thus, we consider a model with one change-point and do not suppose "gluing" in the sequence Ξ. On the one hand, these assumptions simplify the subsequent asymptotical analysis. On the other hand, they do not influence subsequent generalizations, because (as we see it below from description of the algorithm of change-point detection), the case of multiple change-points is reduced to the case of only one change-point.

Consider the main statistic $Y_N(n, \delta)$ (see (2.6)) constructed by the sequence X^N. We associate with $Y_N(n, \delta)$ the process $y_N(t, \delta) \stackrel{\text{def}}{=} Y_N([Nt], \delta)$, $t \in [a, b]$, with trajectories from Skorokhod space $D[a, b]$.

The function $\mathbf{E}_\vartheta y_N(t, \delta)$ uniformly converges in $[a, b]$ (with the rate $O(N^{-1})$) to the function

$$m(t) = \begin{cases} h(1-\vartheta)t^\delta(1-t)^{\delta-1} & \text{if } t \le \vartheta \\ h\vartheta t^{\delta-1}(1-t)^\delta & \text{if } t \ge \vartheta, \end{cases} \quad (2.12)$$

where $h = a_2 - a_1$.

An arbitrary point of maximum of the process $|y_N(t, \delta)|$ in the interval $[a, b]$ is assumed to be the estimator of the change-point. We accept the hypothesis of a change-point if $\sup_{a\le t\le b} |y_N(t, \delta)| > C$, where $C > 0$ is a certain threshold. The choice of C is described in the following section in the algorithm of change-point detection.

The following values are characteristics of a method of change-point detection:

$$\alpha(N) = \mathbf{P}_0\{ \sup_{a \le t \le b} |y_N(t, \delta)| > C\},$$

which is the probability of the 1st-type error ("false alarm");

$$\beta(N) = \mathbf{P}_\vartheta\{ \sup_{a \le t \le b} |y_N(t, \delta)| \le C\},$$

the probability of the 2nd-type error ("false tranquility");

$$\gamma(\epsilon, N) = \mathbf{P}_\vartheta\{|\hat{\vartheta}_N - \vartheta| > \epsilon\},$$

the probability of the error of estimation.

Define the process

$$\eta_N(t, \delta) \stackrel{\text{def}}{=} y_N(t, \delta) - \mathbf{E}_\vartheta y_N(t, \delta).$$

From Theorem 2.4.1 and Conditions (2.11), it follows that the process $\sqrt{N}\eta_N(t, \delta)$ weakly converges in $D[a, b]$ to the process $z(t) \stackrel{\text{def}}{=} \sigma[t(1-t)]^{\delta-1}W^0(t)$, where $W^0(t)$ is the standard Brownian bridge and σ is the parameter from Conditions (2.11).

For the asymptotical analysis of the values $\alpha(N), \beta(N), and \gamma(\epsilon, N)$, we need to obtain the asymptotical (as $N \to \infty$) estimate of the probability

$$\mathbf{P}_0\{ \sup_{t \in [a,b]} |\eta_N(t, \delta)| > x\} = \mathbf{P}_0\{\|\eta_N(t, \delta)\| > x\},\ x > 0.$$

Denote

$$F(x) = \mathbf{P}_0\{ \max_{t \in [a,b]} |z(t)| > x\} = \mathbf{P}_0\{\|z(t)\| > x\}. \tag{2.13}$$

Evidently, $F(x)$ is a continuous and monotonous function. Therefore, there exists the inverse function to it. So, in virtue of the weak convergence under the null: for every $\epsilon > 0$, $\eta > 0$ there exists the number $N_0(\epsilon, \eta)$, such that for $N > N_0(\cdot)$ the following relationship holds:

$$\left| \mathbf{P}_0\left\{ \|\eta_N(t, \delta)\| > \frac{F^{-1}(\epsilon)}{\sqrt{N}} \right\} - \epsilon \right| < \eta, \tag{2.14}$$

which can be rewritten as follows:

$$\mathbf{P}_0 \left\{ \|\eta_N(t, \delta)\| > A(\epsilon, N) \right\} \sim \epsilon, \tag{2.15}$$

where $A(\epsilon, N) = \frac{F^{-1}(\epsilon)}{\sqrt{N}}$.

But $A(\epsilon, N)$ is the root (w.r.t. A) of the equation $F(A\sqrt{N}) = \epsilon$. Therefore, from (2.15), it follows that (as $N \to \infty$):

$$\mathbf{P}_0 \left\{ \|\eta_N(t, \delta)\| > A \right\} \sim F(A\sqrt{N}). \tag{2.16}$$

Now we consider the asymptotical estimates.

Theorem 2.4.2. *The following equality holds:*

$$\lim_{N\to\infty} \frac{|\ln \alpha(N)|}{N} = \begin{cases} (1/2)\,(C/\sigma)^2\,\Delta^2 & \text{if } 0 \le \delta \le 1/2 \\ (1/2)\,(C/\sigma)^2\,2^{4\delta-2} & \text{if } 1/2 < \delta \le 1, \end{cases} \tag{2.17}$$

where $\Delta = \Delta(a,b) = \min\left((a(1-a))^{0.5-\delta}, (b(1-b))^{0.5-\delta}\right)$.

Proof.

Let $0 \le \delta \le 1/2$. By definition of $\alpha(N)$ and from (2.16), we can write as $N \to \infty$

$$\begin{aligned} \alpha(N) &= \mathbf{P}\{\tfrac{\sigma}{\sqrt{N}} \max_{a\le t\le b} \frac{|W^0(t)|}{[t(1-t)]^{1-\delta}} > C\}(1+o(1)) \le \\ &\le \mathbf{P}\{\max_{a\le t\le b} \frac{|W^0(t)|}{\sqrt{t(1-t)}} > \frac{C}{\sigma}\sqrt{N}\Delta(a,b)\}(1+o(1)). \end{aligned} \tag{2.18}$$

Making time transformations $t \to \frac{u}{u+1}$, $u \to \frac{a}{1-a}z$ and taking into account the equality (with respect to the distribution) $W^0(\frac{t}{t+1}) \overset{d}{=} \frac{1}{1+t} W(t)$, we obtain from (2.18)

$$\alpha(N) \le \mathbf{P}\{\max_{1\le z\le d(1-c)/c(1-d)} \frac{|W(z)|}{\sqrt{z}} > \frac{C}{\sigma}\sqrt{N}\Delta\}(1+o(1)). \tag{2.19}$$

Now we use the following relationship (Vostrikova, 1981)

$$\begin{aligned} &\mathbf{P}\{\max_{1\le t\le T} \frac{|W(t)|}{\sqrt{t}} > x\} = \\ &= \frac{1}{\sqrt{\pi}}\left((1-x^{-2})\ln\sqrt{T} + 2x^{-2} + o(x^{-4})\right) x\exp(-x^2/2), \end{aligned} \tag{2.20}$$

which is true as $x \to \infty$.

Then, from (2.19), we obtain

$$\liminf_{N} \frac{|\ln \alpha(N)|}{N} \ge \left(\frac{C}{\sigma}\right)^2 \frac{\Delta^2}{2}. \tag{2.21}$$

Using change of time and the equality in distribution, after taking the upper estimate, we can write

$$\begin{aligned} \alpha(N) &= 1 - \mathbf{P}\left\{\max_{a\le t\le b} \frac{|W^0(t)|}{[t(1-t)]^{1-\delta}} \le \sqrt{N}\frac{C}{\sigma}\right\} \\ &= 1 - \mathbf{P}\left\{\max_{1\le z\le \frac{b(1-a)}{a(1-b)}} \frac{|W(z)|}{z^{1-\delta}} \frac{1}{(1+\dfrac{a}{1-a}z)^{2\delta-1}} \le \sqrt{N}\frac{C}{\sigma}\left(\frac{a}{1-a}\right)^{0.5-\delta}\right\} \\ &= 1 - \frac{1}{\sqrt{2\pi}}\int_{-A}^{A} \mathbf{P}\left\{\max_{0\le y\le \frac{b-a}{a(1-b)}} \frac{|W(y)+x|}{(1+y)^{1-\delta}\,(1+(a(1+y)/(1-a))^{2\delta-1}}\right. \\ &\left.\le \sqrt{N}\frac{C}{\sigma}\left(\frac{a}{1-a}\right)^{0.5-\delta}\right\}\exp(-\frac{x^2}{2})\,dx \ge 1 - \frac{1}{\sqrt{2\pi}}\int_{-A}^{A} \exp(-x^2/2)\,dx, \end{aligned} \tag{2.22}$$

where $A = \sqrt{N}\,\frac{C}{\sigma}\,(a(1-a))^{0.5-\delta}$.

Hence,

$$\limsup_N \frac{|\ln \alpha(N)|}{N} \le \frac{C^2}{2\sigma^2}\,(a(1-a))^{1-2\delta}.$$

The same considerations for the reversed time yield

$$\limsup_N \frac{|\ln \alpha(N)|}{N} \le \frac{C^2}{2\sigma^2}\,(b(1-b))^{1-2\delta}.$$

Therefore,

$$\limsup_N \frac{|\ln \alpha(N)|}{N} \le \frac{C^2}{2\sigma^2}\,\Delta^2,$$

and, hence, for $0 \le \delta \le 1/2$,

$$\lim_N \frac{|\ln \alpha(N)|}{N} = \left(\frac{C}{\sqrt{2}\sigma}\Delta(a,b)\right)^2. \tag{2.23}$$

Consider now the case $1/2 < \delta \le 1$. Since

$$\frac{1}{(t(1-t))^{1-\delta}} = \frac{(t(1-t))^{\delta-1/2}}{\sqrt{t(1-t)}} \le \frac{1}{4^{\delta-1/2}\sqrt{t(1-t)}},$$

we have as before

$$\liminf_N \frac{|\ln \alpha(N)|}{N} \ge \left(\frac{C(4^{\delta-1/2})}{\sqrt{2}\sigma}\right)^2 = \left(\frac{C}{\sigma}\right)^2 2^{4\delta-3}.$$

For the upper estimate, we have

$$\alpha(N) \ge \mathbf{P}\left\{\frac{|W^0(1/2)|}{(1/4)^{1-\delta}} \ge \frac{C}{\sigma}\sqrt{N}\right\}.$$

Hence,

$$\limsup_N \frac{|\ln \alpha(N)|}{N} \le \frac{C^2}{\sigma^2}\,2^{4\delta-3},$$

and, therefore,

$$\lim_N \frac{|\ln \alpha(N)|}{N} = 2^{4\delta-3}\,\frac{C^2}{\sigma^2}. \tag{2.24}$$

The theorem is proved. ■

Corollary 2.4.1. *The asymptotically best method that furnishes the minimum of the 1st-type error probability in the considered family (2.6) is the method with $\delta = 1$. For this method,* $\lim_{N\to\infty} (|\ln \alpha(N)|/N) = 2(C/\sigma)^2$.

Indeed, it is easy to see from (2.17) that the function $r(\delta) \triangleq \lim_{N\to\infty}(|\ln\alpha(N)|/N)$ monotonously increases on $0 \le \delta \le 1$. This property implies the required result.

From (2.16), it follows that for the asymptotical analysis of the process $y_N(t,\delta)$, the following model can be used:

$$y_N(t,\delta) = m(t) + N^{-1/2}\{\sigma[t(1-t)]^{\delta-1}W^0(t) + o_p(1)\}, \tag{2.25}$$

where $o_p(1)$ denotes the process converging to zero by distribution, and $m(t)$ is defined by (2.12).

We proceed from model (2.25) for the asymptotical analysis of $\beta(N)$ and $\gamma(\epsilon, N)$.

Now begin from the analysis of $\beta(N)$.

It is well known (see, for example, Aubin, 1984) that the maximum type functional is differentiable in any direction in the space of continuous functions. Hence,taking into account that the function $|m(t)|$ has a single maximum on $[a,b]$ at $t^\star = \vartheta$, we obtain the following relation as $\epsilon \to +0$, $g(\cdot) \in C[a,b]$:

$$\max_{a\le t\le b} |m(t) + \epsilon g(t)| = |m(\vartheta)| + \epsilon\, g(\vartheta)\operatorname{sign} m(\vartheta) + o(\epsilon). \tag{2.26}$$

The trajectories of the process $W^0(t)$ are almost surely bounded in the finite segment $[a,b]$. Therefore, it follows from (2.25) and (2.26) that almost surely

$$\max_{a\le t\le b} |y_N(t,\delta)| = |h|\,p^\delta + \frac{\sigma}{\sqrt{N}}\,\xi + o(\frac{1}{\sqrt{N}}), \tag{2.27}$$

where $p = p(\vartheta) \triangleq \vartheta(1-\vartheta)$, $\xi = (\pm W^0\,(\vartheta)/p^{1-\delta})$ is a Gaussian random variable with zero mean and dispersion $p^{2\delta-1}$, and the random variable $\sqrt{N}\cdot o(1/\sqrt{N})$ weakly converges to zero as $N \to \infty$.

As a matter of fact, we assume that the threshold C fulfills the inequality

$$|h|\,p^\delta > C \quad \forall\vartheta \in [a,b]. \tag{2.28}$$

Otherwise, the 2nd-type probability $\beta(N)$ is positive for all N and tends to 1 as $N \to \infty$.

Let $C = \lambda\,|h|$, $\lambda > 0$. Then, (2.28) is equivalent to the following condition:

$$\lambda < \min_{a\le\vartheta\le b} (p(\vartheta))^\delta = \min\left((a(1-a))^\delta,\, (b(1-b))^\delta\right). \tag{2.29}$$

Now, (2.27) yields the following.

Theorem 2.4.3. *Suppose condition (2.27) is fulfilled. Then*

$$\lim_{N\to\infty} \mathbf{P}\{\sqrt{N}\max_{a\le t\le b}(|y_N(t,\delta)| - |h|\,p^\delta) < C - |h|\,p^\delta\} = \Phi\left(\frac{|h|(\lambda - p^\delta)}{\sigma\,p^{\delta-1/2}}\right),$$

where Φ is a standard function of Gaussian distribution.

Corollary 2.4.2. *The following relation holds:*

$$\beta(N) \sim \Phi\left(\frac{\sqrt{N}\,|h|\,(\lambda - p^\delta)}{\sigma\, p^{\delta-1/2}}\right) \sim \exp\left(-\frac{Nh^2(p^\delta - \lambda)^2}{2\sigma^2 p^{2\delta-1}}\right). \qquad (2.30)$$

Corollary 2.4.3. *The asymptotically best method that furnishes the minimum of the 2nd-type error probability $\beta(N)$ in the considered family of change-point detection methods is the method with $\delta = 0$, such that*

$$\beta(N) \sim \exp\left(-\frac{Nh^2}{2\sigma^2}\, p(\vartheta)\,(1-\lambda)^2\right). \qquad (2.31)$$

Indeed, it follows from (2.30) that

$$\frac{|\ln \beta(N)|}{N} \sim \frac{h^2}{2\sigma^2}\, \psi\,(p,\delta),$$

where $\psi\,(p,\delta) = (p^\delta - \lambda)^2/p^{2\delta-1}$.

Therefore, the best method in this case is the method with $\delta^\star$, such that

$$\min_{a\le\vartheta\le b} \psi(p(\vartheta),\delta^\star) = \max_{0\le\delta\le1}\ \min_{a\le\theta\le b} \psi(p(\vartheta),\delta)\ .$$

Since $1/4 \ge p(\vartheta) \ge \min_{a\le\vartheta\le b} p(\vartheta) \triangleq p_m = \min\,(a(1-a)),\, b(1-b))$, we have

$$\min_{a\le\vartheta\le b} \psi(p(\vartheta),\delta) = \min_{p_m\le p\le 1/4} \psi(p,\delta).$$

Then, (2.29) yields

$$\psi_p^{'}\,(p,\delta) = \frac{p^\delta - \lambda}{p^{2\delta}}\,(p^\delta - \lambda + 2\lambda\delta) > 0, \quad 0 \le \delta \le 1, \quad p_m \le p \le 1/4.$$

Therefore,

$$\min_{a\le\vartheta\le b} \psi(p(\vartheta),\delta) = \psi(p_m,\delta) = \frac{(p_m^\delta - \lambda)^2}{p_m^{2\delta-1}} \triangleq g(\delta).$$

Computing the derivative of the function $g(\delta)$, we obtain by virtue of (2.29)

$$g^{'}\,(\delta) = 2\lambda\, p_m^{1-\delta} \ln p_m\,(1 - \lambda\, p_m^{-\delta}) < 0 \quad \forall\delta\ .$$

Thus, the maximum of $g(\delta)$ is achieved for $\delta = 0$. Corollary **2.4.3** is proved.

Note that the above considerations imply the relation

$$\min_{0\le\delta\le1}\ \max_{a\le\vartheta\le b} \beta(N) \sim \exp\left(-\frac{Nh^2}{2\sigma^2}\,(1-\lambda)^2\, p_m\right).$$

Now, consider the error of estimation $\gamma(\epsilon, N)$. Since $W^0(\vartheta+u) - W^0(\vartheta) \overset{d}{=} W^0(u)$, then, for $u \to +0$, we have using (2.25)

$$\eta_N(u) \triangleq |y_N(\vartheta+u,\delta)| - |y_N(\vartheta,\delta)| = -|h|\,\rho_1(\delta,\vartheta)u +$$

$$+o(u) + \frac{\sigma}{\sqrt{N}}\left(\frac{W^0(u)}{p^{1-\delta}(\vartheta)} + \left(\frac{(2\vartheta-1)(1-\delta)}{p^{2-\delta}(\vartheta)}u + o(u)\right) W^0(\vartheta+u)\right), \quad (2.32)$$

where

$$\rho_1(\delta,\vartheta) = (p(\vartheta))^{\delta-1}\left(\delta\vartheta + (1-\delta)(1-\vartheta)\right).$$

Making a change of time $u \to z/(z+1)$ and taking into account that $W^0(z/(z+1)) \overset{d}{=} W(z)/(z+1)$, we obtain from (2.32)

$$\eta_N(z) = -|h|\,\rho_1(\delta,\vartheta)z + o(z) + \frac{1}{\sqrt{N}}\left(\frac{\sigma}{p^{1-\delta}}W(z) + z\,\lambda(z)\right), \quad (2.33)$$

where $\lambda(z)$ is the function bounded with probability 1 on any bounded interval.

Again, making the time transformation $z \to s/N$ and taking into account that $W(s/N) \overset{d}{=} N^{-1/2}W(s)$, we obtain from (2.33):

$$\eta_N(s) = \frac{1}{N}\left(-|h|\rho_1(\delta,\vartheta)s + \frac{\sigma}{p^{1-\delta}}W(s) + Ng(s)\cdot o(N^{-1})\right), \quad (2.34)$$

where $g(s)$ is the function bounded almost surely on every bounded interval.

By virtue of the same considerations for $u \le 0$, we conclude that on every bounded interval, the process $N\,\eta_N(\cdot)$ weakly converges to the process $\xi(\cdot)$

$$\xi(t) = \begin{cases} \xi^+(t) \triangleq -|h|\rho_1(\delta,\vartheta)t + \sigma p^{\delta-1}W(t) & \text{if } t \ge 0 \\ \xi^-(t) \triangleq |h|\rho_2(\delta,\vartheta)t + \sigma p^{\delta-1}\tilde{W}(t) & \text{if } t \le 0, \end{cases} \quad (2.35)$$

where $W, \tilde{W}$ are independent standard Wiener processes, $\rho_2(\delta,\vartheta) = p^{\delta-1} - \rho_1(\delta,\vartheta)$.

Let

$$\xi_1 = \max_{t\ge 0}\left(-\frac{|h|\rho_1(\delta,\vartheta)}{\sigma}p^{1-\delta}t + W(t)\right),$$

$$\xi_2 = \max_{t\le 0}\left(\frac{|h|\rho_2(\delta,\vartheta)}{\sigma}p^{1-\delta}t + \tilde{W}(t)\right).$$

Here, we use the following relationship for the process $u(t) = bt + W(t)$, $-\infty < b < \infty$ in the segment $[0,T]$, $T > 0$ (Robbins and Siegmund (1970)):

$$\mathbf{P}\{\max_{0\le t\le T} y(t) \le x\} = \Phi\left(\frac{-bT+x}{T^{-1/2}}\right) - \Phi\left(\frac{-bT-x}{T^{-1/2}}\right)\exp(2bx), \quad x \ge 0, \quad (2.36)$$

where Φ is the standard normal distribution function.

From (2.36), we obtain

$$\mathbf{P}\{\xi_i < x\} = 1 - \exp\left(-\frac{2|h|\rho_i(\delta,\vartheta)}{\sigma}\, p^{1-\delta}\, x\right), \; i = 1, 2 \tag{2.37}$$

Since ξ_1 and ξ_2 are independent r.v.'s, from (2.37), we have

$$\mathbf{P}\{\xi_1 < \xi_2\} = \int_0^\infty \mathbf{P}(\xi_1 < x)\, dP(\xi_2 < x) =$$

$$= 2\tfrac{|h|\rho_2(\delta,\vartheta)}{\sigma} p^{1-\delta} \int_0^\infty [1-$$

$$-\exp\left(-2\tfrac{|h|\rho_1(\delta,\vartheta)}{\sigma} p^{1-\delta} x\right)\Big] \cdot \exp\left(-2\tfrac{|h|\rho_2(\delta,\vartheta)}{\sigma} p^{1-\delta} x\right)\, dx =$$

$$\delta\vartheta + (1-\delta)(1-\vartheta) \triangleq \mu(\delta,\vartheta).$$

By analogy,

$$\mathbf{P}\{\xi_2 < \xi_1\} = \delta(1-\vartheta) + (1-\delta)\vartheta = 1 - \mu(\delta,\vartheta)$$

and, therefore,

$$\mathbf{P}\{\xi_1 = \xi_2\} = 0. \tag{2.38}$$

The process $\eta(t) = \alpha t - W(t)$, $\alpha > 0$ achieves its maximum at a single point almost surely on each finite interval. Therefore, from (2.38), we conclude that the process $\xi(t)$ also achieves its maximum at a single point (almost surely) on each finite interval.

Let $x(t)$ be a continuous function, with a single point of maximum $t^\star$ on a compact. Suppose that $x_n(t) \to x(t)$ in the metric of the space of continuous functions. If $\mathcal{M}_n$ is the set of maximum points of the function $x(t)$ on the compact, then from Corollary 1.6.1 of Lemma 1.4.6, we obtain:

$$\sup_{t \in \mathcal{M}_n} |t - t^\star| \to 0 \quad \text{as } n \to \infty. \tag{2.39}$$

Therefore the random variable $N(\hat{\vartheta}_N - \vartheta)$ weakly converges to a single (almost surely) point of maximum τ of the process $\xi(t)$ as $N \to \infty$.

Let $\tau^+\,(\tau^-)$ be a moment of time when the maximum of $\xi^+\,(\xi^-)$ is achieved, and suppose that $F^+\,(F^-)$ is the distribution function of this moment.

Now, let us use the following relationship (see, e.g., Khakhubia, 1986). Suppose τ is the point of maximum of the process $z(t) = bt + W(t)$, $-\infty < b < \infty$ on the segment $[0, T]$ and $F_T(t) = \mathbf{P}\{\tau \le t\}$. Then,

$$\begin{aligned}\lim_{T\to\infty} F_T(t) &== 2\Phi\,(b\sqrt{t}) - 1 - 2b^2\, t\, \Phi(-b\sqrt{t}) + \\ &+ 2b\sqrt{\tfrac{t}{2\pi}}\exp(-\tfrac{b^2 t}{2}) \triangleq G(b,t).\end{aligned} \tag{2.40}$$

From (2.40), we obtain

$$F^+(t) = \begin{cases} G(\frac{|h|\rho_1}{\sigma}\, p^{1-\delta}, t) & \text{if } t \ge 0 \\ 0 & \text{if } t < 0 \end{cases}$$

$$F^-(t) = \begin{cases} 1 & \text{if } t > 0 \\ 1 - G(\frac{|h|\rho_2}{\sigma}\, p^{1-\delta}, t) & \text{if } t \leq 0. \end{cases}$$

Thus,

$$\mathbf{P}\{\tau < t\} = (1 - \mu(\delta, \vartheta))\, F^+(t) + \mu(\delta, \vartheta)\, F^-(t)\;,$$

i.e., the following theorem holds true.

Theorem 2.4.4. *For any* $z > 0$

$$\begin{aligned} &\lim_{N\to\infty} \mathbf{P}\{N\,|\hat{\vartheta}_N - \vartheta| > z\} \\ &= 1 - (1 - \mu(\delta, \vartheta)) \cdot G(\frac{|h|\rho_1(\delta, \vartheta)}{\sigma}\, p^{1-\delta}, z) - \\ &-\mu(\delta, \vartheta)\, G\left(\frac{|h|\rho_2(\delta, \vartheta)}{\sigma} p^{1-\delta}, z\right). \end{aligned} \tag{2.41}$$

Corollary 2.4.4. *As* $N \to \infty$,

$$\begin{aligned} \gamma(\epsilon, N) \sim 2\epsilon N\, \tfrac{h^2\, p^{2-2\delta}}{\sigma^2}\, (1 - \mu(\delta, \vartheta))\, \rho_1^2\,(\delta, \vartheta) \cdot \exp\left(-\tfrac{h^2\, \rho_1^2\,(\delta,\vartheta)}{2\sigma^2}\, p^{2-2\delta}\, \epsilon N\right) + \\ +\mu(\delta, \vartheta)\rho_2^2(\delta, \vartheta) \cdot \exp\left(-\tfrac{h^2\, \rho_2^2(\delta,\vartheta)}{2\sigma^2}\, p^{2-2\delta}\, \epsilon N\right). \end{aligned} \tag{2.42}$$

From (2.42), we obtain as $N \to \infty$

$$\begin{aligned} \frac{\ln|\gamma(\epsilon, N)|}{N} \sim \frac{h^2\, p^{2-2\delta}}{2\sigma^2}\, \epsilon\, \min(\rho_1^2\,(\delta, \vartheta), \rho_2^2(\delta, \vartheta)) = \\ = \tfrac{h^2\, \epsilon}{2\sigma^2}\, \min\left((\delta\vartheta + (1-\delta)(1-\vartheta))^2,\, (\delta(1-\vartheta) + (1-\delta)\vartheta)^2\right) \triangleq R(\delta, \vartheta). \end{aligned} \tag{2.43}$$

It easily follows from (2.43) that

$$\max_{0\leq\delta\leq 1}\; \min_{a\leq\vartheta\leq b}\; R(\delta, \vartheta) = \frac{h^2\, \epsilon}{8\sigma^2}, \tag{2.44}$$

and the *asymptotically optimal method is the method with* $\delta = 1/2$.

Note that for $\vartheta = 1/2$, all methods of this family are asymptotically equivalent, and for $\delta = 0$ and $\delta = 1$,

$$\ln \frac{|\gamma(\epsilon, N)|}{N} \sim \frac{h^2\, \epsilon}{2\sigma^2}\, \min\left(\vartheta^2, (1-\vartheta)^2\right),$$

i.e., for these methods, the ratio $\gamma(\epsilon, N)/(\gamma(\epsilon, N))_{\min}$ is asymptotically equal to

$$\exp \frac{h^2\, \epsilon N}{2\sigma^2}\, \left(1/4 - \min(\vartheta^2, (1-\vartheta)^2)\right).$$

In conclusion, we give here results of the asymptotical analysis of Statistic (2.6): the best values of the prameter δ in this family of statistics.

Asymptotically best value in the sense of minimum of false alarm probability (type-1 error) is $\delta = 1$.

Asymptotically best value in the sense of minimum of type-2 error probability (false tranquility) is $\delta = 0$.

Asymptotically best value in the sense of minimum of the error of estimation of the parameter ϑ is $\delta = 1/2$.

2.5 Three-Step Procedure of Change-Point Estimation

In this section, we consider the main ideas of the algorithm of change-point estimation in an observed (diagnostic) random sequence. This algorithm is for long years used for proceeding of the real data in various applications (see, e.g., Brodsky and Darkhovsky, 2000).

Remember, that according to the first idea we suppose that the diagnostic sequence $X^N = \{x^N(n)\}$ is written as follows:

$$x^N(n) = \varphi(\vartheta, n/N) + \xi^N(n), \; n = 1, \ldots, N, \tag{2.45}$$

where $\xi^N = \{\xi^N(n)\}_{n=1}^{\infty}$ is the "glued" sequence (according to the scheme of section 2.1, i.e., $\Xi \stackrel{\text{def}}{=} \{\Xi^{(1)}, \ldots, \Xi^{(k+1)}\}$, $\Xi^{(i)} = \{\xi^{(i)}(n)\}_{n=1}^{\infty}$ п»„ $\xi^N(n) = \xi^{(i)}(n)$ if $[\vartheta_{i-1}N] \le n < [\vartheta_i N]$, $i = 1, \ldots, k+1$), $\mathbf{E}_\theta \xi^N(n) \equiv 0$, and the function $\varphi(\vartheta, t)$, $t \in [0, 1]$ is written as follows:

$$\varphi(\vartheta, t) = \sum_{i=1}^{k} a_i \mathbb{I}(\vartheta_{i-1} \le t < \vartheta_i) + a_{k+1}\mathbb{I}(\vartheta_k \le t \le 1),$$

with $|a_i - a_{i+1}| \stackrel{\text{def}}{=} h_i \ge h > 0$.

First stage of the algorithm: obtaining a preliminary list of change-points.

1. Find the global maximum of the statistic $|Y_N(n, 1)|$ and fix the point of maximum n_1.
2. Divide the sample into two parts (left and right from n_1), and find the points of global maximum of $|Y_N(n, 1)|$ at each two sub-samples. Therefore, we get points n_2, n_3.
3. Continue that division process up to the subsamples become rather small. In such a way we have the list of *preliminary change-points* $n_1, n_2, \ldots, n_s$.

We use at this stage the statistic $Y_N(n, 1)$, since for $\delta = 1$, we get the asymptotically best method in the sense of false alarm probability (see Section 2.4).

For an arbitrary segment $[a,b] \subseteq [0,1]$, put

$$\psi(a,b) = (b-a)^{-1} \int_a^b \varphi(\vartheta,t)dt.$$

Suppose that the following condition

$$\varphi(t,\vartheta) - \psi(a,b) \neq 0 \tag{2.46}$$

is satisfied on every nonzero segment $[a,b]$ at all stages of the procedure.

(2.46) is, evidently, a *general situation* condition. Any violation of this condition is eliminated by a "small change" of initial data. Thus, we can suppose that (2.46) is always satisfied in real situations.

From (2.45) and (2.46), it follows that the mathematical expectation of the statistic $|Y_N(n,1)|$ is a piecewise-linear function *without horizontal segments.* From here, it follows that the point of the global maximum of this statistic is the consistent estimator for one of the points ϑ.

Second stage of the procedure: rejection of points from the preliminary list

1. Take the first point from the preliminary list and consider the sub-sample around this point.
2. Calculate the threshold C_1 for chosen sub-sample using the relation

   $$\sqrt{N} \cdot Y([Nt],1) \to \sigma W^\circ(t), \qquad N \to \infty, \tag{2.47}$$

   and given value of the false alarm probability, where W° is the standard Brownian bridge, $t \in [0,1]$, and σ is the empirical estimate.

   The threshold C_1 is computed from the well-known d.f. of the maximum absolute value of the Brownian bridge and the limit relationship (2.47) (see formulas (2.13) and (2.16) for $\delta = 1$; in these formulas, N is the size of the corresponding subsample). For estimation of the parameter σ, the basis relationships from Lemma 2.2.2 are used with empirical estimates of the correlation function for the corresponding subsample.
3. If $\max |Y_N(n,1)| > C_1$, then point n_1 remains in the list of *confirmed change-points*; if not — we delete this point from the list. Here maximum is taken over the subsample.
4. Continue the previous procedure for all preliminary change-points. For every point $n_i, i = 2, ...$ from the preliminary list of change-points, the corresponding threshold $C_i, i = 2, ...$ is computed using the same relationships as the precious item.

As a result of the second stage, we get the *list of confirmed change-points* (LCCP).

Third stage of the procedure: refining the change-points and calculation of confidence intervals

- Take point n_1 from the LCCP and consider the subsample around this point.
- Find $\max|Y_N(n,0)| = |Y_N(\tilde{n}_1,0)| = B_1$ (the maximum over the subsample). The point $\tilde{n}_1$ is called the *refined change-point.*

  Let us explain this. From (2.12), it follows that the limit value of the mathematical expectation of the statistic $|Y_N(n,0)|$ in the neighborhood of a unique change-point n_1 (at the second stage of the algorithm, we consider a unique change-point and its neighborhood) has the following form,

$$m(t) = \begin{cases} h_1 \frac{(1-\vartheta)}{1-t} & \text{if } t \le \vartheta \\ h_1 \vartheta \frac{1}{t} & \text{if } t \ge \vartheta, \end{cases} \tag{2.48}$$

  where h_1 is the absolute value of a change in the mathematical expectation of the sequence $x^N(n)$ at the change-point n_1. From (2.48), it follows that at the change-point $m(\vartheta) = h_1$. Besides, from Section 2.4, it follows that the statistic $Y_N(n,0)$ is the asymptotically best choice in the sense of probability of false tranquility. Therefore, the use of the statistic $Y_N(n,0)$ at this stage helps to refine the coordinate of a change-point and to estimate the value h_1.

- At this stage, we compute the confidence interval for a refined change-point. As it is demonstrated in Section 2.4, the asymptotically best statistic for this problem is $Y_N(n,1/2)$. For computation of the confidence interval, we need first to center this statistic. Taking into account the estimate $h_1 \approx B_1$ found at the previous step, we compute the function

$$g(n) = \begin{cases} \sqrt{\frac{n}{N-n}}(1-\tilde{n}_1/N)B_1, & \text{if} n \le \tilde{n}_1 \\ \sqrt{\frac{N-n}{n}}\frac{\tilde{n}_1}{N}B_1, & \text{if} n > \tilde{n}_1 \end{cases} \tag{2.49}$$

  and begin to center the statistic.

- Calculate sequence $V(n) \stackrel{\text{def}}{=} |Y_N(n,1/2)| - g(n)$ over the subsample.
- Now the confidence interval around $\tilde{n}$ is computed (for a given confidence probability) from the limit relationship for the process $V(n)$ (see (2.40 – 2.42)).
- Repeat previous items for each point from the LCCP.

The procedure of change-point estimation is finished now.

Let us formulate the limit theorem characterizing the quality of change-point estimation by the proposed method.

Put

$$\Theta_\delta = \{\vartheta \in \mathbb{R}^k : 0 \equiv \vartheta_0 < \delta \leq \vartheta_1 < \vartheta_2 < ..$$
$$< \vartheta_k \leq (1-\delta) < \vartheta_{k+1} \equiv 1, \min_i |\vartheta_{i+1} - \vartheta_i| \geq \delta > 0\}.$$

Suppose for the "glued" vector-valued sequence Ξ in Scheme (2.45) the following conditions are satisfied:

i) for each component of this sequence there exists the limit

$$\sigma_i^2 \stackrel{\text{def}}{=} \lim_{n\to\infty} \mathbf{E}_\vartheta \left(\sum_{k=1}^n \xi^{(i)}(k) \right)^2 ;$$

ii) for the vector-valued sequence Ξ, the unified Cramer and ρ-mixing conditions are satisfied; and

iii) $\sum\limits_{k=1}^{\infty} \max\limits_{(i,j)} \sqrt{\rho_{ij}(k)} < \infty$,

where $\rho_{ij}(\cdot)$ be a ρ-mixing coefficient between i-th and j-th components of the vector sequence $\{\Xi\}$, and п»„ $\rho_{ii}(\cdot) \stackrel{\text{def}}{=} \rho_i(\cdot)$ is the coefficient of - ρ-mixing for the sequence $\{\xi^{(i)}(n)\}$).

Theorem 2.5.1. *Suppose conditions (i)–(iii) are satisfied. Then*

a) the estimate $\hat{\vartheta}_N \to \vartheta$ as $N \to \infty$ $\mathbf{P}_\vartheta$-a.s., and for any (sufficiently small) $\epsilon > 0$ there exist $N(\epsilon), A(\epsilon) > 0, B(\epsilon) > 0$, such that as $N > N(\epsilon)$, the following inequality holds:

$$\sup_{\vartheta\in\Theta_\delta} \mathbf{P}_\vartheta\{\|\hat{\vartheta}_N - \vartheta\| > \epsilon\} \leq A(\epsilon)\exp(-B(\epsilon)N). \tag{2.50}$$

b) process $N^{1/2}(Y_N([Nt],1) - \mathbf{E}_\vartheta Y_N([Nt],1))$ weakly converges in the space $D[0,1]$ to the process $\tilde{\sigma}(t)W^0(t)$, where

$$\tilde{\sigma}^2(t) = t^{-1} \int_0^t \sigma^2(s)ds, \ \sigma^2(s) = \sum_{i=1}^{k+1} \sigma_i^2 \mathbb{I}(\vartheta_{i-1} \leq s < \vartheta_i).$$

Now, let us give here the sketch of the proof of this theorem.

Item b) follows from Theorem 1.3.2, if we take into account that the moment condition of this theorem follows from the unified Cramer's condition.

Considering item a), we first analyze the situation when the vector ϑ is one-dimensional, i.e., there exists a unique change-point ϑ, $0 < a \leq \vartheta \leq b < 1$.

Consider the random process $y_N(t) \stackrel{\text{def}}{=} Y_N([Nt],0)$ in the Skorokhod space $D[a,b]$ (remember that this process is used for refining change-points). The function $m_N \stackrel{\text{def}}{=} \mathbf{E}_\vartheta Y_N([Nt],0)$ uniformly converges in $[a,b]$ (with the rate $O(1/N)$) to the function

$$f(t) = \begin{cases} h\frac{(1-\vartheta)}{1-t} & \text{if } t \leq \vartheta \\ h\frac{\vartheta}{t} & \text{if } t \geq \vartheta, \end{cases} \tag{2.51}$$

where $h = (a_1 - a_2)$ is the size of a change (for this case) of the function $\varphi(\vartheta, t)$ (see (2.5.1) at the point ϑ.

Denote $\eta_N(t) \stackrel{\text{def}}{=} y_N(t) - \mathbf{E}_\vartheta y_N(t)$. Then, from (2.5.7), it follows that

$$y_N(t) = f(t) + \eta_N(t) + O(1/N). \tag{2.52}$$

From (2.51), it follows that the unique maximum of the function $|f(t)|$ is attained at the point $t = \vartheta$ and

$$|f(\vartheta)| - |f(t)| \geq |h||t - \vartheta|. \tag{2.53}$$

Trajectories of the process $y_N(t)$ are piecewise constant and for each N have a finite number of jumps. Therefore, the set of maximum points of the process $|y_N(t)|$ is not empty. By definition, the estimate $\hat{\vartheta}_N$ of the parameter ϑ is an arbitrary point of the set $\mathfrak{M}_N \stackrel{\text{def}}{=} \arg\max\{|y_N(t)|, t \in [a, b]\}$.

From Remark 1.6.3 to Lemma 1.6.2, and from (2.52), it follows that

$$\text{Dist}(\vartheta, \mathfrak{M}_N) \leq \frac{2}{|h|}\|\eta_N(t)\| + O(1/N). \tag{2.54}$$

Now, remark that the process η_N has the following form:

$$\begin{aligned} \eta_N(t) &= ([Nt])^{-1} \textstyle\sum_{k=1}^{[Nt]} \xi^N(k) - (N - [Nt])^{-1} \displaystyle\sum_{k=[Nt]+1}^{N} \xi^N(k) = \\ &= N(N - [Nt])^{-1} \left(([Nt]^{-1} \sum_{k=1}^{[Nt]} \xi^N(k) - N^{-1} \sum_{k=1}^{[N]} \xi^N(k) \right) \end{aligned}$$

Therefore,

$$\|\eta_N(t)\| \leq 2/b \|([Nt])^{-1} \sum_{k=1}^{[Nt]} \xi^N(k)\|. \tag{2.55}$$

But for the sums of type (2.55), the exponential upper estimate is satisfied. Therefore, from (2.54), we obtain the exponential rate of convergence of the estimator of the parameter ϑ to its true value.

In its turn, the exponential rate of convergence guaranteers a.s. convergence of the estimator $\mathbf{P}_\vartheta$ to its true value. This fact follows from here: Let $\mathbf{P}\{|\xi_n - \xi| \geq \epsilon\} \leq A(\epsilon)\exp(-B(\epsilon)n)$, where $A(\epsilon), B(\epsilon)$ are some positive functions. Then, ξ_n tends to ξ a.s.

In fact,

$$\begin{aligned} &\mathbf{P}\{\sup_{k \geq 0} |\xi_{n+k} - \xi_n| \geq \epsilon\} \leq 2\mathbf{P}\{\sup_{k \geq n} |\xi_k - \xi| \geq \epsilon/2\} \leq \\ &\leq \sum_{k=n}^{\infty} \mathbf{P}\{\xi_k - \xi| \geq \epsilon/2\} \leq A(\epsilon/2) \textstyle\sum_{k=n}^{\infty} \exp(-B(\epsilon/2)k) \to 0, \end{aligned}$$

and this is a necessary and sufficient condition for a.s. convergence.

Now, let us consider the general case. From the algorithm's description, it follows that the process of splitting the whole sample into subsamples is carried out in a sequential way, and, as a result of this process we arrive to the situation (with the probability exponentially close to unit) when it is at maximum one change-point in each subsample. Therefore, for a fixed coordinates of change-points, we obtain the strongly consistent estimator of the vector ϑ.

In order to obtain a uniform (w.r.t. the set Θ_δ) and consistent estimator, we need to find the worst change-point estimate for one change-point. Suppose Φ denotes the set of functions $\varphi(\vartheta, t)$, which are piecewise constant, and the interval of their constancy is no less than δ, and the module of jumps is no less than h. Then the "worst" function is the realisator of the infimum (and this infimum is attainable):

$$\inf_{\varphi(\cdot)\in\Phi} \max_{t\in[0,1]} \varphi(\vartheta, t) \stackrel{\text{def}}{=} A.$$

It can be proved that $A = 3^{-1}h\delta$. Therefore, we estimate the rate of convergence of $\hat{\vartheta}_N$ to ϑ. In analogy, we can demonstrate that this estimate is uniform, i.e., to prove the item a) of the theorem.

2.6 A Priori Lower Estimates of Performance Efficiency in Retrospective Disorder Problems

In this section, we obtain the a priori lower bounds for performance efficiency in retrospective change-point detection problems. These bounds help in conclusion about the asymptotic optimality of the basis method of change-point detection considered in Section 2.5.

2.6.1 One Change-Point

Consider the problem at a maximum of one change-point (i.e., the parameter $0 < \vartheta < 1$ is one-dimensional; see Subsection 2.1.3). We suppose that $x^N(n) \equiv \xi^N(n)$ (i.e., the function $\varphi(\vartheta, t) \equiv 0$), the sequence ξ^N is "glued" from two independent, stationary, and strongly ergodic (Borovkov, 1998) k-step Markov chains with stationary densities and transitional density functions w.r.t. a certain σ-finite measure (for the sake of simplicity, we suppose below that these are densities w.r.t. the Lebesgue measure), $f_1(y), f_2(y)$, $y \in \mathbb{R}^k$, and $\varphi_1(x|y)$, $\varphi_2(x|y)$.

Besides, we suppose that condition *iii*) is satisfied. This condition guarantees the summability of correlation functions and, consequently, existence of limits *i*) in the same theorem.

For detection of a change-point, we need to distinguish d.f. of observations before and after this change-point. A natural (and probably minimal) condition of such distinguishing can be formulated as follows:

There exists the set $M \subseteq \mathbb{R}^k$ of a positive measure, such that

$$\int_M f_1(x)dx \neq \int_M f_2(x)dx. \tag{2.56}$$

Below, we suppose that condition (2.56) holds.

Besides, we assume existence of the following integrals:

$$\begin{aligned} &\int dx \int_{\mathbb{R}^k} \Big| \left(\ln \frac{\varphi_1(x|y)}{\varphi_2(x|y)} \right) \varphi_1(x|y) f_1(y) \Big| dy < \infty \\ &\int dx \int_{\mathbb{R}^k} \Big| \left(\ln \frac{\varphi_2(x|y)}{\varphi_1(x|y)} \right) \varphi_2(x|y) f_2(y) \Big| dy < \infty. \end{aligned} \tag{2.57}$$

Condition (2.56) guarantees that the set of consistent estimators of the parameter ϑ is not empty. In fact, consider the diagnostic sequence $Y^N = \{y^N(n)\}$,

$$y^N(n) = \mathbb{I}(x^N(n) \in M),$$

where M is the set from condition (2.56).

Then,

$$\mathbf{E}_\vartheta y^N(n) = \begin{cases} A \triangleq \int_M f_1(x)dx & \text{if } n \leq [\vartheta N] \\ B \triangleq \int_M f_2(x)dx & \text{if } n > [\vartheta N], \end{cases}$$

and in virtue of Condition (2.56), $A \neq B$. Therefore, the sequence Y^N satisfies all conditions of our problem from Subsection (2.5) (Cramer's condition is satisfied automatically, because this sequence takes only two values, and ψ-mixing condition is required only for strong consistency), and from **Theorem 2.5.1**, it follows that the estimator of the parameter ϑ obtained from the method of 2.5 is consistent.

Now, we introduce the following objects:

$$T_N(\Delta) : \mathbb{R}^N \to \Delta \subset \mathbb{R}^1,$$

the Borel function on $\mathbb{R}^N$ with values in Δ;

$$\mathfrak{M}_N(\Delta) = \{T_N(\Delta)\},$$

the set of all Borel functions $T_N(\Delta)$;

$$\mathfrak{M}(\Delta) = \{T : T = \{T_N(\Delta)\}_{N=1}^\infty\},$$

the set of all sequences of elements $T_N(\Delta) \in \mathfrak{M}_N(\Delta)\}$; and

$$\tilde{\mathfrak{M}}(\Delta) = \{T(\Delta) \in \mathfrak{M}(\Delta) : \lim_N \mathbf{P}_\vartheta(|T_N - \vartheta| > \epsilon) = 0 \quad \forall \vartheta \in \Delta, \forall \epsilon > 0\},$$

the set of all consistent estimates of the parameter $\vartheta \in \Delta$.

Theorem 2.6.1. *Let $0 < \epsilon < 1$ be fixed. Then, under above assumptions,*

$$\liminf_N N^{-1} \ln \inf_{\vartheta_N \in \mathfrak{M}_N(0,1)} \sup_{0<\vartheta<1} \mathbf{P}_\vartheta\{|\vartheta_N - \vartheta| > \epsilon\} \geq -\epsilon \min(\rho_1, \rho_2), \tag{2.58}$$

where

$$\begin{aligned} \rho_1 &= \int dx \int_{\mathbb{R}^k} \left(\ln \frac{\varphi_1(x|y)}{\varphi_2(x|y)}\right) \varphi_1(x|y) f_1(y) dy, \\ \rho_2 &= \int dx \int_{\mathbb{R}^k} \left(\ln \frac{\varphi_2(x|y)}{\varphi_1(x|y)}\right) \varphi_2(x|y) f_2(y) dy. \end{aligned} \tag{2.59}$$

Proof.
Fix the numbers $0 < a < b < 1$ and denote

$$A(a,b) = \liminf_N N^{-1} \ln \inf_{\vartheta_N \in \mathfrak{M}_N([a,b])} \sup_{a \leq \vartheta \leq b} \mathbf{P}_\vartheta\{|\vartheta_N - \vartheta| > \epsilon\}.$$

Taking into account the definition of the sets $\mathfrak{M}([a,b])$ п»„ $\tilde{\mathfrak{M}}([a,b])$, we have (the symbol $\{\vartheta_N\}$ below denotes the sequence (by N) of estimates of the parameter ϑ):

$$A(a,b) \geq \inf_{\{\vartheta_N\} \in \mathfrak{M}([a,b])} \liminf_N N^{-1} \ln \sup_{a \leq \vartheta \leq b} \mathbf{P}_\vartheta\{|\vartheta_N - \vartheta| > \epsilon\} =$$

$$= \inf_{\{\vartheta_N\} \in \tilde{\mathfrak{M}}([a,b])} \liminf_N N^{-1} \ln \sup_{a \leq \vartheta \leq b} \mathbf{P}_\vartheta\{|\vartheta_N - \vartheta| > \epsilon\} \equiv B(a,b). \tag{2.60}$$

Equality in (2.60) is explained by the fact that nonconsistent estimates $\mathbf{P}_\vartheta\{|\vartheta_N - \vartheta| > \epsilon\}$ do not converge to zero, and therefore, the minimum for such estimates cannot be less than for consistent estimates.

Suppose ϑ_N is a certain consistent estimate of the parameter $\vartheta \in [a,b]$, and $\delta > 0$ and $\epsilon\prime > \epsilon > 0$ are certain fixed numbers. Define the following random value:

$$\lambda_N = \lambda_N\left(x^N(1), \ldots, x^N(N)\right) = \mathbb{I}\{|\vartheta_N - \vartheta| > \epsilon\}.$$

For any $d > 0$, we have:

$$\mathbf{P}_\vartheta\{|\vartheta_N - \vartheta| > \epsilon\} =$$

$$= \mathbf{E}_\vartheta \lambda_N \geq \mathbf{E}_\vartheta \left(\lambda_N \mathbb{I}(f(X^N, \vartheta + \epsilon\prime)/f(X^N, \vartheta) < e^d)\right),$$

where $f(X^N, \vartheta)$ is the likelihood function of the sample X^N.

Since the sequences before and after the change-point are independent and k is the finite number (remember that we consider the case of k-step Markov sequences), subsequent considerations are the same as in Theorem 1.5.1. As a result, we obtain the required lower estimate for $B(a,b)$.

Remark 2.6.1. The estimate that furnishes the lower limit in (2.58) is called the *asymptotically minimax.* We do not know whether the lower limit obtained in **Theorem 2.6.1** is attainable. However, we know precisely that the order of this estimate is true. It follows from the work of Korostelev (1997) where for the Gaussian case and the continuous time is demonstrated that the left side of (2.58) is equal to $(-1/4\epsilon(\nu)^2)$, where ν is the signal/noise ratio. At the same time, from Theorem 2.6.1, we obtain that this left side is greater or equal to $(-1/2\epsilon(\nu)^2)$.

However, in the same paper it was remarked that this asymptotically minimax estimate is unknown and it can be biased. Therefore the lower limit (2.58) determines (with the accuracy of a constant) the order of the asymptotically minimax estimate in one change-point problem.

2.6.2 Multiple Change-Point

Consider the problem of a multiple change-point (i.e., the parameter $\vartheta \in \mathbb{R}^k$, $k > 1$, $0 \equiv \vartheta_0 < \vartheta_1 < \vartheta_2 < \ldots < \vartheta_k < \vartheta_{k+1} \equiv 1$).

We suppose that $x^N(n) \equiv \xi^N(n)$ (i.e., the function $\varphi(\vartheta, t) \equiv 0$), the sequence ξ^N is "glued" from $(k+1)$ independent, stationary, and strongly ergodic m-step Markov chains with stationary densities and transition d.f. w.r.t. a certain σ-finite measure (below we assume that these are d.f. w.r.t. the Lebesgue measure), respectively, and $f_1(y), f_2(y), \ldots, f_{(k+1)}(y) \in \mathbb{R}^m$ п»„ $\varphi_1(x|y)$, $\varphi_2(x|y), \ldots, \varphi_{(k+1)}(x|y)$.

As before, we suppose that condition *iii*) of Theorem 2.5.1 is satisfied.

We suppose that the following conditions are satisfied:

For all $i = 1, \ldots, k$

a) there exist the sets of a positive measure $M_i \subseteq \mathbb{R}^m$, such that

$$\int_{M_i} f_i(x)dx \neq \int_{M_i} f_{(i+1)}(x)dx, \text{and} \tag{2.61}$$

b)

$$\begin{aligned}
&\int dx \int_{\mathbb{R}^m} \Big| \left(\ln \frac{\varphi_i(x|y)}{\varphi_{(i+1)}(x|y)} \right) \varphi_i(x|y) f_i(y) \Big| dy < \infty \\
&\int dx \int_{\mathbb{R}^m} \Big| \left(\ln \frac{\varphi_{(i+1)}(x|y)}{\varphi_i(x|y)} \right) \varphi_{(i+1)}(x|y) f_{(i+1)}(y) \Big| dy < \infty.
\end{aligned} \tag{2.62}$$

As before. Condition (2.61) guarantees that set of consistent estimators of the parameter ϑ is not empty. For the proof of this fact it suffices to consider the diagnostic sequence $Y^N = \{y^N(n)\}$,

$$y^N(n) = \mathbb{I}(x^N(n) \in M),$$

where $M = \bigcup_{i=1}^k M_i$.

Abrupt changes in the mathematical expectation at the instants $[N\vartheta_i]$, $i = 1, \ldots, k$ occur in such a diagnostic sequence. Therefore, the main algorithm applied to this sequence gives consistent estimates of the parameter ϑ.

We consider the vector of parameters $\vartheta = (\vartheta_1, \dots, \vartheta_k)$ in the space $\mathbb{R}^k$, with the norm $\|\cdot\|_\infty$, п»„.п»„. $\|\vartheta\| = \max_i |\vartheta_i|$. Besides, if necessary, we provide the symbol of this norm with the upper index showing the space dimension, e.g., $\|\cdot\|^{(k)}$.

In the problem of a multiple change-point, we estimate the vector ϑ and the number of change-points k. Therefore, we define the following objects:

For any $s = 1, \dots, s^\star \equiv [1/\delta]$ ($\delta > 0$ is the minimal distance between $\vartheta_{(i+1)}$ and ϑ_i; we suppose that the value δ is a priori known).

Put

$$\mathfrak{D}_s = \{x \in \mathbb{R}^s : \delta \le x_i \le 1 - \delta,\ x_{i+1} - x_i \ge \delta, x_0 \equiv 0, x_{s+1} \equiv 1\}$$

$$\mathfrak{D}^\star = \bigcup_{i=1}^{s^\star} \mathfrak{D}_i, \mathfrak{D}^\star \subset \mathbb{R}^{s^\star} \equiv \mathbb{R}^\star.$$

The vector ϑ is any point of the set $\mathfrak{D}_k$ by construction.

$$T_N(D_s) : \mathbb{R}^N \to \mathfrak{D}_s \subset \mathbb{R}^s$$

is a Borel function on $\mathbb{R}^N$ with values in $\mathfrak{D}_s$;

$$\mathfrak{M}_N(\mathfrak{D}_s) = \{T_N(\mathfrak{D}_s)\},$$

the set of all Borel functions $T_N(\mathfrak{D}_s)$;

$$\mathfrak{M}(\mathfrak{D}_s) = \{T : T = \{T_N(\mathfrak{D}_s)\}_{N=1}^\infty\},$$

the set of all sequences of elements $T_N(\mathfrak{D}_s) \in \mathfrak{M}_N(\mathfrak{D}_s)$; and

$$\begin{gathered}\tilde{\mathfrak{M}}(\mathfrak{D}_k) = \{T(\mathfrak{D}_k) \in \mathfrak{M}(\mathfrak{D}_k) : \\ \lim_N \mathbf{P}_\vartheta(\|T_N - \vartheta\|^{(k)} > \epsilon) = 0 \quad \forall \vartheta \in \mathfrak{D}_k, \quad \forall \epsilon > 0\},\end{gathered}$$

the set of all consistent estimates of the parameter $\vartheta \in \mathfrak{D}_k$.

Suppose $x \in \mathbb{R}^p, y \in \mathbb{R}^q$, п»„ $m = \max(p, q)$. Define the inclusions

$$\mathrm{im}_x : \mathbb{R}^p \to \mathbb{R}^m, \quad \tilde{x} = \mathrm{im}_x\, x, \quad \mathrm{im}_y : \mathbb{R}^q \to \mathbb{R}^m, \quad \tilde{y} = \mathrm{im}_y\, y$$

(all lacking components are substituted by zeros in inclusions) and denote

$$dist(x, y) = \|\tilde{x} - \tilde{y}\|^{(m)}.$$

Theorem 2.6.2. *Let* $0 < \epsilon < \delta$. *Then,*

$$\begin{aligned}\liminf_N N^{-1} \ln \inf_{\vartheta_N \in \mathfrak{M}_N(\mathfrak{D}^\star)} \sup_{\vartheta \in \mathfrak{D}_k} &\Big\{\mathbf{P}_\vartheta(\vartheta_N \in \mathfrak{D}_k, \|\vartheta_N - \vartheta\|^{(k)} > \epsilon) \\ &+ \mathbf{P}_\vartheta(\vartheta_N \notin \mathfrak{D}_k)\Big\} \ge -\epsilon \min_{0 \le i \le k-1} \rho_{i,i+1},\end{aligned} \tag{2.63}$$

where

$$\rho_{i,i+1} = \min\left(\int dx \int_{\mathbb{R}^m} \left(\ln \frac{\varphi_i(x|y)}{\varphi_{(i+1)}(x|y)}\right)\varphi_i(x|y)f_1(y)dy,\right.$$
$$\left.\int dx \int_{\mathbb{R}^m} \left(\ln \frac{\varphi_{(i+1)}(x|y)}{\varphi_i(x|y)}\right)\varphi_{(i+1)}(x|y)f_i(y)dy\right). \tag{2.64}$$

Proof. We use the same idea as in the proof of Theorem 2.6.1 and, therefore, bound ourselves with the sketch of it.

For $\epsilon < \delta$, for any estimate $\vartheta_N \in \mathfrak{M}_N(\mathfrak{D}^*)$, and every $\vartheta \in \mathfrak{D}_k$, the following relationships hold:

$$(dist(\vartheta_N, \vartheta) > \epsilon) = \left(\vartheta_N \in \mathfrak{D}_k, \|\vartheta_N - \vartheta\|^{(k)} > \epsilon\right) \bigcup (\vartheta_N \notin \mathfrak{D}_k, dist(\vartheta_N, \vartheta) > \epsilon)$$
$$= \left(\vartheta_N \in \mathfrak{D}_k, \|\vartheta_N - \vartheta\|^{(k)} > \epsilon\right) \bigcup (\vartheta_N \notin \mathfrak{D}_k). \tag{2.65}$$

Here, we used the definition of *dist* and condition $(\vartheta_N \notin \mathfrak{D}_k)$. It follows from here that $(dist(\vartheta_N, \vartheta) > \delta)$, and for $\epsilon < \delta$, we obtain $dist(\vartheta_N, \vartheta) > \epsilon)$.

Therefore, the probability that is estimated in (2.63), i.e., the probability of the event that we estimate correctly the number of change-points (k) but $\mathbb{R}^k$ differs from the true vector of parameters for more than ϵ, or we do not estimate correctly the number of change-points (i.e., some of the points ϑ_i are missed or there are some "extra" points ϑ_i), coincides with the probability $(\mathbf{P}_\vartheta\, dist(\vartheta_N, \vartheta) > \epsilon)$. This value, namely, is estimated below.

First, we demonstrate nonemptiness of the set $\tilde{\mathfrak{M}}(\mathfrak{D}_k)$ of all consistent estimators of the parameter $\vartheta \in \mathfrak{D}_k$. It follows from Condition (2.61), as in the case of one change-point.

Second, remark that the infimum in the left hand of (2.63) can be found only in the set $\mathfrak{M}_N(\mathfrak{D}_k)$. This is demonstrated from considerations analogous to those presented in Brodsky, and Darkhovsky (2000).

On the set $\mathfrak{M}_N(\mathfrak{D}_k)$ by definition of *dist*, we have:

$$dist(\vartheta_N, \vartheta) = \|\vartheta_N - \vartheta\|^{(k)}.$$

Further, for any $i = 1, \ldots, k$, the following inclusion holds:

$$\{\|\vartheta_N - \vartheta\|^{(k)} > \epsilon, \vartheta_N \in \mathfrak{D}_k\} \supseteq \{|\vartheta_N^i - \vartheta^i| > \epsilon, \vartheta_N \in \mathfrak{D}_k\},$$

where ϑ_N^i is the i-th component of the vector ϑ_N.

Therefore,

$$\mathbf{P}_\vartheta\{\|\vartheta_N - \vartheta\|^{(k)} > \epsilon, \vartheta_N \in \mathfrak{D}_k\} \geq$$
$$\geq \max_{1 \leq i \leq k} \mathbf{P}_\vartheta\{|\vartheta_N^i - \vartheta^i| > \epsilon, \vartheta_N \in \mathfrak{D}_k\}.$$

But the problem of estimation of the value

$$\liminf_N N^{-1} \ln \inf_{\vartheta_N \in \mathfrak{M}_N(\mathfrak{D}_k)} \sup_{\vartheta \in \mathfrak{D}_k} \mathbf{P}_\vartheta\{|\vartheta_N^i -$$

$$-\vartheta^i| > \epsilon, \vartheta_N \in \mathfrak{D}_k\} \triangleq A_i$$

is analogous to those already considered in the proof of **Theorem 2.6.1** (estimation of $B(a,b)$ for the case of one change-point). Therefore, from the same considerations. we obtain (2.63).

From this theorem, as before the asymptotic minimax property, follows for the problem of estimation of the parameter ϑ.

2.7 Simulations

The nonparametric method proposed in this chapter is flexible enough to detect changes in the mean value of observations, in dispersion, and the correlation function of an observed sequence. Moreover, we can consider non-Gaussian random errors and multiple change-points in observations. Therefore, we are planning this section on simulations as follows. First, we formulate a model of observations that can exhibit different cases of possible changes:

1) Change-in-mean
2) Change-in-dispersion
3) Change-in-correlation

This model can be considered first for the Gaussian case:

$$x_{n+1} = \rho x_n + \sigma\sqrt{1-\rho^2}u_{n+1}, \qquad n = 1, \ldots, N-1,$$

where $u_{n+1} \sim \mathcal{N}(0,1) \to \mathcal{N}(a,1),\ a \neq 0.$

For the first change-in-mean model, the parameter a is varying, while parameters $\sigma > 0, -1 \leq \rho < 1$, and $0 < \theta < 1$ are fixed.

Second, we consider the case of non-Gaussian observations, and, third, the case of multiple change-points. The obtained results are given below.

1. Change-in-mean

For homogenous samples with $a = 0$ (without change-points), we put $\sigma = 1, \rho = 0$ and computed 95 percent and 99 percent quantiles of the proposed statistic (2.6) for different sample volumes. Here and below, each cell in the tables is filled with the average in 1000 independent trials of the Monte Carlo tests. The obtained results are given in Table 2.1.

Then we considered dependent observations: $\sigma = 1, \rho = 0.7$. The obtained results are presented in Tables 2.4–2.6.

2. Change-in-dispersion

For the change-in-dispersion model, the following diagnostic sequence was considered: $x_i = y_i^2$, where

$$y_{n+1} = \rho y_n + \sigma\sqrt{1-\rho^2}u_{n+1}, \qquad n = 1, \ldots, N-1,$$

where $u_{n+1} \sim \mathcal{N}(0,1) \to \mathcal{N}(a,1),\ a \neq 0.$

TABLE 2.1
Change-In-Mean Problem

N	100	200	300	500	700	1000	1500
$\alpha = 0.95$	0.1418	0.1012	0.0860	0.0590	0.0483	0.0405	0.0340
$\alpha = 0.99$	0.1507	0.1410	0.0930	0.0653	0.0508	0.0487	0.0414

Note: Problem = α-quantiles of the decision statistic; the case of independent observations $\sigma = 1, \rho = 0$; homogenous samples.

TABLE 2.2
Change-In-Mean Problem

N		300	500	700
$a = 0 \to 0.5$	C	0.0860	0.0590	0.0483
	β_N	0.107	0	0
	$\hat{\theta}_N$	0.3079	0.3273	0.3101

Note: Problem = $a = 0 \to a = 0.5$. C – decision threshold; β_N – estimate of type 2 error; θ_N – change-point estimate of $\theta = 0.3$. 1000 replications.

TABLE 2.3
Change-In-Mean Problem

N		100	200	300
$a = 0 \to 1.0$	C	0.1418	0.1012	0.0860
	β_N	0.035	0	0
	$\hat{\theta}_N$	0.0380	0.3126	0.3007

Note: Problem = $a = 0 \to a = 1.0$. C – decision threshold; β_N – estimate of type 2 error; θ_N – change-point estimate of $\theta = 0.3$. 1000 replications.

TABLE 2.4
Change-In Mean Problem

N	200	300	500	700	1000
$\alpha = 0.95$	0.1932	0.1662	0.1302	0.1182	0.1007
$\alpha = 0.99$	0.2464	0.1789	0.1603	0.1393	0.1206

Note: Problem = α-quantiles of the decision statistic; the case of dependent observations $\sigma = 1, \rho = 0.7$; homogenous samples.

TABLE 2.5
Change-In-Mean Problem

N		200	300	500
$a = 0 \to 1.0$	C	0.1932	0.1662	0.1302
	β_N	0.40	0	0
	$\hat{\theta}_N$	0.1982	0.3114	0.3112

Note: Problem = dependent observations: $a = 0 \to a = 1.0$. C – decision threshold; β_N – estimate of type 2 error; θ_N – change-point estimate of $\theta = 0.3$. 1000 replications.

TABLE 2.6
Change-In-Mean Problem

N		300	500	700
$a = 0 \to 0.5$	C	0.1662	0.1302	0.1182
	β_N	0.205	0.014	0
	$\hat{\theta}_N$	0.2826	0.3218	0.3241

Note: Problem = Dependent Observations: $a = 0 \to a = 0.5$. C – decision threshold; β_N – estimate of type 2 error; θ_N – change-point estimate of $\theta = 0.3$. 1000 replications.

In this model, σ is varying, but $\rho = 0, a = 0, \theta = 0.3$ are assumed to be fixed. The obtained results are given in Tables 2.7–2.9.

Change-in-correlation

For the change-in-correlation model, the parameter ρ is varying, but parameters $\sigma = 1, a = 0$, and $\theta = 0.3$ are assumed to be fixed. The following diagnostc sequence was used: $x_i = y_{i+1}y_i$, $i = 1, \ldots, N-1$.

The obtained results are given in Tables 2.10–2.11.

For homogenous samples, the regime $\rho = 0, \sigma = 1, a = 0, \theta = 0.3$ was assumed. α-quantiles of the proposed statistic are given in Table 2.10.

TABLE 2.7
Change-In-Dispersion Problem

N	200	300	500	700	1000
$\alpha = 0.95$	0.2272	0.1711	0.1380	0.0317	0.0287
$\alpha = 0.99$	0.2639	0.2729	0.1720	0.0409	0.0345

Note: Problem = α-quantiles of the decision statistic; the case of independent Observation $a = 0, \rho = 0; \theta = 0.3$; homogenous samples.

TABLE 2.8
Change-In-Dispersion Problem

N		300	500	700
	C	0.1711	0.1380	0.0317
$\sigma = 1.0 \to \sigma = 0.5$	β_N	0.159	0	0
	θ_N	0.2459	0.2457	0.2618

Note: Problem = $\sigma = 1.0 \to \sigma = 0.5$. C – decision threshold; β_N – estimate of type 2 error; θ_N – change-point estimate of $\theta = 0.3$. 1000 replications.

TABLE 2.9
Change-In-Dispersion Problem

N		500	700	1000
	C	0.1380	0.0317	0.0287
$\sigma = 1.0 \to \sigma = 2.0$	β_N	0.077	0.005	0
	θ_N	0.3089	0.3439	0.3152

Note: Problem = $\sigma = 1.0 \to \sigma = 2.0$. C – decision threshold; β_N – estimate of type 2 error; θ_N – change-point estimate of $\theta = 0.3$. 1000 replications.

Non-Gaussian noises

The proposed method is nonparametric by its nature. Therefore, we can consider non-Gaussian d.f.'s of noises, e.g., in the above model:

$$y_{n+1} = \rho y_n + \sigma\sqrt{1-\rho^2}(u_{n+1} + a), \qquad n = 1, \ldots, N-1.$$

Here, noises u_n are uniformely distributed on $[0, 1]$.

For homogenous samples, the following parameters of this model were used: $\rho = 0, \sigma = 1.0, \theta = 0.3, a = 0$. The obtained α-quantiles (averages in 1000 replications of the experiment) of the proposed statistic (2.6) are reported in Table 2.12.

TABLE 2.10
Change-In-Correlation Problem

N	200	300	500	700
$\alpha = 0.95$	0.1799	0.1040	0.0914	0.0668
$\alpha = 0.99$	0.2120	0.1416	0.1112	0.0897

Note: Problem = α-quantiles of the decision statistic; the case of independent observations $a = 0, \rho = 0; \sigma = 1.0, \theta = 0.3$; homogenous samples. 1000 replications.

TABLE 2.11
Change-In-Correlation Problem

N		300	500	700
$\rho = 0 \to \rho = 0.7$	C	0.1040	0.0914	0.0668
	β_N	0.332	0.060	0
	θ_N	0.0749	0.3355	0.3187

Note: Problem $= \rho = 0 \to \rho = 0.7 = 0.5$. C – decision threshold; β_N – estimate of type 2 error; θ_N – change-point estimate of $\theta = 0.3$. 1000 replications.

TABLE 2.12
Non-Gaussian Noises

N	100	200	300	500	700
$\alpha = 0.95$	0.0088	0.0072	0.0059	0.0044	0.0038
$\alpha = 0.99$	0.0117	0.0090	0.0071	0.0053	0.0045

Note: Problem $=$ α-quantiles of the decision statistic; the case of independent observations $a = 0, \rho = 0; \sigma = 1.0, \theta = 0.3$; homogenous samples. 1000 replications.

TABLE 2.13
Non-Gaussian Noises: The Case of Independent Observations

N		300	500	700
$a = 0 \to 0.1$	C	0.0059	0.0044	0.0038
	β_N	0.338	0.156	0.049
	θ_N	0.22	0.2897	0.3138

Note: Problem $= \rho = 0; \sigma = 1.0, \theta = 0.3;$. 1000 replications; change-in-mean problem: $a = 0 \to a = 0.1$; β_N – estimate of type 2 error; θ_N – change-point estimate of $\theta = 0.3$.

TABLE 2.14
Non-Gaussian Noises: The Case of Independent Observations

N		100	200	300
$a = 0 \to 0.3$	C	0.0086	0.0072	0.0059
	β_N	0.042	0	0
	θ_N	0.3115	0.3080	0.3098

Note: Problem $= \rho = 0; \sigma = 1.0, \theta = 0.3;$. 1000 replications; change-in-mean problem: $a = 0.0 \to a = 0.3$; β_N – estimate of type 2 error; θ_N – change-point estimate of $\theta = 0.3$.

TABLE 2.15
Multiple Change-Points

N		300	500	700	1000	1500
$\sigma = 1.0$	w_N	0.149	0.110	0.111	0.107	0.100
$\rho = 0.$	θ_1^N	0.3020	0.2871	0.2980	0.2999	0.3002
$a = 0 \to b_i$	θ_2^N	0.5033	0.4920	0.50.14	0.5020	0.4917
	θ_3^N	0.7233	0.7000	0.7000	0.7010	0.6993

Multiple changes

In many applications, multiple change-points are observed in obtained samples. To detect these multiple change-points, the method proposed in this chapter can be used.

First, we describe the simulated model. Suppose the concrete parameters of change-points are as follows: $\theta_1 = 0.3$, $\theta_2 = 0.5$, $\theta_3 = 0.7$. At the points, $n_i = [\theta_i N]$ the d.f. of observations in the sample $y_1, \ldots, y_N$, where N is the sample size changes. In the sequel, we assume the following model of observations:

$$y_{i+1} = \rho y_i + \sigma\sqrt{1-\rho^2}(u_{i+1} - b_{i+1}), \qquad i = 1, \ldots, N-1,$$

where $u_i \sim \mathcal{N}(0,1)$ and b_i is the profile of changes at the points $[\theta_i N], i = 1, 2, 3$: $b_i = (0\ -1\ 0\ 0.5)$. It means that

$$b_i = \begin{cases} 0 & 0 \leq i \leq [\theta_1 N] \\ -1 & [\theta_1 N] < i \leq [\theta_2 N] \\ 0 & [\theta_2 N] < i \leq [\theta_3 N] \\ 0.5 & [\theta_3 N] < i \leq N. \end{cases}$$

In simulations, we used the decision thresholds from Table 2.1 for the change-in-mean problem. Then, for different volumes of the sample size N, the following values were computed:

w_N — the estimate of the probability of the estimation error for the number of change-points $k = 3$: $P\{\hat{k}_N \neq k\}$

- θ_1^N — the estimate of the value $E(\theta_1 | k_N = k)$

- θ_2^N — the estimate of the value $E(\theta_2 | k_N = k)$

-θ_3^N — the estimate of the value $E(\theta_3 | k_N = k)$

In words, θ_i^N are the estimates of change-points θ_i on condition that we correctly estimate the number of these change-points.

The obtained results are given in Table 2.15.

2.8 Conclusion

In this chapter change-points problems for univariate models were considered. For the scheme of "gluing" from several stationary parts, I proposed Statistic (2.6) for detection of the number of stationary regimes and proved the result about the almost sure convergence of change-point estimators to their true values. Then, I considered the a priori estimates of quality for retrospective change-point detection methods, including the case of one change-point and the case of multiple change-points. From these a priori lower bounds for performance criteria, it follows that the proposed method is asymptotically optimal by the order of convergence of change-point estimators. Then, I considered the general three-step procedure of change-point estimation with the help of the main statistic with optimally chosen parameters and proved its properties for the case of dependent observations satisfying Cramer's and ψ-mixing conditions.

3

Retrospective Detection and Estimation of Stochastic Trends

3.1 Introduction

The development of methods for detection and estimation of structural changes has parallel development in the analysis of stochastic trend and unit root models. For example, most tests that attempt to distinguish between a unit root and a (trend) stationary process will favor the unit root model when the true process is subject to structural changes.

Consider a univariate time series $\{y_t; t = 1, \ldots, T\}$, which under the null hypothesis is identically and independently distributed with mean μ and finite variance. Under the alternative hypothesis, y_t is subject to a one-time change in mean at some unknown date T_b, i.e.,

$$y_t = \mu_1 + \mu_2 I(t > T_b) + \epsilon_t,$$

where $\epsilon_t \sim i.i.d.(0, \sigma_\epsilon^2)$ and I denotes the indicator function.

Following earlier works of Chernoff and Zacks (1964) and Kander and Zacks (1966), Gardner (1969) proposed the test for a structural change that is Bayesian in nature and assigns weights p_t as a prior probabilities that a change occurs at date $t, 1 \leq t \leq T$. Assuming Gaussian errors and an unknown value of σ_ϵ^2, this strategy leads to the test

$$Q = \hat{\sigma}_\epsilon^2 T^{-2} \sum_{t=1}^{T} [\sum_{j=t+1}^{T} (y_j - \bar{y})]^2,$$

where $\bar{y} = T^{-1} \sum_{t=1}^{T} Y_t$ is the sample average, and $\hat{\sigma}_\epsilon^2 = T^{-1} \sum_{j=1}^{T} (y_j - \bar{y})]^2$ is the sample variance of the data. The limit distribution of the statistic Q was analyzed by MacNeill (1974) under the null and alternative hypotheses of a polynomial trend function with changing coefficients.

With a prior that assigns equal weights to all observations (i.e., $p_t = 1/T$), the test of no change under the alternative of a polynomial trend has the following form (Anderson and Darling (1952), Antoch et al. (1997)):

$$Q_p = \hat{\sigma}_\epsilon^2 T^{-2} \sum_{t=1}^{T} [\sum_{j=t+1}^{T} \hat{e}_j]^2,$$

where $\hat{\sigma}_\epsilon^2 = T^{-1} \sum_{j=1}^{T} \hat{e}_t^2$ and $\hat{e}_t$ are the residuals from a regression of y_t on $\{1, t, \ldots, t^p\}$.

The statistic Q_p is well known to applied economists. It is the so-called KPSS test for testing the null hypothesis of stationarity vs. the alternative of a unit root (see Kiatkowski et al. (1992)).We see that the same statistic can be used for detection of structural changes and stochastic trends in real data. And this is the general situation: a presence of a structural change in data substantially worsens power of all known tests for unit roots (e.g., the ADF (augmented Dickey-Fuller test(1979, 1981)), KPSS, and vice versa, testing for a structural change for models with (possible) unit roots is extremely difficult.

In Perron's opinion, the reason for this situation is that most tests for unit roots and structural changes are statistically "sensitive" to all of these types of nonstationarity. The real problem is to propose statistical tests that can discriminate between unit roots and structural changes (see Perron (2005)) in nonstationary time series models.

In the econometric context, this problem was actual after the Nelson and Plosser (1982) study of 14 macroeconomic time series of the US economy. The authors demonstrated that the ADF test cannot reject the stochastic trend hypothesis for most of these series. After that Rappoport, Reichlin (1989) and especially Perron (1989, and 1994) showed that the mere presence of a structural change in the data precludes effective testing of a stochastic trend hypothesis. For example, if a structural shift in level is present in data, then the MLE estimate of the first autoregressive coefficient is asymptotically biased to 1. Therefore, the ADF test with the modified threshold (MacKinnon (1980)) confirms the hypothesis of a stochastic trend in data.

In subsequent papers by Montanes and Reyes (1999, 2000), it was demonstrated that the Phillips-Perron test (1988) for unit roots has similar problems.

The situation becomes even more complicated if instances of structural changes are unknown. Christiano (1992), Banerjee et al. (1992), Zivot, and Andrews (1992), Perron, and Vogelsang (1992), and Perron (1997) showed that the power of all known tests for unit roots drops significantly for samples with structural changes at unknown instants. Some preliminary ideas, how to deal with these problems were proposed in Perron, and Yabu (2005), and Kim, and Perron (2005).

I conclude that the problem of testing the unit root and the structural change (change-point) hypotheses is very actual for nonstationary time series. Below, we present the most used methods for detection of stochastic trends.

Parametric tests

The most well-known test for unit roots proposed by Dickey and Fuller (1979, 1981) uses a simple AR(1) model of a time series

$$y_t = \gamma y_{t-1} + \epsilon_t, \quad t = 1, \dots, T,$$

where ϵ_t is a centered "white noise" process.

For $|\gamma| < 1$, the ordinary least square (OLS) estimate of the parameter γ is

$$\hat{\gamma}_T = \frac{\sum_{t=1}^{T} y_t y_{t-1}}{\sum_{t=2}^{T} y_{t-1}^2}.$$

This estimate has the following properties:

$$\hat{\gamma}_T \to \gamma, \quad \text{in probability as } T \to \infty,$$

and

$$\sqrt{T}(\hat{\gamma}_T - \gamma) \to \mathcal{N}(0, 1 - \gamma^2).$$

Does this result still hold for $\gamma = 1$? This case is called "unit root," since the characteristic equation for the process y_t: $1 - \gamma z = 0$ has only one unit root. Mann and Wald (1943) began to study this case. However, modern research is based upon Dickey, and Fuller (1979, 1981) who proved that in the case of a unit root,

$$T(\hat{\gamma}_T - \gamma) \to v \quad \text{in probability as } T \to \infty,$$

where v is a certain r.v., $Dv < \infty$, and for finite sample sizes, $E(\hat{\gamma}_T) < 1$.

We can make two conclusions from these results. The first conclusion: the estimate $\hat{\gamma}_T$ is biased for $\gamma = 1$. The second conclusion: the estimate $\hat{\gamma}_T$ converges to the true value of γ with a rate that is much higher than in the case of stationarity (the estimate $\hat{\gamma}_T$ is called *superconsistent*).

In practice, the first conclusion means that if we apply OLS for estimation of autoregression parameter in the case of a unit root, then the OLS estimate will mask the true hypothesis of nonstationarity of a time series. To overcome this drawback, the Dickey-Fuller (DF) test was proposed.

This test is based upon the idea of usual t-statistic for the coefficient γ. However, the critical value of the student statistic in the nonstationary case is different from its standard value. These critical values were computed by Dickey and Fuller via simulation tests.

Thus, in the simplest case, the following statistic is used:

$$DF = \frac{\hat{\gamma}_T - 1}{std.error(\hat{\gamma}_T)},$$

where "std.error" denotes the standard error in parameter estimation.

The critical values of this statistic were computed by Dickey and Fuller for the three specifications of the AR(1) model:

— model without constant and trend: $y_t = \gamma y_{t-1} + \epsilon_t$,
— model with constant and without trend: $y_t = \mu + \gamma y_{t-1} + \epsilon_t$,
— model with constant and trend: $y_t = \mu + \beta t + \gamma y_{t-1} + \epsilon_t$.

The DF test can be used only for the AR(1) model of a time series. For generalizations of it to serially correlated processes, ADF (augmented Dickey-Fuller) test was proposed. This test can be used for the following model:

$$y_t = \mu + \beta t + \gamma y_{t-1} + \gamma_1 \Delta y_{t-1} + \cdots + \gamma_p \Delta y_{t-p} + \epsilon_t.$$

The ADF statistic is analogous to the above-considered case of the AR(1) process:

$$ADF = \frac{\hat{\gamma} - 1}{std.err.(\hat{\gamma})}.$$

Fuller (1979) demonstrated that the asymptotic distribution of this statistic does not depend on lagged differences Δy_{t-i}. Moreover, Said, and Dickey (1984) proved that the ADF test can be used in specifications of time series models with MA(q) terms.

We may elect to include a constant, or a constant and a linear time trend, in ADF test regression. For these two cases, Elliott, Rothenberg, and Stock (ERS) (1996) propose a modification of the ADF test in which the data are detrended so that explanatory variables are removed from the data before running the test regression.

The ERS point optimal test is based on the quasidifferencing regression defined in the previous section. Define the residuals as $\tilde{\eta}_t(a) = d(y_y|a) - d(x_t|a)^{'}\delta(a)$, and let $SSR(a) = \sum \tilde{\eta}_t^2(a)$ be the sum-of-squared residuals function. The ERS point optimal test statistic of the null $\alpha = 1$ against the alternative $\alpha = \tilde{a}$ is then defined as

$$P_T = (SSR(\tilde{a}) - \tilde{a}SSR(1))/f_0,$$

where f_0 is an estimator of the residual spectrum at frequency zero.

Critical values for the ERS statistic are computed by interpolating the simulation results provided by ERS (1996) for $T = \{50, 100, 200, \infty\}$.

Nonparametric tests for unit roots

In order to augment power of tests and to expand the circle of specifications of time series models, several alternatives to the ADF test were proposed. One of the most well-known tests for unit roots was proposed by Phillips and Perron (1988). Specification of the time series model has the following form:

$$y_t = \delta_t + \gamma y_{t-1} + \gamma_1 \Delta y_{t-1} + \cdots + \gamma_p \Delta y_{t-p} + \epsilon_t,$$

where δ_t can be zero, constant, and constant and trend $(0,\ \mu,\ \mu + \beta t)$.

The Phillips-Perron test has the following form:

$$Z = \sqrt{\frac{c_0}{a}}(\frac{\hat{\gamma} - 1}{v}) - \frac{1}{2}(a - c_0)\frac{Tv}{\sqrt{as^2}},$$

where $e_t, t = 1, \ldots, T$ are regression residuals;

$$s^2 = \frac{\sum_{t=1}^{T} e_t^2}{T-K}$$

$$v^2 -$$

is the estimated asymptotic dispersion $\hat{\gamma}$

$$c_j = \frac{1}{T} \sum_{s=j+1}^{T} e_t e_{t-s}, \quad j = 0, \ldots, p$$

$$c_0 = [(T-K)/T]s^2$$

$$a = c_0 + 2 \sum_{j=1}^{L} (1 - \frac{j}{L+1}) c_j; \text{ and}$$

K is equal to the number of degrees of freedom of the considered model, i.e., $K = p+1, p+2, p+3$ in dependence on the type of the model (without constant and trend, with constant without trend, with constant and trend, respectively).

The same tables for computing critical threshold are used for the Phillips-Perron test, as for the ADF test.

We already mentioned the KPSS test. This method is often used for testing regression residuals in OLS procedures for unit roots. Its remarkable feature is the assumption about the absence of the deterministic trend in data. The method uses the sequence of regression residuals e_t, $t = 1, 2, \ldots, T$.

The CUSUM (cumulative sums) statistic for residuals is

$$S_t = \sum_{i=1}^{t} e_i, \quad t = 1, 2, \ldots, T.$$

The decision statistic of this test can be written as follows:

$$LM = \sum_{t=1}^{T} S_t^2 / (T^2 \hat{\sigma}_\epsilon^2),$$

where $\hat{\sigma}_\epsilon^2$ is a certain consistent estimate of dispersion of regression residuals

$$\hat{\sigma}_\epsilon^2 = T^{-1} \sum_{t=1}^{T} e_t^2 + 2T^{-1} \sum_{s=1}^{l} w(s,l) \sum_{t=s+1}^{T} e_t e_{t-s},$$

and $w(s,l)$ is the weight function, depending on the choice of the spectral window. Concretely, the Bartlett window is used: $w(s,l) = 1 - s/(l+1)$, where l has the order $o(T^{1/2}$, e.g., $l = [T^{1/3} + 1]$.

Ng and Perron (2001) construct test statistics that are based upon the GLS detrended data y_y^d constructed by ERS (1996). These test statistics are modified forms of Phillips and Perron statistics and ERS point optimal statistic.

3.2 A Priori Informational Inequalities and Quality Criteria

In this section the we consider theoretical informational lower estimates for performance criteria in problems of estimation of unknown parameters in models with unit roots. These a priori theoretical estimates can be used for conclusions about the asymptotic optimality of concrete estimates for nonstationary models of data with a unit root.

Let us consider some examples.

Random walk

Suppose the model of observations has the following form:

$$x_i = x_{i-1} + u_i, \qquad u_i \sim \mathcal{N}(0, \sigma^2).$$

We ask a question about accuracy of estimation of the parameter σ in this model. Concretely, suppose the sample $x_1, \ldots, x_N$ is obtained, where N of the sample size. Which is the asymptotically optimal (as $N \to \infty$) rate of convergence of estimates $\hat{\sigma}_N$ to the true value of σ?

Regression with a nonstationary predictor

Suppose a regression model has the following form:

$$\begin{aligned} y_i &= \theta x_i + \xi_i, \\ x_i &= x_{i-1} + u_i, \end{aligned}$$

where u_i are i.r.v.'s.

We consider the problem of estimation of the parameter θ by an obtained sample $y_1, \ldots, y_N$. Which is the asymptotically optimal (as $N \to \infty$) rate of convergence of estimates $\hat{\theta}_N$ to the true value of θ?

To answer these questions, we prove Theorem 3.2.1, in which we establish the a priori informational lower bound for the error probability in estimation of an unknown parameter in such problems.

Theorem 3.2.1. *Suppose all assumptions of Theorem 2.6.1 are satisfied. For all* $0 < \epsilon < \theta$, *we assume that there exists*

$$I(\epsilon, \theta) = \lim_{N \to \infty} N^{-1} \sum_{i=1}^{N} E_{\theta+\epsilon} \ln(f_{\theta+\epsilon}(x_i)/f_\theta(x_i)).$$

Then,

$$\liminf_N N^{-1} \ln \inf_{\hat{\theta}_N \in \mathcal{M}_N} \mathbf{P}_\theta\{|\hat{\theta}_N - \theta| > \epsilon\} \geq I(\epsilon, \theta).$$

The proof of this result is analogous to the proof of Theorem 2.6.1 from Chapter 2.

Now, let us demonstrate that this result holds for two above-considered examples (a random walk on the line and the regression model with a random predictor).

For the first example (a random walk), we have:

$$E_{\epsilon+\sigma} \ln(f(X^N, \epsilon+\sigma)/f(X^N, \sigma))$$
$$E_{\epsilon+\sigma} \sum_{i=1}^{N} (\ln(\frac{i\sigma}{i(\sigma+\epsilon)}) + \frac{x^2}{2i}(\frac{1}{\sigma^2} - \frac{1}{(\sigma+\epsilon)^2}))$$
$$= (\frac{1}{\sigma^2} - \frac{1}{(\sigma+\epsilon)^2})) \sum_{i=1}^{N} E_{\sigma+\epsilon}(\frac{x^2}{2i}).$$

Therefore,

$$P_\sigma\{|\hat{\sigma}_N - \sigma| > \epsilon\} \geq \exp(-\frac{\epsilon}{\sigma^3} \ln N).$$

From here, we obtain

$$E_\sigma|\hat{\sigma}_N - \sigma| \geq \frac{\sigma^3}{\ln N}$$
$$E_\sigma(\hat{\sigma}_N - \sigma)^2 \geq \frac{\sigma^6}{\ln^2 N}.$$

Now, let us consider the second example: We are interested in estimation of the regression parameter θ in the following model:

$$y_i = \theta x_i + \xi_i$$
$$x_i = x_{i-1} + u_i,$$

where $\xi_i \sim \mathcal{N}(0, \eta^2)$, $u_i \sim \mathcal{N}(0, \sigma^2)$.

Here, we obtain $x_i \sim \mathcal{N}(0, i\sigma^2)$, $y_i \sim \mathcal{N}(0, i\sigma^2\theta^2 + \eta^2)$, and, therefore,

$$P_\theta\{|\hat{\theta}_N - \theta| > \epsilon\} \geq \exp(-\frac{\sigma^2(2\theta\epsilon + \epsilon^2)}{2} \sum_{i=1}^{N} \frac{1}{(i\sigma^2\theta^2 + \eta^2)(i\sigma^2(\theta+\epsilon)^2 + \eta^2)}).$$

After certain calculations, we obtain

$$E_\theta|\hat{\theta}_N - \theta| \geq \frac{\theta\eta^2}{\ln N}$$
$$E_\theta(\hat{\theta}_N - \theta)^2 \geq \frac{\theta^2\eta^4}{\ln^2 N}.$$

3.3 Method for Retrospective Detection of a Stochastic Trend

Motivation

For motivation of subsequent assumptions, let us consider the following model of a time series that is often used in practice. Suppose that a one-dimensional time series $\{Y_t\}$ is described by the model $AR(p)$, i.e.,

$$Y_t = \alpha_1 Y_{t-1} + \cdots + \alpha_p Y_{t-p} + u_t,$$

where $\{u_t\}$ is the sequence of i.i.d.r.v.'s, $\mathbf{E}u_t = 0$, $\sigma_u^2 = \mathbf{E}u_t^2$, $0 < \sigma_u^2 < \infty$ (here and below we denote by $\mathbf{E}(\mathbf{P})$ the mathematical expectation (probability measure)). Suppose that the characteristic polynomial

$$p(z) = 1 - \alpha_1 z - \cdots - \alpha_p z^p, \quad z \in \mathbb{C}$$

has one unit root $p(1) = 0$ and all other roots of this polynomial lie outside the unit circle. Then $p(z) = p^*(z)(1 - z)$, and the polynomial $p^*(z)$ has no roots inside the unit circle.

It follows from here that $1/p^*(z)$ exists for $|z| \leq 1$. Then, $p(L) = p^*(L)\Delta Y_t = u_t$, where L is the delay operator. Since the polynomial $p^*(L)$ can be inverted, we have the following representation,

$$Y_t = Y_{t-1} + \sum_{j \geq 0} \beta_j u_{t-j},$$

for some coefficients $\{\beta_j\}$.

This means that the considered process Y_t can be described by a random walk model with *dependent* errors. Remark that an analogous representation is valid for the process $\{Y_t\}$ described by the $ARMA(p, q)$ model with one unit root.

Problem statement

Based on the previous subsection, we conclude that most situations of unit roots and structural changes in one-dimensional time series can be described by the following model:

$$e_i = \rho e_{i-1} + u_i, \quad e_0 \equiv 0, \, i = 1, \ldots, N, \tag{3.1}$$

where $\{u_i\}$ is a sequence of *dependent* i.d.r.v.'s with zero mean and finite variance.

The problem statement is as follows. Suppose the sample of observations $\{y_i\}_{i=1}^N$ is given (the model of this sample depends on the hypothesis considered and is presented below). It is necessary to test the following hypotheses:

— The hypothesis of stationarity $\mathbf{H}_0 : y_i = e_i$, $i = 1, \ldots, N$, $|\rho| \leq 1 - \delta$, where δ is known.

— The hypothesis of a structural change $\mathbf{H}_1$: $y_i = h\mathbb{I}(i \geq [\theta N]) + e_i$, $i = 1, \ldots, N$, $|\rho| \leq 1 - \delta$ (here and below, $\mathbb{I}(A)$ is the indicator function of a set A, $[a]$ is the integer part of a number a).

In this case, θ is an unknown parameter (the relative change-point), $0 < a \leq \theta \leq 1 - a < 1$, where a is known constant, h is unknown parameter (the size of a structural change), such that $0 < b \leq |h|$, and b is known constant.

— The unit root hypothesis $\mathbf{H}_2$: $\rho = 1$, $y_i = e_i, i = 1, \ldots, N$. Under hypothesis $\mathbf{H}_1$, the family (under parameters (θ, h)) of probability measures $\{\mathbf{P}_{\theta,h}\}$ is considered. In this case, we consider also the problem of the change-point $\hat{n} \stackrel{\text{def}}{=} [\theta N]$ estimation.

Assumptions

Everywhere below, the following assumptions will be used.

A1. There exists the limit $\sigma^2 = \lim_{n\to\infty} n^{-1}\mathbf{E}V^2(n)$, where $V(n) = \sum_{i=1}^{n} u_i$.

A2. The sequence $\{u_i\}_{i=1}^{\infty}$ is uniformely integrable.

A3. The sequence $\{u_i\}_{i=1}^{\infty}$ is ρ-mixing and $\sum_i \sqrt{\rho(2^i)} < \infty$.

Let us make some comments on the assumptions.

Assumption A1 holds if, for example, the sequence $\{u_i\}$ is a stationary one and has integrable correlation function. Assumption A2 is usually satisfied in most practical cases; a sufficient condition for A2 is existence of the 4-th moment for u_i. Assumption A3 is usually satisfied for Markov chains $\{u_i\}$. Theorem 3.3, in Bradley (2005) states that if $\rho(n) < 1$ for a certain $n \geq 1$ then $\rho(n) \to 0$ at least exponentially as $n \to \infty$.

Result

The method of hypothesis testing is based upon the nonparametric statistic that was first proposed by Brodsky and Darkhovsky in 1979–1989s. This statistic is the modification of the classic Kolmogorov-Smirnov test and has the following form:

$$Z_N(n) = \frac{1}{N^2}\left(N\sum_{j=1}^{n} y_j - n\sum_{j=1}^{N} y_j\right), \quad n = 1, \ldots, N.$$

Together with the statistic $Z_N(n)$, we consider the following random process in the Skorokhod space $D([0,1])$: $Z_N(t) \stackrel{\text{def}}{=} Z_N([Nt])$. Remark that statistical testing based on $Z_N(n)$ and $Z_N(t)$ gives the same results.

The proposed method consists of two steps. At the first step, the hypothesis $\mathbf{H_0}$ is tested against the alternative $\mathbf{H_1}$ or $\mathbf{H_2}$. If the alternative $\mathbf{H_1}$ or $\mathbf{H_2}$ is accepted at the first step, then the hypothesis $\mathbf{H_1}$ against $\mathbf{H_2}$ is tested at the second step.

The first step of the procedure

Denote by $\alpha_0(N)$ the probability to reject the hypothesis $\mathbf{H}_0$ by the observed sample $\{y_i\}_{i=1}^{N}$ on condition that this hypothesis is true, and by $\beta_{0j}(N)$ the probability to accept the hypothesis $\mathbf{H_0}$ on condition that the hypothesis $\mathbf{H}_j$, $j = 1, 2$ holds.

Suppose that hypothesis $\mathbf{H}_0$ is true. Then, from (3.1), we obtain

$$\sum_{i=1}^{N} y_i \stackrel{\text{def}}{=} S_N = \rho S_N - \rho y_N + \sum_{i=1}^{N} u_i. \tag{3.2}$$

Consider the random process $Y_N(t) \stackrel{\text{def}}{=} N^{-1/2} S_{[Nt]}$ in Skorokhod space $D([0,1])$. Then, from (3.2), we get

$$Y_N(t) = \frac{1}{(1-\rho)\sqrt{N}} \sum_{i=1}^{[Nt]} u_i - \frac{\rho}{(1-\rho)\sqrt{N}} y_N. \tag{3.3}$$

In Peligrad (1992), it was proved that the process $\frac{1}{\sqrt{N}} \sum_{i=1}^{[Nt]} u_i$ weakly converges in $D([0,1])$ to the process $\sigma W(t)$ under our assumptions, where $W(t)$ is the standard Wiener process (below we denote weak convergence by the symbol $\Rightarrow$).

Using (3.3), it is easy to prove that $\mathbf{E} y_N^2 \leq \frac{\sigma^2}{1-\rho}$. Therefore, the sequence of random variables $\{\frac{\rho}{(1-\rho)\sqrt{N}} y_N\}$ tends to zero in probability, as $N \to \infty$.

Taking this into account, using (3.3), and Theorem 4.4 from Billingsly (1968), we get that under $\mathbf{H_0}$, the process $Y_N(t)$ weakly converges in $D([0,1])$ to the process $\frac{\sigma}{1-\rho} W(t)$. Therefore, under $\mathbf{H_0}$,

$$\sqrt{N} Z_N([Nt]) \Longrightarrow \frac{\sigma}{1-\rho} W^0(t), \tag{3.4}$$

as $N \to \infty$, where $W^0(t)$ is the standard Brownian bridge.

Put

$$T_N \stackrel{\text{def}}{=} \max_{1 \leq n \leq N} |Z_N(n)|.$$

The solution at the first step is as follows:
if $T_N \leq C$, then $\mathbf{H_0}$ is accepted; and
if $C < T_N$, then the alternative is accepted.
Using (3.4), we get

$$\max_{|\rho| \leq 1-\delta} \lim_{N \to \infty} \mathbf{P}\{\sqrt{N} T_N > C | \mathbf{H}_0\} = 1 - \mathcal{K}\left(\frac{\delta}{\sigma} C\right), \tag{3.5}$$

where $\mathcal{K}(z) = 1 + 2 \sum_{k=1}^{\infty} (-1)^k \exp\left(-2(kz)^2\right)$ is the Kolmogorov function.

Therefore, the following asymptotic relationship holds:

$$\mathbf{P}\{\text{reject } \mathbf{H}_0 | \mathbf{H}_0\} = \alpha_0(N) \sim 1 - \mathcal{K}\left(\frac{\delta\sqrt{N}}{\sigma} C\right). \tag{3.6}$$

Now suppose that hypothesis $\mathbf{H_1}$ is true. Consider the function

$$f(t) = \begin{cases} ht(1-\theta), & \text{if } 0 \leq t \leq \theta \\ h\theta(1-t), & \text{if } \theta \leq t \leq 1. \end{cases}$$

Remark that

$$\min_{a\leq\theta\leq 1-a,\ b\leq|h|}\ \max_{0\leq t\leq 1}|f(t)| = b\,a(1-a) \stackrel{\text{def}}{=} \Delta.$$

Denote $f_N(t) \stackrel{\text{def}}{=} \mathbf{E}Z_N(t)$. It is easy to prove that $\max_{0\leq t\leq 1}|f_N(t)-f(t)| = O(N^{-1})$ under $\mathbf{H_1}$.

Therefore, under $\mathbf{H_1}$, we obtain

$$\sqrt{N}\,(Z_N(t)-f(t)) \Longrightarrow \frac{\sigma}{1-\rho}W^0(t), \tag{3.7}$$

as $N\to\infty$.

Besides, under $\mathbf{H_1}$,

$$\max_{0\leq t\leq 1}|f(t)| - \sup_{0\leq t\leq 1}|Z_N(t)| \leq \sup_{0\leq t\leq 1}|Z_N(t)-f(t)|. \tag{3.8}$$

Suppose that

$$C = \lambda\Delta,\ 0<\lambda<1. \tag{3.9}$$

This condition is necessary for the opportunity to detect the minimal structural change. Then, from (3.8), we obtain

$$\beta_{01}(N) = \mathbf{P}\{T_N \leq C|\mathbf{H}_1\} \leq \mathbf{P}\{\sup_{0\leq t\leq 1}|Z_N(t)-f(t)| \geq (1-\lambda)\Delta\}.$$

Hence, taking into account (3.5) and (3.7), we obtain the following asymptotic relationship:

$$\beta_{01}(N) \sim\leq 1-\mathcal{K}\left(\frac{\delta\sqrt{N}}{\sigma}(1-\lambda)\Delta\right). \tag{3.10}$$

Let us rewrite (3.6) with account of (3.9):

$$\alpha_0(N) \sim 1-\mathcal{K}\left(\frac{\delta\sqrt{N}}{\sigma}\lambda\Delta\right). \tag{3.11}$$

Now suppose that hypothesis $\mathbf{H_2}$ holds. Then, $y_i = \sum_{k=1}^{i} u_k$ and $N^{-1/2}\sum_{k=1}^{[Nt]} u_k \Rightarrow \sigma W(t)$ as $N\to\infty$ due to Peligrad (1992) (as before). Therefore,

$$\frac{Z_N([Nt])}{\sqrt{N}} \Longrightarrow \sigma\left(\int_0^t W(s)ds - t\int_0^1 W(s)ds\right) \stackrel{\text{def}}{=} U(t), \tag{3.12}$$

as $N\to\infty$.

The process $U(t)$ is Gaussian with zero mean. Write

$$\mathbf{E}U^2(t) = \sigma^2 \left((1-2t) \int_0^t \int_0^t \min(s,v) ds dv - (1-t)t^3 + t^2 \int_0^1 \int_0^1 \min(s,v) ds dv \right) \tag{3.13}$$

Calculating the integrals, we obtain

$$\mathbf{E}U^2(t) = \frac{\sigma^2}{3} \left(t^4 - 2t^3 + t^2 \right) \stackrel{\text{def}}{=} D^2(t). \tag{3.14}$$

Therefore, we have the following equality by distribution

$$|U(t)| \stackrel{\text{law}}{=} D(t)|\xi|, \tag{3.15}$$

where ξ is the standard Gaussian r.v.

From (3.15), we obtain

$$\mathbf{P}\{|U(t)| \leq z\} = \frac{2}{\sqrt{2\pi}} \int_0^{z/D(t)} \exp(-x^2/2) dx. \tag{3.16}$$

Remark that

$$\max_{0 \leq t \leq 1} D^2(t) = D^2(t^*) \stackrel{\text{def}}{=} (D^*)^2, \; t^* = 1/2.$$

Therefore,

$$\mathbf{P}\{\max_{0 \leq t \leq 1} |U(t)| \leq C\} \leq \frac{2}{\sqrt{2\pi}} \int_0^{C/D^*} \exp(-x^2/2) dx. \tag{3.17}$$

By the definition,

$$\mathbf{P}\{\text{accept}\, \mathbf{H}_0 | \mathbf{H}_2\} = \beta_{02} = \mathbf{P}\{T_N \leq C | \mathbf{H}_2\}.$$

Therefore, we obtain from (3.17) and (3.9) the following asymptotic inequality:

$$\beta_{02}(N) \sim\leq \frac{2}{\sqrt{2\pi}} \int_0^{\lambda\Delta/D^*\sqrt{N}} \exp(-x^2/2) dx. \tag{3.18}$$

Thus, we proved the following:

Theorem 3.3.1. *Suppose* $0 < C < \Delta$. *Then the error probabilities of the first stage of the proposed method satisfy asymptotic relationships (3.10), (3.11), and (3.18).*

3.4 Discrimination between a Stochastic Trend and a Structural Change Hypotheses

The second step of the procedure

If the hypothesis $\mathbf{H_0}$ was rejected at the first step, then we test the hypothesis $\mathbf{H_1}$ against $\mathbf{H_2}$ at the second step. Preliminarily, consider a method of a structural change estimation (this corresponds to the case that hypothesis $\mathbf{H_1}$ is true). The method is based on the formalism of the scheme of series, i.e., we suppose that the instant of a structural change n^* has the form $n^* = [\theta N]$, where $0 < a \le \theta \le 1 - a < 1$ is an unknown parameter. So we consider the parametric family of the probability measures $\{\mathbf{P}_{\theta,h}\}$ and are interested in estimation of an unknown parameter θ.

Suppose $\hat{n}$ is an arbitrary maximum point of $|Z_N(n)|$, $n = [aN], \dots, [(1 - a)N]$. Then, the value $\hat{\theta} = \hat{n}/N$ can be considered as the estimate of the parameter θ. In other terms, if $\hat{t}$ is an arbitrary maximum point of the process $|Z_N([Nt])|$ on the segment $t \in [a, 1-a]$, then $\hat{\theta} = \hat{t}$ (remark that the process $|Z_N([Nt])|$ has piecewise constant trajectories and, therefore, attains its maximum; hence, there is an equivalence of two definitions of the estimate of the parameter θ).

For the proof of properties of the estimate $\hat{\theta}$, we need the result of Lemma 1.6.1.

From Remark 1.6.3 to Lemma 1.6.1, we obtain

Theorem 3.4.1. *The following asymptotical estimate holds for the normalized instant of the structural change:*

$$\max_{a \le \theta \le 1-a,\ b \le |h|} \mathbf{P}_{\theta,h}\{|\hat{\theta} - \theta| > \epsilon\} \sim\le 1 - \mathcal{K}\left(\frac{\epsilon\delta ab}{2\sigma}\sqrt{N}\right). \tag{3.19}$$

Corollary 3.4.1. *Taking into account the exponential rate of convergence $\hat{\theta} \to \theta$ (see (3.19)), we obtain that $\hat{\theta} \to \theta$ $\mathbf{P}_{\theta,h}$-a.s. under any fixed $a \le \theta \le 1 - a$, $|h| \ge b$.*

Now describe the second step of the procedure. Testing of hypotheses $\mathbf{H_1}$ and $\mathbf{H_2}$ is carried out in the following way:

1. Suppose at the first step, the inequality $T_N > C(N)$ is satisfied (we give formula for $C(N)$ in next section), and the maximum of $|Z_N(n)|$ is attained at the point $\tilde{n}$ (in the case of several points of maximum of $|Z_N(n)|$, we choose the minimal of these points).

2. Fix the parameter $0 < \varkappa < a/2$ and form two subsamples: $Y_1(\tilde{n}) = \{y_i : i = 1, \dots, \tilde{n} - [\varkappa N]\}$ and $Y_2(\tilde{n}) = \{y_i : i = \tilde{n} + [\varkappa N], \dots, N\}$. Denote by N_1, N_2 the sizes of corresponding subsamples.

3. For these subsamples, we compute the values $T_{N_1}(Y_1)$ and $T_{N_2}(Y_2)$,

i.e., the maximums of the statistic $|Z_N(\cdot)|$ designed for each of these samples, respectively.

4. The decision rule at the second step is formulated as follows:
Accept hypothesis $\mathbf{H_1}$ if $T_{N_1}(Y_1) < C(N_1)$ or $T_{N_2}(Y_2) < C(N_2)$, and accept hypothesis $\mathbf{H_2}$ otherwise.

From Corollary 3.4.1, it follows that the two subsamples $Y_1(\tilde{n})$ and $Y_2(\tilde{n})$, formed at the second stage of the proposed method, are almost surely (as $N \to \infty$) classified as statistically homogenous under $\mathbf{H_1}$. However, under the unit root hypothesis (the hypothesis $\mathbf{H_2}$), these two subsamples $Y_1(\tilde{n})$ and $Y_2(\tilde{n})$ are almost surely classified as non-stationary. This is the raw idea of the second stage of our method: to compute the maximums $T_{N_1}(Y_1)$ and $T_{N_2}(Y_2)$ of the statistic $|Z_N(\cdot)|$ for two subsamples and to compare these values with the decision boundary $C(N_i)$, $i = 1, 2$.

The quality of hypothesis testing at the second step can be characterized by the following conditional error probabilities:

$$\delta_{12} = \frac{\mathbf{P}\{\mathbf{H_2}|\mathbf{H_1}\}}{1 - \mathbf{P}\{\mathbf{H_0}|\mathbf{H_1}\}}, \quad \delta_{21} = \frac{\mathbf{P}\{\mathbf{H_1}|\mathbf{H_2}\}}{1 - \mathbf{P}\{\mathbf{H_0}|\mathbf{H_2}\}}.$$

The sense of these conditional error probabilities is as follows: They estimate classification errors at the second step on condition that the true decision was made at the first step of the proposed method.

By definition, $\mathbf{P}\{\mathbf{H_0}|\mathbf{H_1}\} = \beta_{\mathbf{01}}(\mathbf{N})$, and, by the idea of the second step of our method:

$$\mathbf{P}\{\mathbf{H_2}|\mathbf{H_1}\} \leq \max(\alpha_{\mathbf{0}}([\theta\mathbf{N}] - [\varkappa\mathbf{N}]), \alpha_{\mathbf{0}}(\mathbf{N} - [\vartheta\mathbf{N}] - [\kappa\mathbf{N}])).$$

Therefore, from Theorem 3.4.1, we conclude that δ_{12} goes to zero exponentially as N increases to infinity.

Further, $\mathbf{P}\{\mathbf{H_0}|\mathbf{H_2}\} = \beta_{\mathbf{02}}(\mathbf{N})$ and

$$\mathbf{P}\{\mathbf{H_1}|\mathbf{H_2}\} \leq \max(\beta_{\mathbf{02}}([\theta\mathbf{N}] - [\varkappa\mathbf{N}]), \beta_{\mathbf{02}}(\mathbf{N} - [\theta\mathbf{N}] - [\varkappa\mathbf{N}])).$$

Therefore, from Theorem 3.4.1, we conclude that δ_{21} goes to zero as $O(1/\sqrt{N})$ as $N \to \infty$. Thus, we proved the following:

Theorem 3.4.2. *Suppose* $0 < C < \Delta$. *Then the conditional error probabilities at the second step* δ_{12}, δ_{21} *tend to zero with an increasing sample size* N.

3.5 Simulations

Algorithm of testing

The algorithm of hypotheses testing consists of the following steps.

a) *Preliminary data processing*

Suppose $\tilde{n}$ is the maximum point of the main statistic. Under $\mathbf{H_1}$ in virtue of Corollary 1, we obtain $\tilde{n}/N \to \theta$-a.s. For $\mathbf{H_0}$ and $\mathbf{H_2}$, this point is of no interest. Therefore, for calculation of the threshold by the observations $\{y_i\}_{i=1}^N$, we must "clear" this sample from a possible shift in mean. So, first, we construct two subsamples, as was proposed in Section 3.2. Each of these two subsamples has constant mean (almost sure as $N \to \infty$) for any of the considered hypotheses. Put

$$\tilde{y}_i = y_i - \frac{1}{N - \tilde{n} - [\varkappa N] + 1} \sum_{i=\tilde{n}+[\varkappa N]}^{N} y_i$$

and consider the joint sample defined as $\{y_i\}_{i=1}^{\tilde{n}-[\varkappa N]} \bigcup \{\tilde{y}_i\}_{i=\tilde{n}+[\varkappa N]}^{N}$. Denote this joint sample (with the size $\tilde{N} \overset{\text{def}}{=} N - 2[\varkappa N] + 1$) as $\mathbb{W} = \{w_i\}_{i=1}^{\tilde{N}}$. The sample $\mathbb{W}$ has zero mean (almost sure as $N \to \infty$), both under $\mathbf{H_1}$ and $\mathbf{H_2}$.

b) *Calculating the threshold*

In this section, we weaken the condition A1 and assume that $\{u_i\}$ is stationary in the wide sense with the summable autocorrelation function. If $|\rho| < 1$, then the sequence $\{e_i\}$ (and, therefore, $\{w_i\}$) is also stationary in the wide sense and has the summable autocorrelation function. In Section 3.1, we demonstrated weak convergence of the process $Y_N(t)$. From here, it follows that for $|\rho| < 1$

$$\lim_{N\to\infty} \tilde{N}^{-1}\mathbf{E}\left(\sum_{i=1}^{\tilde{N}} w_i\right)^2 = \frac{\sigma^2}{(1-\rho)^2}. \tag{3.20}$$

On the other hand, we have

$$\lim_{N\to\infty} \tilde{N}^{-1}\mathbf{E}\left(\sum_{i=1}^{\tilde{N}} w_i\right)^2 = R(0) + 2\sum_{k=1}^{\infty} R(k) \overset{\text{def}}{=} \mathcal{R}, \tag{3.21}$$

where $R(k)$ is the autocorrelation function of $\{w_i\}$.

For estimation of the series in (3.21), we use the relationships

$$\hat{R}(k) = \frac{1}{\tilde{N} - k} \sum_{i=1}^{\tilde{N}-k} w_i w_{i+k}. \tag{3.22}$$

We take so many terms in (3.21) in order to obtain the difference between two consecutive partial sums of the series less than some given number $\epsilon \ll 1$.

Now, let $0 < \alpha < 1$ be the given significance level for testing hypothesis $\mathbf{H_0}$. We propose the following algorithm for computation of the decision threshold. Denote

$$\hat{\mathcal{R}}_k = \hat{R}(0) + 2\sum_{s=1}^{k} \hat{R}(s),$$

where $\hat{R}(s)$ is computed in (3.22).

We choose the number of terms l in this series as follows:

$$l = \min\{1 \le s \le 0.1\tilde{N} : \ |\hat{\mathcal{R}}_{s+1} - \hat{\mathcal{R}}_s| \le \epsilon\},$$

if this set is nonempty, and $l = 0.1\tilde{N}$ otherwise.

Then, we compute the decision threshold:

$$C(N) = \sqrt{\frac{\hat{\mathcal{R}}_l \, |\ln \alpha/2|/2}{N}}. \tag{3.23}$$

If there is no unit root, this threshold has the order $O(N^{-1/2})$ and, therefore, makes it possible to accept the stationarity hypothesis or to detect a structural change larger than this threshold. If there is a unit root, then this threshold grows as $aN^{1/2}$ with $0 < a < 1/2$, but the decision statistic T_N grows as $N^{1/2}$. Therefore, this statistic exceeds the threshold.

c) *Testing* $\mathbf{H_0}$

For testing hypothesis $\mathbf{H_0}$, we compare the maximum of the module of the main statistic with the threshold computed by (3.23). If this maximum is lower than the threshold, then we accept $\mathbf{H_0}$; otherwise, we go to the following step.

d) *Testing* $\mathbf{H_1}$ or $\mathbf{H_2}$.

This step was described in Section 3.4.

Simulation results

We considered the following model for computer (Monte-Carlo) simulation:

$$x_i = \rho x_{i-1} + \sigma v_i, \quad v_i \sim \mathcal{N}(0,1),$$

where N is the sample size, and $\sigma > 0$, $\mathcal{N}(0,1)$ is the standard Gaussian distribution.

In case of hypotheses $\mathbf{H_0}$ and $\mathbf{H_1}$, we use $-0.99 \le \rho \le 0.9$, and in case of hypothesis $\mathbf{H_2}$, we use $\rho = 1$.

The observations have the form $y_i = x_i, i = 1, \dots, N$ in case of hypotheses $\mathbf{H_0}$ and $\mathbf{H_2}$, and $y_i = x_i + h\mathbb{I}(i \le [\theta N])$, $i = 1, \dots, N$, $0.1 \le \theta \le 0.9$, $h > 0$ in case of hypothesis $\mathbf{H_1}$.

The following characteristics were estimated in Monte-Carlo tests:

For the first step:

— the first-type error: $\alpha_0(N) = \mathbf{P}_0\{T_N > C(N)\}$.

— the second-type error: $\beta_2(N) = \mathbf{P}\{T_N \le C(N)|\mathbf{H_1} \,\text{or}\, \mathbf{H_2}\}$ (note that $\beta_2(N) \le \max(\beta_{01}(N), \beta_{02}(N)))$

For the second step:

— the error probabilities: δ_{12}, δ_{21}.

The estimates of probabilities α_0, β_2, and δ_{12}, δ_{21} in Table 3.1 were obtained by averaging in $k = 5000$ trials of each test. The simulation was done according to Section 3.4.

The obtained results are presented in Table 3.1.

TABLE 3.1
Discrimination between $H_0, H_1,$ and H_2 Hypotheses: Performance Characteristics of the Proposed Test.

N		100	200	300	500	700	1000
$h = 1$	α_0	0.114	0.056	0.04	0.072	0.056	0.038
$\rho = 0.3$	β_2	0.412	0.282	0.106	0.03	0	0
$\sigma = 0.5$	δ_{12}	0.05	0.054	0.06	0.04	0.04	0.028
$\theta = 0.5$	δ_{21}	0.496	0.56	0.52	0.34	0.172	0.090
$h = 0.5$	α_0	0.082	0.062	0.052	0.040	0.038	0.048
$\rho = 0.3$	β_2	0.64	0.23	0.082	0.026	0.006	0
$\sigma = 0.5$	δ_{12}	0.05	0.052	0.058	0.044	0.03	0.024
$\theta = 0.5$	δ_{21}	0.50	0.566	0.568	0.354	0.168	0.072
$h = 0.5$	α_0	0.26	0.124	0.064	0.072	0.038	0.044
$\rho = 0.7$	β_2	0.95	0.86	0.52	0.35	0.136	0.04
$\sigma = 0.5$	δ_{12}	0.044	0.026	0.028	0.012	0.018	0.020
$\theta = 0.5$	δ_{21}	0.468	0.524	0.518	0.308	0.19	0.090
$h = 0.5$	α_0	0.046	0.038	0.058	0.032	0.026	0.030
$\rho = -0.7$	β_2	0.37	0.29	0.088	0.026	0.004	0
$\sigma = 0.5$	δ_{12}	0.024	0.032	0.058	0.024	0.024	0.032
$\theta = 0.5$	δ_{21}	0.494	0.528	0.502	0.346	0.176	0.084

The obtained results require some comments.

First, the type-1 error α_0 was chosen at the level of 5 percentage. The type-2 error estimate quickly goes to zero as the sample size N increases. In Table 3.1, we analyze the samples with the noise/signal ratio of about 1–2, which is rather low for testing. However, even in these difficult conditions, a satisfactory quality of testing is observed for the sample size more than 500. Remark that in most practical problems, a noise/signal ratio is usually much higher (10–100). So, effective testing of hypotheses is practically observed for sample size of 100–150 observations.

Second, the quality of testing strongly depends on the degree of data correlation. A satisfactory quality of testing for strongly correlated observations (the autocorrelation coefficient 0.7–0.8) is attained for rather large sample size (800–1000).

At the second stage of the proposed method, we discriminate between the hypothesis of an unknown structural change and a unit root. From Table 3.1, it follows that the estimates of the error probability δ_{12} are rather low (0.03–0.05). The estimates of the error probability δ_{21} are higher but rapidly go to zero as N increases.

3.6 Conclusions

In this chapter, methods for retrospective detection of stochastic trends in realizations of stochastic processes were considered. After a short review of methods designed for detection of a unit root in data, the problem statement was formulated. I tried to emphasize the main idea: The unit root and the structural change hypotheses are closely interrelated, and, therefore, the problem of their distinguishing from each other is actual.

Then the method of detection and testing nonstationarity of a unit root and a structural change in realizations of dependent time series is proposed. This method is based upon the modified Kolmogorov-Smirnov statistic and consists of two stages. At the first stage I test the hypothesis of stationarity of an obtained time series against the unified alternative of a unit root and a structural change. At the second stage (on condition of a rejected stationarity hypothesis), I discriminate between alternatives of a unit root and a structural change.

The asymptotic (as the sample size N tends to infinity) estimates of error probabilities are given. Empirically, I study statistical characteristics of the proposed method in Monte-Carlo tests for dependent time series.

4

Retrospective Detection and Estimation of Switches in Univariate Models

4.1 Introduction

In this chapter, the problem of the retrospective detection of changes in stochastic models with switching regimes is considered. The main goal is to propose asymptotically optimal methods for detection and estimation of possible "switches", i.e., random and transitory departures from prevailing stationary regimes of observed stochastic models.

First, let us mention previous important steps in this field. Models with switching regimes have a long prehistory in statistics (see, e.g., Lindgren (1978)). A simple switching model with two regimes has the following form:

$$\begin{aligned} Y_t &= X_t\beta_1 + u_{1t} \quad \text{for the 1st regime} \\ Y_t &= X_t\beta_2 + u_{2t} \quad \text{for the 2nd regime} . \end{aligned}$$

For models with endogenous switchings, usual estimation techniques for regressions are not applicable. Goldfeld and Quandt (1973) proposed *regression models with Markov switchings.* In these models, probabilities of sequential switchings are supposed to be constant. Usually, they are described by the matrix of probabilities of switchings between different states.

Another modification of the regression models with Markov switchings was proposed by Lee and Porter (1984). The following transition matrix was studied:

$$\Lambda = [p_{ij}]_{i,j=0,1}, \quad p_{ij} = P\{I_t = j | I_{t-1} = i\}.$$

Lee and Porter (1984) consider an example with railway transport in the US from 1880–1886 which were influenced by the cartel agreement. The following regression model was considered:

$$logP_t = \beta_0 + \beta_1 X_t + \beta_2 I_t + u_t,$$

where $I_t = 0$ or $I_t = 1$ in dependence of "price wars" in the concrete period.

Cosslett and Lee (1985) generalized the model of Lee and Porter to the case of serially correlated errors u_t.

Many economic time series occasionally exhibit dramatic breaks in their behavior, assocoated with events, such as financial crises (Jeanne and Mason,

2000; Cerra, 2005; Hamilton, 2005) or abrupt changes in government policy (Hamilton, 1988; Sims and Zha, 2004; Davig, 2004). Abrupt changes are also a prevalent feature of financial data and empirics of asset prices (Ang and Bekaert, 2003; Garcia, Luger, and Renault, 2003; Dai, Singleton, and Wei, 2003).

The functional form of the "hidden Markov model" with switching states can be written as follows:

$$y_t = c_{s_t} + \phi y_{t-1} + \epsilon_t, \tag{4.1}$$

where s_t is a random variable that takes the values $s_t = 1$ and $s_t = 2$ obeying a two-state Markov chain law:

$$Pr(s_t = j | s_{t-1} = i, s_{t-2} = k, \ldots, y_{t-1}, y_{t-2}, \ldots) = Pr(s_t = j | s_{t-1} = i) = p_{ij}. \tag{4.2}$$

A model of the form (4.1–4.2) with no autoregressive elements ($\phi = 0$) appears to have been first analyzed by Lindgren (1978) and Baum et al. (1980). Specifications that incorporate autoregressive elements date back in the speech recognition literature to Poritz (1982), Juang and Rabiner (1985), and Rabiner (1989). Markov-switching regressions were first introduced in econometrics by Goldfeld and Quandt (1973), the likelihood function for which was first calculated by Cosslett and Lee (1985). General characterizations of moment and stationarity conditions for Markov-switching processes can be found in Tjostheim (1986), Yang (2000), Timmermann (2000), and Francq and Zakoian (2001).

A useful review of modern approaches to estimation in Markov-switching models can be found in Hamilton (2005).

However, the mechanism of Markov chain modeling is far from unique in statistical description of dependent observations. Besides Markov models, we can mention martingale and copula approaches to dealing with dependent data, as well as description of statistical dependence via different coefficients of "mixing". All of these approaches are interrelated and we must choose the most appropriate method for the concrete problem. In this chapter, the mixing paradigm for description of statistical dependence is chosen.

Remark that the ψ-mixing condition is imposed below in this paper in order to obtain the exponential rate of convergence to zero for type-1 and type-2 error probabilities (see Theorems 4.3.1 and 4.3.2 below). Another alternative was to assume α-mixing property, which is always satisfied for periodic and irreducible countable-state Markov chains (see Bradley (2005)). Then, we can obtain the hyperbolic rate of convergence to zero for type-1 and type-2 error probabilities. For the majority of practical applications, it is enough to assume r-dependence (for a certain finite number of lags $r \geq 1$) of observations and state variables. Then all proofs become much shorter.

Now, let us mention some important problems that lead to stochastic models with switching regimes.

Splitting mixtures of probabilistic distributions

In the simplest case, we suppose that the d.f. of observations has the following form:

$$F(x) = (1 - \epsilon)F_0(x) + \epsilon F_1(x),$$

where $F_0(x)$ is the d.f. of ordinary observations; $F_1(x)$ is the d.f. of abnormal observations; and $0 < \epsilon < 1$ is the probability of obtaining an abnormal observation.

We need to test the hypothesis of statistical homogeneity (no abnormal observations) of an obtained sample $X^N = \{x_1, x_2, \ldots, x_N\}$. If this hypothesis is rejected, then we need to estimate the share of abnormal observations (ϵ) in the sample and to classify this sample into subsamples of ordinary and abnormal observations.

Estimation for regression models with abnormal observations

The natural generalization of the previous model is the regression model with abnormal observations

$$Y = X\beta + \epsilon,$$

where Y is the $n \times 1$ vector of dependent observations; X is the $n \times k$ matrix of predictors; β is $k \times 1$ vector of regression coefficients; and ϵ id the $n \times 1$ vector of random noises with the d.f. of the following type:

$$f_\epsilon(x) = (1 - \delta)f_0(x) + \delta f_1(x),$$

where $0 \leq \delta < 1$ is the probability to obtain an abnormal observation; $f_0(x)$ is the density function of ordinary observations; and $f_1(x)$ is the density function of abnormal observations. For example, in the model with Huber's contamination, $f_0(\cdot) = \mathcal{N}(0, \sigma^2)$, $f_1(\cdot) = \mathcal{N}(0, \Lambda^2)$.

Estimation for regression models with changing coefficients

Regression models with changing coefficients is another generalization of the contamination model. Suppose a baseline model is described by the following regression:

$$Y = X\beta + \epsilon.$$

However, in an abnormal regime, this model changes into

$$Y = X\gamma + \epsilon,$$

where $\beta \neq \gamma$.

The mechanism of a change is purely random:

$$\alpha = \begin{cases} \beta \text{ with the probability } 1 - \epsilon \\ \gamma \text{ with the probability } \epsilon. \end{cases}$$

We again need to test the hypothesis of statistical homogeneity of an obtained sample and to divide this sample into subsamples of ordinary and abnormal observations if the homogeneity hypothesis is rejected.

The goal of this chapter is to propose methods that can solve these problems effectively. Theoretically, we mean estimation of type-1 and type-2 errors in testing the statistical homogeneity hypothesis, and with estimation of contaminations parameters in the case of rejectiong this hypothesis. Practically, we propose procedures for implementation of these methods for univariate and multivariate models.

The structure of this chapter is as follows. For binary mixtures of probabilistic distributions, I prove Theorem 4.3.1 about exponential convergence to zero of type 1 error in classification (to detect switches for a statistically homogenous sample) as the sample size N tends to infinity; Theorem 4.3.2 about exponential convergence to zero of type-2 error (vice versa, to accept stationarity hypothesis for a sample with switches); and Theorem 4.6.1, which establishes the lower bound for the error of classification for binary mixtures. From Theorems 4.3.2 and 4.6.1 I, conclude that the proposed method is asymptotically optimal by the order of convergence to zero of the classification error.

Results of a detailed Monte-Carlo study of the proposed method for different stochastic models with switching regimes are presented.

4.2 Model

Binary mixtures

We begin from the following simplest case. Suppose the binary mixture is

$$f(x) = (1-\epsilon)f_0(x) + \epsilon f_0(x-h),$$

where $E_0 x_i = 0$, $h \neq 0$, ϵ is unknown.

An ad hoc method of estimation of these parameters is as follows: ordinary and abnormal observations are heuristically classified to two subsamples, and the estimate $\hat{\epsilon}$ is computed as the share of the size of the subsample of abnormal observations in the whole sample size. Clearly, this method is correct only for large values of h. However, this idea of two subsamples can be used in construction of more subtle methods of estimation.

So the problem is:

1. to check the hypothesis H_0 $(\epsilon = 0)$ against the alternative H_1 $(\epsilon \neq 0)$;

2. if the null hypothesis H_0 is rejected, we need to classify observations by different classes (ordinary observations and outliers); and

3. to estimate the parameter ϵ by the sample $X^N = \{x_i\}_{i=1}^N$ on condition that certain additional information is available (see below; we need to know the functional form of the d.f.'s $f_0(\cdot), f_1(\cdot)$).

For these purposes, we do the following:

1) From the initial sample X^N, compute the estimate of the mean value:

$$\theta_N = \frac{1}{N}\sum_{i=1}^{n} x_i.$$

2) Fix the numbers $0 < \kappa < B$ and the parameter $b \in \mathbb{B} = [\kappa, B]$ and classify observations as follows: If an observation falls into the interval $(\theta_N - b, \theta_N + b)$, then we place it into the subsample of ordinary observations, otherwise to the subsample of abnormal observations.

3) Then, for each $b \in \mathbb{B}$, we obtain the following decomposition of the sample X^N into two subsamples,

$$X_1 = \{\tilde{x}_1, \tilde{x}_2, \ldots, \tilde{x}_{N_1}\}, \quad |\tilde{x}_i - \theta_N| < b,$$
$$X_2 = \{\hat{x}_1, \hat{x}_2, \ldots, \hat{x}_{N_2}\}, \quad |\hat{x}_i - \theta_N| \geq b$$

Denote by $N_1 = N_1(b)$, $N_2 = N_2(b)$, $N = N_1 + N_2$ the sizes of the sub-samples X_1 and X_2, respectively.

The parameter b is chosen so that the subsamples X_1 and X_2 are separated in the best way. For this purpose, consider the following statistic:

$$\Psi_N(b) = \frac{1}{N^2}(N_2 \sum_{i=1}^{N_1} \tilde{x}_i - N_1 \sum_{i=1}^{N_2} \hat{x}_i).$$

4) Define the decision threshold $C > 0$ and compare it with the value

$$J = \max_{b \in \mathbb{B}} |\Psi_N(b)|.$$

If $J \leq C$, then we accept the hypothesis H_0 about the absence of outliers; if, however, $J > C$, then the hypothesis H_0 is rejected. Then, our primary goal, is to separate ordinary observations and outliers in the sample.

5) Define the number b_N^*:

$$b_N^* \in \arg\max_{b \in \mathbb{B}} |\Psi_N(b)|.$$

Then,

$$\epsilon_N^* = N_2(b_N^*)/N, \quad h_N^* = \theta_N/\epsilon_N^*$$

are the nonparametric estimate for ϵ, h. In the general case for construction of an unbiased and consistent estimators of ϵ, h, we can use the following relationships:

$$\frac{1 - \hat{\epsilon}_N}{\hat{\epsilon}_N} = \frac{f_0(\theta_N - b_N^* - \hat{h}_N) - f_0(\theta_N + b_N^* - \hat{h}_N)}{f_0(\theta_N + b_N^*) - f_0(\theta_N - b_N^*)},$$
$$\hat{h}_N = \theta_N/\hat{\epsilon}_N.$$

The estimates $\hat{\epsilon}_N$ and $\hat{h}_N$ tend almost surely to the true values ϵ and h as $N \to \infty$. The subsample of abnormal observations is $X_2(b_N^*)$.

This method helps to solve the problem of separating binary mixtures of probabilistic distributions. The next section deals with the formal analysis of the proposed method.

4.3 Main Results

The main results are formulated in the following theorems. Below, we denote by the same symbols C, N_0, L_1, and L_2 probably different constants that do not depend of N.

Theorem 4.3.1. *Let $\epsilon = 0$. Suppose the d.f. $f_0(\cdot)$ is symmetric w.r.t. zero and bounded. For observed random sequence, Cramer's and ψ-mixing conditions are satisfied.*

Then, for any $0 < \kappa < B$, there exist $C > 0, N_0 > 1, L_1 > 0$, and $L_2 > 0$, such that for $N > N_0$, the following estimate holds:

$$\sup_n \mathbf{P}_0\{\sup_{b\in\mathbb{B}} |\Psi_N(b)| > C\} \leq L_1 \exp(-L_2 N).$$

In the proofs given below, we often use the result of Subsection 1.4 (the exponential estimate for the maximum of sums of dependent r.v.'s).

Proof of theorem 4.3.1

For the statistic $\Psi_N(b)$, we can write

$$\Psi_N(b) = \left(N \sum_{i=1}^{N_1(b)} \tilde{x}_i - N_1(b) \sum_{i\in\mathcal{N}} x_i \right) /N^2. \tag{4.3}$$

We have

$$\begin{aligned} \mathbf{P}_0\{\sup_{b\in\mathbb{B}} |\Psi_N(b)| > C\} &\leq \mathbf{P}_0\{\sup_{b\in\mathbb{B}} |\sum_{i=1}^{N_1(b)} \tilde{x}_i| > \frac{CN}{2}\} \\ &+ \mathbf{P}_0\{|\sum_{i\in\mathcal{N}} x_i| > \frac{CN}{2}\}. \end{aligned} \tag{4.4}$$

Further,

$$\mathbf{P}_0 \left\{ \sup_{b\in\mathbb{B}} |\sum_{i=1}^{N_1(b)} \tilde{x}_i| > \frac{C}{2} N \right\} \leq \sum_{n=1}^{N} \mathbf{P}_0 \left\{ \sup_{b\in\mathbb{B}} \{|\sum_{i=1}^{n} \tilde{x}_i| > \frac{C}{2} n\} \cap \{N_1(b) = n\} \right\}. \tag{4.5}$$

Consider the function

$$\Delta(b) = \int_{-b}^{b} f_0(x)dx.$$

Remark that the function $\Delta(b)$ is continuous (in virtue of bounded values of $f_0(\cdot)$), and $\min_{b\in\mathbb{B}} \Delta(b) \geq \int_{-\kappa}^{\kappa} f_0(x)dx \stackrel{\text{def}}{=} u$.

Now, we split the segment $\mathbb{B} = [\kappa, B]$ into equal parts with the interval of splitting such that $|\Delta(b_i) - \Delta(b_{i+1})| \leq u/2$. In virtue of the uniform continuity of $\Delta(b)$, such splitting is possible (here $\{b_i\}$ is the set if bounds of this splitting).

Denote the number of subsegments as R and subsegments of splitting as $\mathbb{B}_s$, $s = 1, \ldots, R$. Then,

$$\begin{aligned}
&\mathbf{P}_0\left\{\sup_{b\in\mathbb{B}}\{|\sum_{i=1}^{n}\tilde{x}_i| > \frac{C}{2}n\} \cap \{N_1(b) = n\}\right\} \\
&= \mathbf{P}_0\left\{\max_s \sup_{b\in\mathbb{B}_s}\{|\sum_{i=1}^{n}\tilde{x}_i| > \frac{C}{2}n\} \cap \{N_1(b) = n\}\right\} \\
&\leq R \max_s \mathbf{P}_0\left\{\sup_{b\in\mathbb{B}_s}\{|\sum_{i=1}^{n}\tilde{x}_i| > \frac{C}{2}n\} \cap \{N_1(b) = n\}\right\}.
\end{aligned} \tag{4.6}$$

Consider a fixed subsegment $\mathbb{B}_i \stackrel{\text{def}}{=} [b_i, b_{i+1}]$. In virtue of the definition of the numbers $N_1(b)$, we have for any $b \in \mathbb{B}_i$:

$$N_1(b_i)/N \leq N_1(b)/N \leq N_1(b_{i+1})/N.$$

Now, we estimate the probabilities of the following events:

$$|N_1(b_i)/N - \Delta(b_i)| \leq u/4, \quad |N_1(b_{i+1})/N - \Delta(b_{i+1})| \leq u/4.$$

First, we estimate the probability of deviation of θ_N from its mathematical expectation $\mathbf{E}_m\theta_N \equiv 0$.

From (1.15), it follows that for any $\gamma > 0$ for large enough N

$$\mathbf{P}_0\{|\theta_N| > \gamma\} \leq A(\gamma)\exp\left(-B(\gamma)N\right). \tag{4.7}$$

By definition, $N_1(b) = \sum_{k\in\mathcal{N}} \mathbb{I}(|x_k - \theta_N| \leq b)$. Then, for any fixed $r > 0$, we have

$$\mathbf{P}_0\{|x_k - \theta_N| \leq b\} \leq \mathbf{P}_0\{|x_k| \leq b + r\} + \mathbf{P}_0\{|\theta_N| > r\}. \tag{4.8}$$

Moreover,

$$\mathbf{P}_0\{|x_k - \theta_N| \leq b\} \geq \mathbf{P}_0\{|x_k| \leq b - r\} - \mathbf{P}_0\{|\theta_N| > r\}. \tag{4.9}$$

It follows from (1.15) that for any point b_i of the splitting of the segment $\mathbb{B}$ for large enough N,

$$\mathbf{P}_0\{|\frac{1}{N}\sum_{k\in\mathcal{N}}(\mathbb{I}(|x_k - \theta_N| \leq b_i) - \mathbf{E}_0(\mathbb{I}(|x_k - \theta_N| \leq b_i))\,| > u/2\} \leq A(u)\exp(-B(u)N). \tag{4.10}$$

From (4.8) and (4.9), it follows that

$$\begin{aligned}
&\Delta(b_i + r) + O\left(\exp(-B(r)N)\right) \geq \mathbf{E}_0(\mathbb{I}(|x_k - \theta_N| \leq b_i) \geq \Delta(b_i - r) \\
&\quad + O\left(\exp(-B(r)N)\right).
\end{aligned}$$

Since the function $\Delta(\cdot)$ satisfies the Lipschitz condition (due to the finiteness of the density function), then combining all these inequalities, we can conclude for a certain r (e.g., $0 < r < u/4$) that for $N > N_0$,

$$\mathbf{P}_0\{|\frac{1}{N}\sum_{k\in\mathcal{N}} \mathbb{I}(|x_k - \theta_N| \leq b_i) - \Delta(b_i)| > u/4\} \leq A(u)\exp(-B(u)N) \stackrel{\text{def}}{=} \gamma(u,N). \tag{4.11}$$

Estimate (4.11) holds for any point b_i.

Then, in virtue of (4.6) with the probability no less than $(1-\gamma(u,N))$ for $N > N_0$, we have for all $b \in \mathbb{B}_i$:

$$(\Delta(b_i) - u/4)N \leq N_1(b) \leq (\Delta(b_i) + u/2)N. \tag{4.12}$$

Now, split the set of all possible values of $N_1(b), b \in \mathbb{B}_i$ into two subsets: $\mathbb{A}_i \stackrel{\text{def}}{=} \{1 \leq n \leq N : [(\Delta(b_i) - u/4)N] \leq n \leq [(\Delta(b_i) + u/2)N]\}$ and its complement. Then, $\mathbb{P}_0(\mathbb{A}_i) \geq (1-\gamma(u,N))$ for $N > N_0$.

Then,

$$\begin{aligned} &\mathbf{P}_0\left\{\sup_{b\in\mathbb{B}_i}\{|\sum_{i=1}^{n}\tilde{x}_i| > \frac{C}{2}n\} \cap \{N_1(b) = n\}\right\} \leq \gamma(u,N) \\ &+\mathbf{P}_0\left\{\max_{n\in\mathbb{A}_i}\{|\sum_{i=1}^{n}\tilde{x}_i| > \frac{C}{2}n\}\right\}. \end{aligned} \tag{4.13}$$

For the probability in the right hand of (4.13), we again use (1.15) and remark that $\Delta(b) \geq u$. Then, we obtain

$$\begin{aligned} &\mathbf{P}_0\left\{\max_{n\in\mathbb{A}_i}\{|\sum_{i=1}^{n}\tilde{x}_i| > \frac{C}{2}n\}\right\} \leq A(C)\exp\left(-N(\Delta(b_i) - u/4)B(C)\right) \leq \\ &\leq A(C)\exp(-NB(C)3/4u). \end{aligned} \tag{4.14}$$

Since all considerations are valid for any subsegment, from (4.13) and (4.14), we obtain for any $s = 1\ldots, R$

$$\mathbf{P}_0\left\{\sup_{b\in\mathbb{B}_s}\{|\sum_{i=1}^{n}\tilde{x}_i| > \frac{C}{2}n\} \cap \{N_1(b) = n\}\right\} \leq A(C)\exp(-NB(C)u/4). \tag{4.15}$$

An analogous estimate holds true for the second term in (4.4).

Taking into account (4.4), (4.5), (4.6), (4.11), and (4.15) we obtain the exponential estimate considered in formulation of Theorem 4.3.1.

Now, consider characteristics of this method in the general case $\epsilon h \neq 0$. Here, we again assume that $E_0\, x_i = 0$, $i = 1, \ldots, N$.

Probabilities of estimation errors can be written as follows:

$$\alpha(\delta) = P\{|\hat{\epsilon}_N - \epsilon| > \delta\},\ \beta(\gamma) = P\{|\hat{h}_N - h| > \gamma\}, \quad \delta, \gamma > 0.$$

These values depend on the mathematical expectation of the statistic $\Psi_N(b)$:

$$E\Psi_N(b) = \Phi(b) + O(\frac{1}{N}) = r(b) - \epsilon h d(b) + O(\frac{1}{N}),$$

where

$$r(b) = \int_{\epsilon h - b}^{\epsilon h + b} f(x) x dx, \quad d(b) = \int_{\epsilon h - b}^{\epsilon h + b} f(x) dx.$$

For $\Phi^{'}(b)$, we obtain

$$\Phi^{'}(b) = (\epsilon h + b) f(\epsilon h + b) + (\epsilon h - b) f(\epsilon h - b) - \epsilon h[f(\epsilon h + b) + f(\epsilon h - b)]$$
$$= b(f(\epsilon h + b) - f(\epsilon h - b)).$$

If $f_0(\cdot) = f_1(\cdot)$ and $f_0(\cdot)$ is symmetric w.r.t. zero, then $\Phi^{'}(b) \equiv 0$ and $\Psi(b) \equiv 0$. In the general case, $\Phi^{'}(b)$ is zero at the roots of the equation

$$f(\epsilon h + b) = f(\epsilon h - b). \tag{4.16}$$

In the following theorem, type-2 error is studied.

Theorem 4.3.2. *Suppose all assumptions of Theorem 4.3.1 are satisfied and $\epsilon \neq 0$. Then,*

1) for type-2 error, we can write for $0 < C < \max_{b \in \mathbb{B}} |\Phi(b)|$:

$$P_1\{\max_{b \in \mathbb{B}} |\Psi_N(b)| \leq C\} \leq 4\phi_0 \exp(-L(\delta)N),$$

where $L(\delta) = \min(\frac{\delta^2}{16\phi_0^2 g}, \frac{H\delta}{8\phi_0})$, $\delta = \max_{b \in \mathbb{B}} |\Psi(b)| - C > 0$.

2) $b_N^* \to b^*$ almost surely as $N \to \infty$, where $b^* \in arg \max_b |\Psi(b)|$.

3) $\hat{\epsilon}_N \to \epsilon$ and $\hat{h}_N \to h$ almost surely as $N \to \infty$.

Proof of Theorem 4.3.2.

For type-2 error, we can write

$$P_1\{\max_{b \in \mathbb{B}} |\Psi_N(b)| \leq C\} \leq P_1\{|\Psi_N(b) - \Phi(b)| \geq |\Phi(b)| - C\},$$

for any $b \in \mathbb{B}$.

So, we need to choose such $b \in \mathbb{B}$ that $|\Phi(b)| - C \geq \delta > 0$. But the function $|\Phi(b)|$ is continuous. Therefore, its maximum is attained at a certain point b^* of the segment $\mathbb{B}$. So, if $\max_{b \in \mathbb{B}} |\Phi(b)| - C \geq \delta > 0$, we need to prove the above inequality for b^*.

Consider the main statistic

$$\Psi_N(b) = (N \sum_{i=1}^{N_1(b)} \tilde{x}_i - N_1(b) \sum_{i=1}^{N} x_i)/N^2.$$

As before, we can write

$$E\Psi_N(b) = \Phi(b) + O(\frac{1}{N}),$$

where $\Psi(b) = r(b) - \epsilon h\, d(b)$.

For any $C > 0$, we can write

$$\begin{aligned} P\{|\Psi_N(b) - \Psi(b)| > C\} &\leq P\{|\sum_{i=1}^{N_1(b)} \tilde{x}_i - Nr(b)| > \frac{C}{2}N\} \\ &+P\{|\frac{N_1(b)}{N}\sum_{i=1}^{N} x_i - N\epsilon hd(b)| > \frac{C}{2}N\}. \end{aligned} \tag{4.17}$$

Consider the first term in the right hand:

$$\begin{aligned} P\{|\sum_{i=1}^{N_1(b)} \tilde{x}_i - Nr(b)| > \frac{C}{2}N\} &= P\{\sum_{i=1}^{N_1(b)} \tilde{x}_i > \frac{C}{2}N + Nr(b)\} \\ &+P\{\sum_{i=1}^{N_1(b)} \tilde{x}_i < -\frac{C}{2}N + Nr(b)\}. \end{aligned} \tag{4.18}$$

As in Theorem 4.3.1, we decompose the sum $\sum_{i=1}^{N_1} \tilde{x}_i$ into ϕ_0 groups of weakly dependent components, and for each of these groups, use Chebyshev's inequality. Then,

$$P\{\sum_{i=1}^{N_1(b)} \tilde{x}_i > \frac{C}{2}N + Nr(b)\} \leq \phi_0 \max_i P\{|\tilde{S}^i_{N_1}| \geq \frac{\frac{CN}{2} + r(b)N}{\phi_0}\},$$

$$\begin{aligned} &P\{|\tilde{S}^i_{N_1}| \geq \frac{\frac{CN}{2} + r(b)N}{\phi_0}\} \\ &\leq (1+\epsilon)^N \exp(-t\frac{\frac{CN}{2} + r(b)N}{\phi_0}) \exp(t\frac{N}{\phi_0}r(b) + \frac{1}{2}t^2Ng)). \end{aligned}$$

Finally, taking the maximum by t of the right hand, we obtain:

$$P\{\sum_{i=1}^{N_1(b)} \tilde{x}_i > \frac{C}{2}N + Nr(b)\} \leq \phi_0 \begin{cases} \exp(-\frac{C^2N}{16\phi_0^2 g}), & 0 < t < gH, \\ \exp(-\frac{CHN}{8\phi_0}), & t > gH. \end{cases}$$

The second term in the right hand of (4.18) is estimated from the above in the same way.

As to the second term in the right hand of (4.17), since $N_1(b) \leq N$ for any ω, we obtain an analogous exponential upper estimate for it.

So, we proved that for $b = b^*$,

$$P\{|\Psi_N(b) - \Psi(b)| > C\} \leq 4\phi_0 \begin{cases} \exp(-\frac{C^2N}{16\phi_0^2 g}), & 0 < C < gH, \\ \exp(-\frac{CHN}{8\phi_0}), & C > gH. \end{cases}$$

This completes the proof of 1).

As to the proof of 2), remark that the function $\Phi(b) = E\Psi_N(b)$ satisfies the reversed Lipschitz condition in a neighborhood of b^*. In fact, we have $\Phi(b^*) = 0$, $\Phi'(b^*) = 0$, and $\Phi''(b^*) = (f(\epsilon h + b^*) - f(\epsilon h - b^*)) + b^*(f'(\epsilon h + b^*) - f'(\epsilon h - b^*)) = 2(b^*)^2 f''(u) \neq 0$, where $0 \leq u = u(b^*) \leq b^*$. Therefore, in a small neighborhood of b^*, we obtain:

$$|\Phi(b) - \Phi(b^*)| = (b^*)^2 \, |f''(u(b^*))|(b - b^*)^2 \geq C(b - b^*)^2,$$

for a certain $C = C(b^*) > 0$ (we assume that $f''(u) \neq 0$ and is continuous for any $u > 0$).

Now, for any $0 < \epsilon < 1$, consider the event $|b_N - b^*| > \epsilon$. Then,

$$\begin{aligned} P\{|b_N - b^*| > \epsilon\} &\leq P\{\max_{b \in \mathbb{B}} |\Psi_N(b_N) - \Phi(b^*)| > \frac{1}{2} C\epsilon^2\} \\ &\leq 4\,\phi_0 \exp(-L(C)N), \end{aligned}$$

where $L(C) = \min(\dfrac{C^2\epsilon^4}{64\phi_0^2 g}, \dfrac{HC\epsilon^2}{16\phi_0})$.

From this inequality, it follows that $b_N \to b^*$ $P-$a.s. as $N \to \infty$.

Then,

$$\epsilon_N = N_2(b_N)/N, \qquad h_N = \theta_N/\epsilon_N$$

are the nonparametric estimates for ϵ and h, respectively.

In general, these estimates are asymptotically biased and nonconsistent. For construction of consistent estimates of ϵ and h, we need information about the d.f.'s $f_0(\cdot), f_1(\cdot)$. These consistent estimates can be obtained from the following relationships:

$$\begin{aligned} &\frac{1 - \hat{\epsilon}_N}{\hat{\epsilon}_N} = \frac{f_0(\theta_N - b_N - \hat{h}_N) - f_0(\theta_N + b_N - \hat{h}_N)}{f_0(\theta_N + b_N) - f_0(\theta_N - b_N)} \\ &\hat{h}_N = \theta_N/\hat{\epsilon}_N. \end{aligned}$$

The estimates $\hat{\epsilon}_N$ and $\hat{h}_N$ are connected with the estimate b_N of the parameter b^* via this system of deterministic algebraic equations. Therefore, the rate of convergence $\hat{\epsilon}_N \to \epsilon$ and $\hat{h}_N \to h$ is determined by the rate of convergence of b_N to b^* (which is exponential w.r.t. N). So, we conclude that $\hat{\epsilon}_N \to \epsilon$ and $\hat{h}_N \to h$ P-a.s. as $N \to \infty$.

Theorem 4.3.2 is proved.

4.4 Example

The proposed method was tested in experiments with different switching models.

TABLE 4.1
Binary Mixture of Distributions; Change-In-Mean Problem; α-Quantiles of the Decision Statistic

N	100	300	500	800	1000	1200	1500	2000
$\alpha = 0.95$	0.1213	0.0710	0.0534	0.044	0.0380	0.037	0.034	0.029
$\alpha = 0.99$	0.1410	0.0869	0.0666	0.050	0.0471	0.0390	0.038	0.035

TABLE 4.2
Binary Mixtue of Distributions; Change-In-Mean Problem; Characteristics of the Proposed Method

$\epsilon = 0.1$	h=2.0				h=1.5			
N	300	500	800	1000	800	1200	2000	3000
C	0.0710	0.0534	0.044	0.038	0.044	0.037	0.029	0.022
w_2	0.26	0.15	0.05	0.02	0.62	0.42	0.16	0.03
$\hat{\epsilon}$	0.104	0.101	0.097	0.099	0.106	0.103	0.102	0.0985

1) In the first series of tests, the following mixture model was studied:

$$f_\epsilon(x) = (1-\epsilon)f_0(x) + \epsilon f_0(x-h), \quad f_0(\cdot) = \mathcal{N}(0,1), \quad 0 \le \epsilon < 1/2.$$

First, the critical thresholds of the decision statistic $\max|\Psi_N(b)|$ by $b > 0$ were computed. For this purpose, for homogenous samples (for $\epsilon = 0$), α-quantiles of the decision statistic $\max_b |\Psi_N(b)|$ were computed ($\alpha = 0.95, 0.99$). The results obtained in 5000 independent trials of each experiment are presented in Table 4.1.

The quantile value for $\alpha = 0.95$ was chosen as the critical threshold C in experiments with non-homogenous samples (for $\epsilon \neq 0$). For different sample sizes in 5000 independent trials of each test, the estimate of type 2 error w_2 (i.e. the frequency of the event $\max_b |\Psi_N(b)| < C$ for $\epsilon > 0$) and the estimate $\hat{\epsilon}$ of the parameter ϵ were computed. The results are presented in table 4.2.

4.5 Recommendations for the Choice of Decision Threshold

For practical applications of obtained results, we need to compute the threshold C.

For computation of this threshold, we begin from homogenous samples (without outliers). For such samples, we can define the threshold C from the following empirical formula, which can be obtained from Theorem 4.3.1:

$$C = C(N) \sim \sigma \sqrt{\frac{\phi_0(\cdot)\,|\ln \alpha|}{N}}, \tag{4.19}$$

where N is the sample size; σ^2 is the dispersion of ϕ_0-dependent observations; and α is the 1st-type error level.

We compute the dispersion σ^2 of observations and the integer ϕ_0 from the argument of the first zero of the autocorrelation function of the sample. Then we compute the threshold C.

Now, let us give a practical example:

Consider the following model of observations without outliers:

$$x(n) = \rho x(n-1) + \sigma\, \xi_n, \qquad n = 1, \ldots, N,$$

where ξ_n are i.i.d.r.v.'s with the Gaussian d.f. $N(0, 1)$.

Here, we choose $\phi_0(\cdot) = [(1-\rho)^{-1}]$, taking into account that the value $(1-\rho)^{-1}$ is the sum of the series with degrees of ρ.

Below, for computation of the relationship for the threshold C, we derived predictors from this formula: the sample size N; the parameter σ; and the autocorrelation coefficient ρ; the confidence probability β.

In our experiment, we used the sample with the volume $n = 234$ which includes different values of these parameters: ($N = 50 \to 2000$; $\sigma = 0.1 \to 5$; $\rho = -0.8 \to 0.8$; and $\beta = 0.95 \to 0.99$).

The obtained regression relationship for C has the following form:

$$\begin{aligned} &log(C) = -0.9490 - 0.4729 * log(N) + 1.0627 * log(\sigma) - 0.6502 * log(1-\rho) \\ &-0.2545 * log(1-\beta). \end{aligned} \tag{4.20}$$

Remark that for this relationship, $R^2 = 0.978$ and the series of regression residuals is stationary for the error level 5%. The elasticity coefficient by the factor N is close to its theoretical level -0.5 (form the above formula).

4.6 Asymptotic Optimality

Now, consider the question about the asymptotic optimality of the proposed method in the class of all estimates of the parameter ϵ. The a priori theoretical lower bound for the estimation error of the parameter ϵ in the model $f_\epsilon(x) = (1-\epsilon)f_0(x) + \epsilon f_1(x)$ is given in the following theorem.

Theorem 4.6.1. *Suppose assumptions of Theorem 2.6.1 are satisfied. Let $\mathcal{M}_N$ be the class of all estimates of the parameter ϵ. Then, for any $0 < \delta < \epsilon$,*

$$\liminf_{N\to\infty} \inf_{\hat{\epsilon}_N \in \mathcal{M}_N} \sup_{0<\epsilon<1/2} \frac{1}{N} \ln P_\epsilon\{|\hat{\epsilon}_N - \epsilon| > \delta\} \geq -\delta^2 J(\epsilon),$$

where $J(\epsilon) = \int [(f_0(x) - f_1(x))^2 / f_\epsilon(x)]\, dx$ *is the generalized* $\varkappa^2$ *distance between densities* $f_0(x)$ *and* $f_1(x)$, *and* P_ϵ *is the measure corresponding to the density* $f_\epsilon(x)$.

The proof of this theorem is analogous to the proof of Theorem 2.6.1 from Chapter 2.

Comparing results of Theorems 4.3.2 and 4.5.1, we conclude that the proposed method is asymptotically optimal by the order of convergence of the estimates of a mixture of parameters to their true values.

4.7 Generalizations. Nonsymmetric Distribution Functions

Results obtained in Theorems 4.3.1 and 4.3.2 can be generalized to the case of nonsymmetric distribution functions. Suppose the d.f. $f_0(\cdot)$ is asymmetric w.r.t. zero. Then, we can modify the proposed method as follows.

1. From the initial sample $X^N = \{x_1, \ldots, x_N\}$, compute the mean value $\theta_N = \frac{1}{N}\sum_{i=1}^{N} x_i$ and the sample $Y^N = \{y_1, \ldots, y_N\}$; $y_i = x_i - \theta_N$. Then we divide the sample Y^N into two subsamples, $I_1(b)$, and $I_2(b)$, as follows:

$$y_i \in \begin{cases} I_1(b), & -\phi(b) \le y_i \le b \\ I_2(b), & y_i > b \text{ or } y_i < -\phi(b), \end{cases}$$

where the function $\phi(b)$ is defined from the following condition: $0 = \int_{-\phi(b)}^{b} y\, f_0(y) dy$ and $f_0(y) = f_0(x - \bar{x})$.

2. As before, we compute the statistic

$$\Psi_N(b) = \frac{1}{N^2}(N_2(b) \sum_{i=1}^{N_1(b)} \tilde{y}_i - N_1(b) \sum_{i=1}^{N_2(b)} \hat{y}_i),$$

where $N = N_1(b) + N_2(b)$ and $N_1(b), N_2(b)$ are sample sizes of $I_1(b)$, and $I_2(b)$, respectively.

3. Then, the value $J = \max_b |\Psi_N(b)|$ is compared with the threshold C. If $J \le C$, then the hypothesis H_0 (no abnormal observations) is accepted; if, however, $J > C$, then the hypothesis H_0 is rejected and the estimate of the parameter ϵ is constructed.

4. For this purpose, define the value b_N^*:

$$b_N^* \in \arg\max_{b>0} |\Psi_N(b)|.$$

Then,

$$\epsilon_N^* = N_2(b_N^*)/N.$$

This method can be used for the study of the classic ϵ-contamination model

$$f_\epsilon(\cdot) = (1-\epsilon)\mathcal{N}(\mu, \sigma^2) + \epsilon\mathcal{N}(\mu, \Lambda^2), \quad \Lambda^2 >> \sigma^2, \quad 0 \le \epsilon < 1/2.$$

For this model, the following method was used:

1. From the sample of observations $X^N = \{x_1, \dots, x_N\}$, the mean value estimate $\hat{\mu} = \sum_{i=1}^N x_i/N$ was computed.

2. The sequence $y_i = (x_i - \hat{\mu})^2$, $i = 1, \dots, N$ and its empirical mean, $\theta_N = \sum_{i=1}^N y_i/N$, are computed.

3. Then for each $b \in [0, B_{max}]$, the sample $Y^N = \{y_1, \dots, y_N\}$ is divided into two sub-samples in the following way: for $\theta_N(1 - \phi(b)) \le y_i \le \theta_N(1+b)$ put $\tilde{y}_i = y_i$ (the size of the subsample $N_1 = N_1(b)$), otherwise put $\hat{y}_i = y_i$ (the size of the subsample $N_2 = N_2(b)$). Here, we choose

$$\phi(b) = 1 - \frac{b}{e^b - 1}.$$

4. For any $b \in [0, B_{max}]$, the following statistic is computed:

$$\Psi_N(b) = \frac{1}{N^2}(N_2 \sum_{i=1}^{N_1} \tilde{y}_i - N_1 \sum_{i=1}^{N_2} \hat{y}_i),$$

where $N = N_1 + N_2$, $N_1 = N_1(b)$, and $N_2 = N_2(b)$ are sizes of subsamples of ordinary and abnormal observations, respectively.

5. Then, as above, the threshold $C > 0$ is chosen and compared with the value $J = \max_b |\Psi_N(b)|$. If $J \le C$ then the hypothesis H_0 (no abnormal observations) is accepted; if, however, $J > C$ then the hypothesis H_0 is rejected and the estimate of the parameter ϵ is constructed.

5) For this purpose, define the value b_N^*:

$$b_N^* \in \arg\max_{b>0} |\Psi_N(b)|.$$

Then,

$$\epsilon_N^* = N_2(b_N^*)/N.$$

In experiments, the critical values of the statistic $\max_b |\Psi_N(b)|$ were computed. For this purpose, as above, for homogenous samples (for $\epsilon = 0$), α-quantiles of the decision statistic $\max_b |\Psi_N(b)|$ were computed ($\alpha = 0.95, 0.99$). The results obtained in 1000 trials of each test are presented in Table 4.3.

The quantile value for $\alpha = 0.95$ was chosen as the critical threshold C in experiments with nonhomogenous samples (for $\epsilon \ne 0$). For different sample sizes in 5000 independent trials of each test, the estimate of type-2 error w_2

TABLE 4.3
Binary Mixture of Distributions; Change-In-Dispersion Problem; α-Quantiles of the Decision Statistic

N	100	300	500	800	1000	1200	1500	2000
0.95	0.2330	0.1570	0.1419	0.1252	0.1244	0.1146	0.1107	0.1075
0.99	0.2862	0.1947	0.1543	0.1436	0.1331	0.1269	0.1190	0.1157

TABLE 4.4
Binary Mixture of Distributions; Change-In-Dispersion Problem; Characteristics of the Proposed Method ($\Lambda = 3.0$)

$\Lambda = 3.0$	$\epsilon = 0.05$			
N	300	500	800	1000
C	0.1570	0.1419	0.1252	0.1244
w_2	0.27	0.15	0.06	0.04
$\hat{\epsilon}$	0.064	0.056	0.052	0.05

(i.e., the frequency of the event $\max_b |\Psi_N(b)| < C$ for $\epsilon > 0$) and the estimate $\hat{\epsilon}$ of the parameter ϵ were computed. The results are presented in Tables 4.4 and 4.5.

TABLE 4.5
Binary Mixture of Distributions; Change-In-Dispersion Problem; Characteristics of the Proposed Method ($\Lambda = 5.0$)

$\Lambda = 5.0$	$\epsilon = 0.01$				
N	1000	1200	1500	2000	3000
C	0.1244	0.1146	0.1107	0.1075	0.1019
w_2	0.25	0.20	0.15	0.10	0.04
$\hat{\epsilon}$	0.0135	0.013	0.012	0.011	0.010

4.8 Conclusions

In this chapter, problems of the retrospective detection/estimation of random switches in univariate models were considered. For the scheme of a mixture of probabilistic distributions, I propose a nonparametric method for the retrospective detection of the number of d.f.'s and classification of observations with different d.f.'s. Beginning from the simple model of a binary mixture, I prove that probabilities of type-1 (Theorem 4.3.1) and type-2 (Theorem 4.3.2) errors of the proposed method converge to zero exponentially as the sample size N tends to infinity.

Then I consider the informational lower bound for performance efficiency of classification methods (Theorem 4.6.1). The asymptotic optimality of the proposed method follows from Theorem 4.6.1. The lower bound for performance efficiency is attained for the proposed method (by the order of convergence to zero of the estimation error).

Then, some results of computer experiments with the proposed methods are given.

5

Retrospective Change-Point Detection and Estimation in Multivariate Stochastic Models

5.1 Introduction

The change-point problem for regression models was first considered by Quandt (1958, 1960). Using econometric examples, Quandt proposed a method for estimation of a change-point in a sequence of independent observations based upon the likelihood ratio test.

Let us describe the change-point problem for the linear regression models considered in the literature. Let $y_1, y_2, \dots, y_n$ be independent random variables (i.r.v.'s). Under the null hypothesis $\mathbf{H}_0$, the linear model is

$$y_i = \mathbf{x}_i^* \beta + \epsilon_i, \quad 1 \le i \le n,$$

where $\beta = (\beta_1, \beta_2, \dots, \beta_d)^*$ is an unknown vector of coefficients, and $\mathbf{x}_i^* = (1, x_{2i},$
$\dots, x_{di})$ are known predictors (here and below, $*$ is the transposition symbol).

The errors ϵ_i are supposed to be independent identically distributed random variables (i.i.d.r.v.'s), with $\mathbf{E}\epsilon_i = 0, \quad 0 < \sigma^2 = var\, \epsilon_i < \infty$.

Under the alternative hypothesis $\mathbf{H}_1$, a change at the instant k^* occurs, i.e.,

$$y_i = \begin{cases} \mathbf{x}_i^* \beta + \epsilon_i, & 1 \le i \le k^* \\ \mathbf{x}_i^* \gamma + \epsilon_i, & k^* < i \le n, \end{cases}$$

where k^* and $\gamma \in \mathbb{R}^d$ are unknown parameters, and $\beta \ne \gamma$.

Denote

$$\bar{y}_k = \frac{1}{k} \sum_{1 \le i \le k} y_i, \bar{\mathbf{x}}_k = \frac{1}{k} \sum_{1 \le i \le k} \mathbf{x}_i,$$
$$Q_n = \sum_{1 \le i \le n} (\mathbf{x}_i - \bar{\mathbf{x}}_n)(\mathbf{x}_i - \bar{\mathbf{x}}_n)^*$$

and $\mathbf{X}_n = (\mathbf{x}_1, \mathbf{x}_2, \dots, \mathbf{x}_n)^*$, $Y_n = (y_1, y_2, \dots, y_n)^*$.

The least square estimate of β is

$$\hat{\beta}_{\mathbf{n}} = (\mathbf{X}_n^* \mathbf{X}_n)^{-1} \mathbf{X}_n^* Y_n.$$

Siegmund with coauthours (James, James, and Siegmund (1989)) proposed to reject $\mathbf{H}_0$ for the large values of $\max_{1\le k\le n} |U_n(k)|$, where

$$U_n(k) = (\frac{k}{1-k/n})^{1/2} \frac{\bar{y}_k - \bar{y}_n - \hat{\beta}_n(\bar{\mathbf{x}}_\mathbf{k} - \bar{\mathbf{x}}_n)^*}{(1 - k(\bar{x}_k - \bar{x}_n)(\bar{x}_k - \bar{x}_n)^*/(Q_n(1-k/n)))^{1/2}}.$$

Earlier, Brown, Durbin, and Evans (1975) used the cumulative sums of regression residuals

$$\sum_{1\le i\le k} (y_i - \bar{y}_n - \hat{\beta}_n(\mathbf{x}_i - \bar{\mathbf{x}}_n)^*), \quad 1\le k\le n.$$

It is easy to see that

$$U_n(k) = w_n(k)\, R_n(k)$$
$$R_n(k) = (\frac{n}{k(n-k)})^{1/2} \sum_{1\le i\le k} (y_i - \bar{y}_n - \hat{\beta}_n(\mathbf{x}_i - \bar{\mathbf{x}}_n)^*)$$
$$w_n(k) = 1 - k(\bar{\mathbf{x}}_k - \bar{\mathbf{x}}_n)(\bar{\mathbf{x}}_k - \bar{\mathbf{x}}_n)^*/(Q_n(1-k/n)))^{-1/2}.$$

The functionals of $U_n(k)$ and $R_n(k)$ were used as the test statistics for detection of change-points in regression relashionships.

Kim and Siegmund (1989) obtained the limit distribution of $\max_{1\le k<n} |U_n(k)|$. Alternatively, Maronna and Yohay (1978) and Worsley (1986) used the maximum likelihood method for testing $\mathbf{H}_0$ against $\mathbf{H}_1$ for Gaussian errors. Later, Gombay and Horváth (1994) studied the limit distributions of statistics $Z_n(i,j) = \max_{i\le k<j} |U_n(k)|$, $T_n(i,j) = \max_{i\le k<j} |R_n(k)|$ for deterministic and stochastic regression plans. The monograph by Cśorgo and Horváth (1997) puts together various results in detection of structural changes in regression models.

Besides change-point detection problems, results in change-point estimation for regressions are of especial practical importance. This theme is considered in papers by Darkhovsky (1995), Huskova (1996), and Horvath, Huskova, and Serbinovska (1997). In the last two papers, the asymptotical characteristics of change-point estimates based upon the maximum likelihood statistics are studied. For the case of contiguous alternatives, the limit distribution of the change-point estimates is obtained, and weak and strong consistency of these estimates is proved. The paper by Darkhovsky (1995) develops the nonparametric approach to retrospective change-point estimation. Here, the limit characteristics of change-point estimates in the functional regression model are studied without the contiguity assumption, and the rate of convergence of these estimates to the "true" change-point parameters is estimated. Some generalizations of these results can be found in the monograph by Brodsky and Darkhovsky (2000).

A new wave of research interest to change-point problems in regressions was formed in the 2000s. Different generalizations to change-point problems for autoregressive time series (Huskova, Praskova, and Steinebach (2007, 2008), Gombay (2008)), for multiple change-point estimation in nonstationary time series (Davis, Lee, and Rodriguez-Yam (2006)), and for testing change-points in covariance structure of linear processes (Berkes, Gombay, and Horvath (2009)) were studied.

However, as a result, we see the multitude of methods proposed for solving different change-point problems in linear relationships and almost no theoretical approaches to their *comparative analyses.* We cannot even estimate the asymptotic efficiency of these methods. All that is empirically observed for "structural breaks" tests in statistics and econometrics can be reduced to the following "vague" statement: The power of these methods is rather low. Let us agree that this "practical conclusion" requires a more serious verification.

In this chapter, I pursue the following main goals:

1) To prove the prior theoretical lower bounds for the error probability in change-point estimation in multivariate models. These bounds provide the theoretical basis for the proofs of the asymptotic optimality of change-point estimates and for the comparative analysis of these estimates.

2) To propose a new nonparametric method for the problem of retrospective change-point detection and estimation in multivariate linear systems. Then we study the main performance characteristics of this method: type-1 and type-2 errors, the error of change-point estimation.

3) For the problem of multiple change-point detection and estimation, to propose a general statement in which both *the number of change-points and their coordinates in the sample are unknown.* For this problem statement, I propose a new asymptotically optimal method that gives consistent estimates of an unknown number of change-points and their coordinates.

The structure of this chapter is as follows. In Section 5.2, the general change-point problem for multivariate linear systems is formulated and general assumptions are given. In Section 5.3, I prove the prior informational inequalities for the main performance characteristic of the retrospective change-point problem, namely, the error of change-point estimation. The lower bounds for the error of estimation are found in different situations of change-point detection (deterministic and stochastic regression plan, multiple change-points). In Section 5.4, I propose a new method for the retrospective change-point detection and estimation in multivariate linear models and study its main performance characteristics (type-1 and type-2 errors, the error of estimation) in different situations of change-point detection and estimation (dependent observations, deterministic and stochastic regression plan, multiple change-points). I prove that this method is asymptotically optimal by the order of convergence of change-point estimates to their true values as the sample size tends to infinity. In Section 5.5, a simulation study of characteristics of the proposed method for finite sample sizes is performed. The main goals of this study are as follows: to compare performance characteristics of the proposed method with characteristics of other well-known methods of change-point detection in linear regression models, and to consider more general multivariate linear models and performance characteristics of the proposed method in these multivariate models. Section 5.6 contains main conclusions.

5.2 Problem Statement

5.2.1 Model

The following basic specification of the multivariate system with structural changes is considered:

$$\mathbf{Y}(n) = \mathbf{\Pi X}(n) + \nu_n, \quad n = 1, \ldots, N, \tag{5.1}$$

where $\mathbf{Y}(n) = (y_{1n}, \ldots, y_{Mn})^*$ is the vector of endogenous variables, $\mathbf{X}(n) = (x_{1n}, \ldots, x_{Kn})^*$ is the vector of predetermined variables, Π is $M \times K$ matrix, and $\nu_n = (\nu_{1n}, \ldots, \nu_{Mn})^*$ is the vector of random errors.

The matrix $\Pi = \Pi(\vartheta, n)$, $\vartheta = (\theta_1, \ldots, \theta_k)$ can change abruptly at some unknown change-points $m_i = [\theta_i N]$, $i = 1, \ldots, k$ (here and below, $[a]$ denotes the integer part of number a), i.e.,

$$\mathbf{\Pi}(\vartheta, n) = \sum_{i=1}^{k+1} \mathbf{a}_i \, \mathbb{I}([\theta_{i-1} N] < n \leq [\theta_i N]),$$

where θ_i are unknown change-point parameters, such that $0 \equiv \theta_0 < \theta_1 < \ldots \theta_k < \theta_{k+1} \equiv 1$, $\mathbf{a}_i \neq \mathbf{a}_{i+1}$, $i = 1, \ldots, k$ are unknown matrices (here and below, $\mathbb{I}(A)$ is the indicator of the set A).

The problem is to estimate the unknown parameters θ_i (and, therefore, the change-points m_i) by observations $\mathbf{Y}(i), \mathbf{X}(i)$, $i = 1, \ldots, N$ (the case $\theta_i \equiv 1, i = 1, \ldots, k$ corresponds to the model without change-points).

Therefore, first, we need to test an obtained dataset of observations for the presence of change-points. Second, in the case of a rejected stationarity hypothesis, we wish to estimate all detected change-points.

Model (5.1) generalizes many widely used regression models, namely:

a) *autoregression model (AR)*

$$y_n = c_0 + c_1 y_{n-1} + \cdots + c_m y_{n-m} + \nu_n,$$

Here, $\mathbf{X}(n) = (1, y_{n-1}, \ldots, y_{n-m})^*$, $\mathbf{\Pi} = (c_0, c_1, \ldots, c_m)$.

b) *autorgression-moving average (ARMA) model*

$$y_n = c_1 y_{n-1} + \cdots + c_k y_{n-k} + d_1 u_{n-\Delta} + \cdots + d_m u_{n-\Delta-m} + \nu_n,$$

where u_n is the input variable, y_n is the output variable at the instant n, and Δ is the delay time. Here, $\mathbf{X}(n) = (y_{n-1}, \ldots, y_{n-m}, u_{n-\Delta}, \ldots, u_{n-\Delta-m})^*$, $\mathbf{\Pi} = (c_1, \ldots, c_k, d_1, \ldots, d_m)$.

c) *multi-factor regression model*

$$y_n = c_1 y_{n-1} + \cdots + c_k y_{n-m} + \sum_{i=1}^{r} \sum_{j=1}^{l_i} d_{ij} x_i(n-j) + \nu_n,$$

where $r, m, l_i \geq 1$. Here, $\mathbf{X}(n) = (y_{n-1}, \ldots, y_{n-m}, x_1(n-1), \ldots, x_1(n-l_1), x_2(n-1), \ldots, x_2(n-l_2), \ldots, x_r(n-1), \ldots, x_r(n-l_r))^*$, $\mathbf{\Pi} = (c_1, \ldots, c_k, d_{11}, \ldots, d_{rl_r})$.

d) *simultaneous equation systems (SES)*

$$B\mathbf{Y}(n) + \Gamma\mathbf{X}(n) = \epsilon_n,$$

where $\mathbf{Y}(n) = (y_{1n}, y_{2n}, \ldots, y_{Mn})^*$ is the vector of endogenous variables, $\mathbf{X}(n) = (x_{1n}, x_{2n}, \ldots, x_{Kn})^*$ is the vector of predetermined variables (all exogenous variables plus lagged endogenous variables), $\epsilon_n = (\epsilon_{1n}, \epsilon_{2n}, \ldots, \epsilon_{Mn})^*$ is the vector of random errors, B is an $M \times M$ nondegenerate matrix ($\det B \neq 0$), and Γ is as $M \times K$ matrix.

This general structural form of the SES can be written in the following *reduced form:*

$$\mathbf{Y}(n) = -B^{-1}\Gamma\mathbf{X}(n) + B^{-1}\epsilon_n = \mathbf{\Pi}\mathbf{X}(n) + \nu_n.$$

This system is usually used for the analysis of change-points (structural changes) in multivariate linear models (see, e.g., Bai, Lumsdaine, and Stock (1998)).

5.3 Prior Inequalities

5.3.1 Unique Change-Point

On a probability space $(\Omega, \mathcal{F}, \mathbf{P}_\theta)$, consider a sequence of i.r.v.'s $x_1, \ldots, x_N$ with the following density function (w.r.t. some σ-finite measure μ):

$$f(x_n) = \begin{cases} f_0(x_n, n/N), & 1 \leq n \leq [\theta N], \\ f_1(x_n, n/N), & [\theta N] < n \leq N. \end{cases} \tag{5.2}$$

Here, $0 < \theta < 1$ is an *unknown change-point parameter.*

Define the following objects:

$$T_N(\Delta) : \mathbb{R}^N \longrightarrow \Delta \subset \mathbb{R}^1 \tag{5.3}$$

is the Borel function on $\mathbb{R}^N$ with the values in the set Δ; and

$$\mathcal{M}_N(\Delta) = \{T_N(\Delta)\} \tag{5.4}$$

is the collection of all Borel functions T_N.

Theorem 5.3.1. *Suppose the following assumption is satisfied:*

The functions $J_0(t) \stackrel{def}{=} \mathbf{E}_0 \ln \frac{f_0(x,t)}{f_1(x,t)}$ *and* $J_1(t) \stackrel{def}{=} \mathbf{E}_1 \ln \frac{f_1(x,t)}{f_0(x,t)}$ *are continuous at* $[0,1]$ *and such that*

$$J_0(t) \geq \delta > 0, \; J_1(t) \geq \delta > 0.$$

Then, for any fixed $0 < \theta < 1$, $0 < \epsilon < \theta \wedge (1-\theta)$, *the following inequality holds:*

$$\liminf_{N\to\infty} N^{-1} \ln \inf_{\hat{\theta}_N \in \mathcal{M}_N((0,1))} \mathbf{P}_\theta\{|\hat{\theta}_N - \theta| > \epsilon\} \geq -\min\left(\int_{\theta}^{\theta+\epsilon} J_0(t)dt, \int_{\theta-\epsilon}^{\theta} J_1(t)dt\right).$$

The proof of this theorem is given below.

Proof of Theorem 5.3.1

Using notations (5.3)–(5.4), put

$$\mathcal{M}(\Delta) = \{T(\Delta) : T(\Delta) = \{T_N(\Delta)\}_{N=1}^{\infty}\}.$$

This is the set of all sequences of the elements $T_N(\Delta) \in \mathcal{M}_N(\Delta)$. Consider also the collection of all consistent estimates of the parameter $\theta \in \Delta$, i.e.,

$$\tilde{\mathcal{M}}(\Delta) = \{T(\Delta) \in \mathcal{M}(\Delta) : \lim_{N\to\infty} \mathbf{P}_\theta(|T_N(\Delta) - \theta| > \epsilon) = 0, \forall \theta \in \Delta, \forall \epsilon > 0\}.$$

Under the assumptions of Theorem 5.3.1, the set $\tilde{\mathcal{M}}([a,b])$ is *nonempty* for any $0 < a < b < 1$. Indeed, consider the sequence $y_n = \ln \frac{f_0(x_n, n/N)}{f_1(x_n, n/N)}$. Due to the assumption, $\mathbf{E}_\theta y_n \geq \delta > 0$ before the change-point θ, $a \leq \theta \leq b$, and less than $(-\delta)$ after the change-point. Now, using the same idea as in Brodsky and Darkhovsky (2000), it is easy to construct the consistent estimate of the change-point.

Further, without loss of generality, we can consider only consistent estimates of the change-point parameter θ, because for nonconsistent estimates, the probability of the error of estimation does not converge to zero, and the considered inequality is satisfied trivially.

Let $\hat{\theta}_N$ be some consistent estimate of the change-point parameter θ constructed by the sample $X^N = \{x_1, \ldots, x_N\}$. Consider the random variable $\lambda_N = \lambda_N(x_1, \ldots, x_N) = \mathbb{I}\{|\hat{\theta}_N - \theta| > \epsilon\}$.

Under the change-point parameter θ, the likelihood function for the sample X^N can be written as follows:

$$f(X^N, \theta) = \prod_{i=1}^{[\theta N]} f_0(x_i, i/N) \cdot \prod_{i=[\theta N]+1}^{N} f_1(x_i, i/N).$$

We have for any $d > 0$ and $0 < \epsilon < \epsilon^{'}$:

$$\mathbf{P}_\theta\{|\hat{\theta}_N - \theta| > \epsilon\} = \mathbf{E}_\theta \lambda_N \geq \mathbf{E}_\theta(\lambda \mathbb{I}(f(X^N, \theta + \epsilon^{'})/f(X^N, \theta) < e^d)) \geq$$
$$\geq e^{-d} \left(\mathbf{E}_{\theta+\epsilon\prime}(\lambda_N \mathbb{I}(f(X^N, \theta + \epsilon\prime)/f(X^N, \theta) < e^d\}\right) \geq$$
$$e^{-d} \left(\mathbf{P}_{\theta+\epsilon\prime}\{|\theta_N - \theta| > \epsilon\} - \mathbf{P}_{\theta+\epsilon\prime}\{f(X^N, \theta + \epsilon\prime)/f(X^N, \theta) \geq e^d\}\right)$$

(here we used the elementary inequality $\mathbf{P}(AB) \geq \mathbf{P}(A) - \mathbf{P}(\Omega \backslash B)$).

Consider the probabilities in the right-hand side of the last inequality. Since θ_N is a consistent estimate of θ, we have $\mathbf{P}_{\theta+\epsilon\prime}\{|\theta_N - \theta| > \epsilon\} \to 1$ as $N \to \infty$. For estimation of the second probability, we take into account that

$$\ln\left(f(X^N, \theta + \epsilon\prime)/f(X^N, \theta)\right) = \sum_{i=[\theta N]+1}^{[(\theta+\epsilon\prime)N]} \ln\left(f_0(x_i, i/N)/f_1(x_i, i/N)\right).$$

Therefore,

$$\mathbf{E}_{\theta+\epsilon\prime} \ln\left(f(X^N, \theta + \epsilon\prime)/f(X^N, \theta)\right) =$$
$$= N \int_{\theta}^{\theta+\epsilon'} \mathbf{E}_0 \ln \frac{f_0(x,t)}{f_1(x,t)} dt + O(1).$$

Then,

$$\mathbf{P}_{\theta+\epsilon'}\{f(X^N, \theta + \epsilon')/f(X^N, \theta) \geq e^d\} =$$
$$= \mathbf{P}_{\theta+\epsilon'} \left\{ \sum_{i=[\theta N]+1}^{[(\theta+\epsilon^{'})N]} (\ln(f_0(x_i, i/N)/f_1(x_i, i/N)) \right.$$
$$-\mathbf{E}_0 \ln\left(f_0(x_i, i/N)/f_1(x_i, i/N)\right)$$
$$\left. \geq d - N \int_{\theta}^{\theta+\epsilon'} \mathbf{E}_0 \ln \frac{f_0(x,t)}{f_1(x,t)} dt + O(1) \right\}.$$

Put $d = d_1(N) = N(\int_{\theta}^{\theta+\epsilon'} \mathbf{E}_0 \ln \frac{f_0(x,t)}{f_1(x,t)} dt + \delta)$ for some $\delta > 0$ and use the law of large numbers, which holds due to existence of $\mathbf{E}_0 \ln \frac{f_0(x,t)}{f_1(x,t)}$. Then, we obtain

$$\mathbf{P}_{\theta+\epsilon'}\{f(X^N, \theta + \epsilon')/f(X^N, \theta) \geq e^{d_1(N)}\} \to 0$$

as $N \to \infty$.

The same considerations for $d = d_2(N) = N(\int_{\theta-\epsilon'}^{\theta} \mathbf{E}_1 \ln \frac{f_1(x,t)}{f_0(x,t)} dt + \delta)$ yield

$$\mathbf{P}_{\theta-\epsilon'}\left\{f(X^N, \theta - \epsilon')/f(X^N, \theta) \geq e^{d_2(N)}\right\} \to 0$$

as $N \to \infty$.

Therefore,

$$\mathbf{P}_\theta\{|\hat{\theta}_N - \theta| > \epsilon\} \geq (1 - o(1)) \max(e^{-d_1(N)}, e^{-d_2(N)}).$$

It follows from here

$$\liminf_{N\to\infty} N^{-1} \ln \inf_{\hat{\theta}_N \in \mathcal{M}_N} \mathbf{P}_\theta\{|\hat{\theta}_N - \theta| > \epsilon\} \geq -\min\left(\int_\theta^{\theta+\epsilon'} J_0(t)dt, \int_{\theta-\epsilon'}^{\theta} J_1(t)dt\right) - \delta.$$

Note that the left-hand side of this inequality does not depend on the parameters δ, ϵ', and the right-hand side exists for each $\delta > 0$, $\theta \wedge (1-\theta) > \epsilon' > \epsilon > 0$. From the continuity assumption for the functions $J_0(\cdot), J_1(\cdot)$, we conclude that our result follows after taking the limits of both sides of this inequality as $\delta \to 0$ and $\epsilon' \to \epsilon$.

Theorem 5.3.1 is proved.

Consider the following particular cases of Model (5.2).

1. A break in the trend function $\phi(t)$ of the mathematical expectation of Gaussian observations

Let

$$\begin{aligned} f_0(x,t) &= h(x)\exp\left(\phi_0(t)x - \phi_0^2(t)/2\right), \quad t \leq \theta \\ f_1(x,t) &= h(x)\exp\left(\phi_1(t)x - \phi_1^2(t)/2\right), \quad t > \theta, \end{aligned}$$

where $h(x) = \frac{1}{\sqrt{2\pi}}\exp(-x^2/2)$, $\phi_0(\cdot) \neq \phi_1(\cdot)$.

In this case, from Theorem 5.3.1, we obtain the following lower bound for the error probability:

$$\mathbf{P}_\theta\{|\hat{\theta}_N - \theta| > \epsilon\} \geq (1 - o(1))\cdot$$
$$\cdot \exp\left(-\frac{N}{2}\min\Big(\int_\theta^{\theta+\epsilon} (\phi_0(t) - \phi_1(t))^2 dt, \int_{\theta-\epsilon}^{\theta} (\phi_0(t) - \phi_1(t))^2\, dt\Big)\right).$$

2. Linear regression with deterministic predictors and Gaussian errors

Let

$$y_n = c_1(n)x_{1n} + \cdots + c_k(n)x_{kn} + \xi_n, \quad n = 1, \dots, N, \tag{5.5}$$

where $\{\xi_n\}$ is a sequence of independent Gaussian r.v.'s with zero mean, $\xi_n \sim \mathcal{N}(0, \sigma^2)$, $\mathbf{c}(n) \overset{\text{def}}{=} (c_1(n), \dots, c_k(n))^* = \mathbf{a}\mathbb{I}(n \leq [\theta N]) + \mathbf{b}\mathbb{I}(n > [\theta N])$, $\mathbf{a} = (a_1, \dots, a_k)^* \neq \mathbf{b} = (b_1, \dots, b_k)^*$, $x_{in} = f_i(n/N)$, $n = 1, \dots, N$, and $f_i(\cdot) \in C[0,1]$, $i = 1, \dots, k$.

In this case from Theorem 5.3.1, applied to the sequence of observations $y_1, \dots, y_N$, we obtain:

$$\mathbf{P}_\theta\{|\hat{\theta}_N - \theta| > \epsilon\} \geq (1 - o(1))\cdot$$
$$\cdot \exp\left(-\frac{N}{2\sigma^2}\min\Big(\int_\theta^{\theta+\epsilon} (\sum_{i=1}^k f_i(t)(a_i - b_i))^2 dt, \int_{\theta-\epsilon}^{\theta} (\sum_{i=1}^k f_i(t)(a_i - b_i))^2 dt\Big)\right).$$

3. Linear stochastic regression model with Gaussian predictors

Consider model (5.5) with $\xi_n \equiv 0$. Suppose that there exist continuous functions $f_i(\cdot), \sigma_i(\cdot)$, $i = 1, \dots, k$, such that x_{in} are Gaussian i.r.v.'s, $x_{in} \sim \mathcal{N}\left(f_i(n/N), \sigma_i^2(n/N)\right), n = 1, \dots, N$. Suppose also that x_{in} and x_{jn} are independent for $i \neq j$, and $\mathbf{c}(n)$ is the same as in Model (5.5).

Then, from Theorem 5.3.1, we obtain:

$$\mathbf{P}_\theta\{|\hat{\theta}_N - \theta| > \epsilon\} \geq (1 - o(1)) \exp\left(-\frac{N}{2} \min\left(\int_\theta^{\theta+\epsilon} J_0(t)dt, \int_{\theta-\epsilon}^{\theta} J_1(t)dt \right) \right),$$

where

$$J_0(t) = \left(\frac{\phi_0(t)}{\Delta_0(t)} - \frac{\phi_1(t)}{\Delta_1(t)} \right)^2 + 2\frac{\phi_0(t)}{\Delta_0(t)} \frac{\phi_1(t)}{\Delta_1(t)} \left(1 - \frac{\Delta_0(t)}{\Delta_1(t)}\right) +$$
$$2 \ln \frac{\Delta_1(t)}{\Delta_0(t)} + \left(1 + \frac{\phi_0^2(t)}{\Delta_0^2(t)}\right) \left(\frac{\Delta_0(t)}{\Delta_1(t)} - 1\right),$$

and

$$\phi_0(t) = a_1 f_1(t) + \cdots + a_k f_k(t),\ \Delta_0^2(t) = a_1^2 \sigma_1^2(t) + \cdots + a_k^2 \sigma_k^2(t),$$
$$\phi_1(t) = b_1 f_1(t) + \cdots + b_k f_k(t),\ \Delta_1^2(t) = b_1^2 \sigma_1^2(t) + \cdots + b_k^2 \sigma_k^2(t).$$

5.3.2 Multiple Change-Points

Theorem 5.3.1 can be generalized to the case of several change-points in the sequence of independent r.v.'s with the following density function:

$$f(x_n) = f_i(x_n, n/N)\mathbb{I}([\theta_{i-1}N] < n \leq [\theta_i N]), \quad n = 1, \dots, N,$$

where $i = 1, \dots, k+1$ and $0 \equiv \theta_0 < \theta_1 < \cdots < \theta_k < \theta_{k+1} \equiv 1$.

Suppose the following assumptions are satisfied:

i) change-points θ_i are such that $\min\limits_{1 \leq i \leq k+1} (\theta_i - \theta_{i-1}) \geq \delta > 0$.

ii) the functions $J_i(t) = \mathbf{E}_i \ln \dfrac{f_i(x,t)}{f_{i-1}(x,t)}$ and $J^{i-1}(t) = \mathbf{E}_{i-1} \ln \dfrac{f_{i-1}(x,t)}{f_i(x,t)}$, $i = 1, \dots, k$ are continuous at $[0, 1]$ and such that

$$J_i(t) \geq \Delta > 0,\ i = 1, \dots, k$$

For the multiple change-point problem, we estimate both the number k and the vector $\vartheta \stackrel{\text{def}}{=} (\theta_1, \dots, \theta_k)$ of change-points' coordinates. Let $s^* \stackrel{\text{def}}{=} [1/\delta]$ and denote $Q = \{1, 2, \dots, s^*\}$.

For any $s \in Q$ define

$$\begin{aligned} &\mathcal{D}_s = \{x \in \mathbb{R}^s : \delta \leq x_i \leq 1 - \delta,\ x_{i+1} - x_i \geq \delta, x_0 \equiv 0, x_{s+1} \equiv 1\} \\ &\mathcal{D}^\star = \textstyle\bigcup_{i=1}^{s^*} \mathcal{D}_i, \mathcal{D}^\star \subset \mathbb{R}^{s^*} \equiv \mathbb{R}^\star. \end{aligned} \tag{5.6}$$

By the construction, an unknown vector ϑ is an arbitrary point of the set $\mathcal{D}_k$, and an unknown number of the change-points k is an arbitrary point of the set Q.

As before, it is reasonable to consider objects (5.3)–(5.4). In this notation, $\mathcal{M}_N(\mathcal{D}^*)$ is the set of all arbitrary estimates of the parameter ϑ, and $\mathcal{M}_N(Q)$ is the set of all arbitrary estimates of the parameter k on the basis of observations with the sample size N.

Let $\hat{k} \in \mathcal{M}_N(Q)$ be an estimate of an unknown number of change-points k, and $\hat{\vartheta} \in \mathcal{M}_N(\mathcal{D}_k)$ is an estimate of unknown change-point coordinates on the condition that the number of the coordinates was estimated correctly.

Theorem 5.3.2. *Suppose assumptions i) and ii) are satisfied. Then, for any fixed* $0 < \epsilon < \delta$, *the following inequality holds:*

$$\liminf_{N\to\infty} N^{-1} \ln \inf_{\hat{\vartheta}\in\mathcal{M}_N(\mathcal{D}_k)} \inf_{\hat{k}\in\mathcal{M}_N(Q)} \sup_{\vartheta\in\mathcal{D}_k} \sup_{k\in Q} \mathbf{P}_\theta\{\{\hat{k} \neq k\} \cup \{(\hat{k} = k) \cap$$

$$\cap (\max_{1\le i\le k} |\hat{\theta}_i - \theta_i| > \epsilon)\} \ge - \min_{1\le i\le k} \min(\int_{\theta_i}^{\theta_i+\epsilon} J^{i-1}(\tau)d\tau, \int_{\theta_i-\epsilon}^{\theta_i} J_i(\tau)d\tau).$$

The idea of the proof of this theorem was given in Chapter 2.

5.4 Main Results

Now, consider Model (5.1). In this section, we assume that the uniform mixing condition (A) and the uniform Cramer condition (B) (see Section 2) are satisfied, and an unknown vector of change-point parameters $\vartheta = (\theta_1, \ldots, \theta_k)$ is such that $0 < \beta \le \theta_1 < \theta_2 < \cdots < \theta_k \le \alpha < 1$, where β, α are known numbers. Everywhere below the measure $\mathbf{P}_\vartheta$ corresponds to a sample with the change-point ϑ ($\mathbf{P}_0$ corresponds to a sample without change-points).

5.4.1 Unique Change-Point

In this subsection, Model (5.1) with unique change-point $0 < \beta \le \theta \le \alpha < 1$ is considered.

5.4.1.1 Deterministic Predictors

Let us formulate assumptions for Model (5.1) in the case of a unique change-point (remember that in Model (5.1) the vector $\mathbf{X}(n)$ has the dimension K and the vector $\mathbf{Y}(n)$ has the dimension M):

a) the vector random sequence $\{\nu_n\}$ satisfies conditions (A) and (B) (see Section 2).

b) there exist functions $f_i(\cdot) \in C[0, 1]$, $i = 1, \ldots, K$, such that $x_{in} = f_i(n/N), n = 1, \ldots, N$.

Denote $F(t) = (f_1(t), \ldots, f_K(t))^*$, $t \in [0, 1]$.

c) for arbitrary $0 \leq t_1 < t_2 \leq 1$, the matrix

$$A(t_1, t_2) \stackrel{\text{def}}{=} \int_{t_1}^{t_2} F(s)F^*(s)ds$$

is positive definite (below, we denote $A(t) \stackrel{\text{def}}{=} A(0,t),\ A(1) \stackrel{\text{def}}{=} I$).

In virtue of our assumptions, the matrix I is symmetric and positive definite.

Define $K \times M$ matrix

$$Z(n_1, n_2) = \sum_{i=n_1}^{n_2} F(i/N)\mathbf{Y}^*(i)$$

and $K \times K$ matrix

$$\mathcal{P}_{n_1}^{n_2} \stackrel{\text{def}}{=} \sum_{k=n_1}^{n_2} F(k/N)F^*(k/N), \quad 1 \leq n_1 < n_2 \leq N.$$

The following matrix statistic is used for estimation of an unknown change-point:

$$\mathcal{Z}_N(n) = N^{-1}\left(Z(1,n) - \mathcal{P}_1^n(\mathcal{P}_1^N)^{-1}\, Z(1,N)\right). \tag{5.7}$$

An arbitrary point $\hat{n}$ of the set $arg \max_{[\beta N] \leq n \leq [\alpha N]} \|\mathcal{Z}_N(n)\|^2$ is assumed to be the estimate of an unknown change-point (here and below, $\|C\|$ denotes the Gilbert norm of a quadratic matrix C, namely, $\|C\| = \sqrt{tr(CC^*)}$).

We define also the value $\hat{\theta}_N = \hat{n}/N$ – the estimate of the change-point parameter θ.

Denote $B \stackrel{\text{def}}{=} B(\theta) = \left(E - I^{-1}A(\theta)\right)(\mathbf{a} - \mathbf{b})^*$.

Theorem 5.4.1. *Suppose assumptions a)–c) are satisfied, and rank$(B) = M$ if $\theta \in [\beta, \alpha]$.*

Then the estimate $\hat{\theta}_N$ converges to the change-point parameter θ $\mathbf{P}_\theta$-almost surely as $N \to \infty$.

Besides, for any fixed $(\alpha - \beta) > \epsilon > 0$, the following inequality is satisfied for $N > N_0(F)$:

$$\sup_{\beta \leq \theta \leq \alpha} \mathbf{P}_\theta\{|\hat{\theta}_N - \theta| > \epsilon\} \leq m_0\left(C(\epsilon, N)/\mathcal{R}\right) \begin{cases} \exp\left(-\dfrac{N\beta\left(C(\epsilon,N)/\mathcal{R}\right)^2}{4gm_0\left(C(\epsilon,N)/\mathcal{R}\right)}\right), \\ \quad \text{if } C(\epsilon,N) \leq \mathcal{R}gT \\ \exp\left(-\dfrac{TN\beta\left(C(\epsilon,N)/\mathcal{R}\right)}{4m_0\left(C(\epsilon,N)/\mathcal{R}\right)}\right), \\ \quad \text{if } C(\epsilon,N) > \mathcal{R}gT. \end{cases} \tag{5.8}$$

where the constants g, T, and $mm_0(\cdot) \geq 1$ are taken from the uniform Cramer's and ψ-mixing conditions, respectively, $C(\epsilon, N) = [\frac{\epsilon\lambda_F}{4\mathcal{M}}\|\mathbf{a}-\mathbf{b}\|^2 - L_F/N]$, $N_0(F)$, λ_F, L_F, and $\mathcal{R}$ are constants that can be exactly calculated for any given family of functions $F(t)$, and the constant $\mathcal{M}$ is given in the proof.

Remark 1. *The assumption rank$B = M$ yields $K \geq M$, i.e., the number M of endogenous variables in (1) cannot exceed the number K of predetermined variables. Note that for one regression equation, this assumption is always satisfied.*

Remark 2. *For independent random errors, $m_0(\epsilon) = 1$.*

Remark 3. *Comparing Theorems 5.4.1 and 5.3.3, we conclude that the order of convergence of the proposed estimate of the change-point parameter to its true value is asymptotically optimal as $N \to \infty$.*

Remark 4. *For any given family of functions $F(t)$, one can calculate the function $f(t) = \|m(t)\|^2$, $m(t) = \lim_{N\to\infty} \mathbf{E}_\theta \mathcal{Z}_N([Nt])$ (see the proof) and investigate this function on the square $(\theta, t) \in [\beta, \alpha] \times [\beta, \alpha]$. Such investigation gives the opportunity to calculate all constants from the formulation.*

The proof of Theorem 5.4.1 is given below.

Proof of Theorem 5.4.1

Due to the assumptions, the matrix $I = \int_0^1 F(t)F^*(t)dt$ is positive definite. Therefore, there exists the matrix $\left[N\left(\mathcal{P}_1^N\right)^{-1}\right]$ for all $N > N_0(F)$. The constant $N_0(F)$ can be exactly estimated for any given family of functions $F(t)$.

Let us consider the matrix random process with continuous time $\mathcal{Z}_N(t) \stackrel{\text{def}}{=} \mathcal{Z}_N([Nt])$, $t \in [0, 1]$.

It is easy to see that the mathematical expectation of the process $\mathcal{Z}_N(t)$ can be written as follows:

$$\mathbf{E}_\theta \mathcal{Z}_N(t) = N^{-1}\left(\sum_{i=1}^{[Nt]} F(i/N)F^*(i/N)\,\mathbf{\Pi}^*(\theta, i)\right.$$
$$\left. -\mathcal{P}_1^{[Nt]}(\mathcal{P}_1^N)^{-1}\sum_{i=1}^{N} F(i/N)F^*(i/N)\,\mathbf{\Pi}^*(\theta, i)\right).$$

After simple transformations, we obtain that $m(t) \stackrel{\text{def}}{=} \lim_{N\to\infty} \mathbf{E}_\theta \mathcal{Z}_N(t)$ has the form

$$m(t) = \begin{cases} A(t)I^{-1}(I - A(\theta))(\mathbf{a}-\mathbf{b})^*, & t \leq \theta \\ (I - A(t))I^{-1}A(\theta)(\mathbf{a}-\mathbf{b})^*, & t > \theta. \end{cases} \tag{5.9}$$

Consider the square of the Gilbert norm of the matrix $m(t)$, i.e., the function $f(t) = \mathrm{tr}(m^*(t)m(t))$, and show that the function $f(t)$ has a unique global maximum on the segment $[0, 1]$ at the point $t = \theta$.

First, for each $t \leq \theta$,

$$f(\theta) - f(t) = \mathrm{tr}(B^*(A^2(\theta) - A^2(t))B),$$

where matrix B was defined above. Consider the matrix

$$A^2(\theta) - A^2(t) = A(\theta)(A(\theta) - A(t)) + (A(\theta) - A(t))A(t).$$

Denote $L = A(\theta)(A(\theta) - A(t))$ and prove that the matrix L is positive definite as $t < \theta$. In fact, since the matrix $A(\theta)$ is symmetric and positive definite, we can write

$$x^* Lx = x^* A^{1/2}(\theta)A^{1/2}(\theta)(A(\theta) - A(t))x = y^* A^{1/2}(\theta)(A(\theta) - A(t))A^{-1/2}(\theta)\, y,$$

where $y = A^{1/2}(\theta)x$.

The matrices $A(\theta) - A(t)$ and $A^{1/2}(\theta)(A(\theta) - A(t))A^{-1/2}(\theta)$ have identical characteristic polynomial and eigenvalues. Besides, $A(\theta) - A(t)$ is positive definite as $t < \theta$. Therefore, the matrix $A^{1/2}(\theta)(A(\theta) - A(t))A^{-1/2}(\theta)$ is also positive definite as $t < \theta$, and, therefore, the matrix L is positive definite.

In analogy, the matrix $(A(\theta) - A(t))A(t)$ is positive definite as $t < \theta$. Therefore, the matrix $A^2(\theta) - A^2(t)$ is positive definite as $t < \theta$.

Now, consider the matrix $D = B(A^2(\theta) - A^2(t))B^*$. The matrix D is positive definite if $\mathrm{rank}(B) = M$, but this is our assumption.

So, we obtain $\mathrm{tr}(B(A^2(\theta) - A^2(t))B^*) > 0$ for $t < \theta$, and, therefore, the function $f(t)$ has a unique global maximum on the segment $[0, \theta]$ at the point $t = \theta$.

The same considerations for $t < \theta$ yield that $f(t)$ monotonically decreases on the segment $[\theta, 1]$. As a result, we obtain that $f(t)$ has a unique global maximum on the segment $[0, 1]$ at the point $t = \theta$.

Further, we are going to show the following: There exists a positive constant c, such that $f(\theta) - f(t) \geq c \cdot |\theta - t|$. This estimate can be obtained as follows: Taking into account the continuity of the functions $f_j(t)$, we obtain

$$A(\theta) - A(t) = \int_t^\theta F(\tau)F^*(\tau)\, d\tau = (\theta - t)U(t, \theta) > 0, \tag{5.10}$$

where the matrix $U(t, \theta)$ is positive definite for $0 \leq t < \theta$ and negative definite for $t > \theta$. Due to the continuity, we can write

$$U(t, \theta) = U(\theta, \theta) + \kappa(t, \theta), \tag{5.11}$$

where $\kappa(t, \theta) \to 0$ as $t \to \theta$.

Then,

$$\begin{aligned} f(\theta) - f(t) &= \operatorname{tr}\left(B^*(A^2(\theta) - A^2(t))B\right) = \\ &= \operatorname{tr}\left(BB^*A(\theta)(A(\theta) - A(t))\right) + \operatorname{tr}\left(BB^*(A(\theta) - A(t))A(t)\right) = \\ &= (\theta - t)\operatorname{tr}\left((\mathbf{a} - \mathbf{b})^*(\mathbf{a} - \mathbf{b})V(t,\theta)\right), \end{aligned} \tag{5.12}$$

where $V(t,\theta) = \left(E - A(\theta)I^{-1}\right)\left(A(\theta)U(t,\theta) + U(t,\theta)A(t)\right)\left(E - I^{-1}A(\theta)\right)$.

Taking into account (5.9) and (5.10), we have

$$\begin{aligned} V(t,\theta) &= \left(E - A(\theta)I^{-1}\right)\left(A(\theta)U(t,\theta) + U(t,\theta)A(t)\right)\left(E - I^{-1}A(\theta)\right) = \\ &= \left(E - A(\theta)I^{-1}\right)\left(A(\theta)U(\theta,\theta) + U(\theta,\theta)A(\theta)\right)\left(E - I^{-1}A(\theta)\right) + \\ &+ \left(E - A(\theta)I^{-1}\right)\left(A(\theta)\kappa(t,\theta) + \kappa(t,\theta)A(\theta)\right)\left(E - I^{-1}A(\theta)\right) + \\ &+ (t - \theta)\left(E - A(\theta)I^{-1}\right)U(t,\theta)U(t,\theta)\left(E - I^{-1}A(\theta)\right). \end{aligned} \tag{5.13}$$

Denote

$$\begin{aligned} G(\theta) &= \left(E - A(\theta)I^{-1}\right)\left(A(\theta)U(\theta,\theta) + U(\theta,\theta)A(\theta)\right)\left(E - I^{-1}A(\theta)\right) \\ R(t,\theta) &= \left(E - A(\theta)I^{-1}\right)\left(A(\theta)\kappa(t,\theta) + \kappa(t,\theta)A(\theta)\right)\left(E - I^{-1}A(\theta)\right) \\ H(t,\theta) &= \left(E - A(\theta)I^{-1}\right)U(t,\theta)U(t,\theta)\left(E - I^{-1}A(\theta)\right) \end{aligned} \tag{5.14}$$

and put

$$\tilde{G}(\theta) = \begin{cases} G(\theta), & \theta > t \\ -G(\theta), & \theta \le t. \end{cases} \tag{5.15}$$

Then, from (5.12), (5.13), (5.14), and (5.15), we get

$$\begin{aligned} f(\theta) - f(t) &= |\theta - t|\operatorname{tr}\left((\mathbf{a} - \mathbf{b})^*(\mathbf{a} - \mathbf{b})\tilde{G}(\theta)\right) + \\ &+ (\theta - t)\operatorname{tr}\left((\mathbf{a} - \mathbf{b})^*(\mathbf{a} - \mathbf{b})R(t,\theta)\right) - \\ &- (\theta - t)^2\operatorname{tr}\left((\mathbf{a} - \mathbf{b})^*(\mathbf{a} - \mathbf{b})H(t,\theta)\right). \end{aligned} \tag{5.16}$$

Since $R(t,\theta) \to 0$ as $t \to \theta$, and $H(t,\theta)$ is positive definite, we conclude that

$$f(\theta) - f(t) \ge |\theta - t|\operatorname{tr}\left((\mathbf{a} - \mathbf{b})^*(\mathbf{a} - \mathbf{b})\tilde{G}(\theta)\right) + o(|t - \theta|),$$

i.e., there exists a positive definite matrix $W(\theta)$, such that

$$\|m(\theta)\|^2 - \|m(t)\|^2 = f(\theta) - f(t) \ge |\theta - t|\operatorname{tr}\big((\mathbf{a} - \mathbf{b})^*(\mathbf{a} - \mathbf{b})W(\theta)\big)$$

for some neighborhood of θ. Therefore, we have got the estimate of sharpness of the maximum for the function $f(t)$:

$$f(\theta) - f(t) \ge |\theta - t|\lambda_F \operatorname{tr}\big[(\mathbf{a} - \mathbf{b})^*(\mathbf{a} - \mathbf{b})\big], \tag{5.17}$$

where

$$\lambda_F \stackrel{\text{def}}{=} \min_{\beta \le \theta \le \alpha} \frac{\operatorname{tr}\left[(\mathbf{a}-\mathbf{b})^*(\mathbf{a}-\mathbf{b})W(\theta)\right]}{\operatorname{tr}\left[(\mathbf{a}-\mathbf{b})^*(\mathbf{a}-\mathbf{b})\right]}.$$

Let us describe how to calculate λ_F. For given family of functions $F(t)$, we can calculate the function $f(t) = \operatorname{tr}\left[m^*(t)m(t)\right]$. Then, it is possible to calculate

$$\lambda_F = \min_{\beta \le t \le \alpha,\, \beta \le \theta \le \alpha} \frac{f(\theta) - f(t)}{|\theta - t| \operatorname{tr}\left[(\mathbf{a}-\mathbf{b})^*(\mathbf{a}-\mathbf{b})\right]}.$$

Due to the condition $0 < \beta \le \theta \le \alpha < 1$, we get $\lambda_F > 0$ (see (5.4.8)). Note that from (5.4.11) and definition of $f(t)$, we have for any $t \in [\beta, \alpha]$

$$\|m(\theta)\|^2 - \|m(t)\|^2 \ge \frac{\lambda_F}{2\|m(\theta\|} |\theta - t| \, \|\mathbf{a}-\mathbf{b}\|^2. \tag{5.18}$$

The process $\mathcal{Z}_N(t)$ can be decomposed into deterministic and stochastic terms:

$$\mathcal{Z}_N(t) = m(t) + \gamma_N(t) + \eta_N(t), \tag{5.19}$$

where the norm of the deterministic function $\gamma_N(t)$ converges to zero with the rate L_F/N) (this term estimates the difference between corresponding integral sum and the integral; the constant L_F depends of the function family $F(t)$ and *can be estimated explicitly for any given family*), and the stochastic term is equal to

$$\eta_N(t) = N^{-1} \left(\sum_{i=1}^{[Nt]} F(i/N)\nu_i^* - \mathcal{P}_1^{[Nt]} (\mathcal{P}_1^N)^{-1} \sum_{i=1}^{N} F(i/N)\nu_i^* \right).$$

The norm of the process $\eta_N(t)$ can be estimated as follows:

$$\begin{aligned}
\sup_{\beta \le t \le \alpha} \|\eta_N(t)\| &\le R \left[\sqrt{K} + \|I\| \cdot \|I^{-1}\| + \tfrac{L_F}{N} \left(\|I\| + \|I^{-1}\| + L_F/N \right) \right] \times \\
&\times \left(\max_{1 \le i \le K} \max_{1 \le l \le M} \max_{[\beta N] \le n \le N} N^{-1} \Big| \sum_{j=1}^{n} f_i(j/N)\nu_{lj} \Big| \right) \stackrel{\text{def}}{=} \\
&= \mathcal{R} \left(\max_{1 \le i \le K} \max_{1 \le l \le M} \max_{[\beta N] \le n \le N} N^{-1} \Big| \sum_{j=1}^{n} f_i(j/N)\nu_{lj} \Big| \right),
\end{aligned} \tag{5.20}$$

where $\mathcal{R} = \mathcal{R}(F, N)$. Here, we used the following relations

$$\max_{t \in [0,1]} \|N^{-1}\mathcal{P}_1^{[Nt]} - A(t)\| \le \tfrac{L_F}{N}, \quad \max_{t \in [0,1]} \|A(t)\| \le \|I\|$$
$$\|N(\mathcal{P}_1^N)^{-1} - I^{-1}\|| \le \tfrac{L_F}{N}$$

and took into account that for any matrix M, we have the relation $\|M\| = \sqrt{\operatorname{tr}(M^*M)} \le R \max_{i,j} |m_{ij}|$, where constant R depends only on the dimensionality.

Denote $\tilde{S}_n = \sum_{j=1}^{n} f_i(j/N)\nu_{lj}$, $\tilde{\xi}(j) = f_i(j/N)\nu_{lj}$ and put $\sigma^2 = \sup_i \sup_{1\le n\le N} \sup_{1\le l\le M} \mathbf{E}_\theta(f_i(n/N)\nu_{ln})^2$. Choose the number $\epsilon(x)$ from the following condition:

$$\ln(1+\epsilon(x)) = \begin{cases} x^2/4g, & x \le gT, \\ xT/4, & x > gT, \end{cases}$$

where the constant T is taken from the uniform Cramer condition and $g > \sigma^2$.

For the chosen $\epsilon(x) = \epsilon$, we choose the number $m_0(x) \ge 1$ from the uniform ψ-mixing condition, such that $\psi(m) \le \epsilon$ for $m \ge m_0(x)$.

Decompose the sum $\tilde{S}_n$ into groups of weakly dependent terms,

$$\tilde{S}_n = \tilde{S}_n^1 + \tilde{S}_n^2 + \cdots + \tilde{S}_n^{m_0(x)},$$

where

$$\tilde{S}_n^i = \tilde{\xi}(i) + \tilde{\xi}(i + m_0(x)) + \cdots + \tilde{\xi}\left(i + m_0(x)[\frac{n-i}{m_0(x)}]\right),$$

and $i = 1, 2, \ldots, m_0(x)$.

The number of summands $k(i)$ in each group is no less than $[n/m_0(x)]$ and no more than $[n/m_0(x)]+1$. The ψ-mixing coefficient between summands within each group is no larger than ϵ. Therefore,

$$\begin{aligned} &\mathbf{P}_\theta\{|\tilde{S}_n|/n \ge x\} \le \sum_{i=1}^{m_0(x)} \mathbf{P}_\theta\{|\tilde{S}_n^i/n| \ge x/m_0(x)\} \le \\ &\le m_0(x) \max_{1\le i\le m_0(x)} \mathbf{P}_\theta\{|\tilde{S}_n^i| \ge (k(i)-1)x\}. \end{aligned} \tag{5.21}$$

From Chebyshev's inequality, we have,

$$\mathbf{P}_\theta\left\{\tilde{S}_k^i = \sum_{j=0}^{k} \tilde{\xi}(i + m_0 j) \ge x\right\} \le e^{-tx}\mathbf{E}_\theta e^{t\tilde{S}_k^i}, \quad \forall t > 0. \tag{5.22}$$

Further, from ψ-mixing condition, it follows that (see Ibragimov, and Linnik, 1971)

$$\mathbf{E}_\theta e^{t\tilde{S}_k^i} \le (1+\epsilon)^k \mathbf{E}_\theta \exp(t\tilde{\xi}(i))\mathbf{E}_\theta \exp(t\tilde{\xi}(i+m_0)) \ldots \mathbf{E}_\theta \exp(t\tilde{\xi}(i+m_0 k)). \tag{5.23}$$

Consider the term $\mathbf{E}_\theta \exp(t\tilde{\xi}(i))$. From the uniform Cramer's condition, it follows that for each $0 < t < T$,

$$\mathbf{E}_\theta e^{t\tilde{\xi}(i)} \le \exp(t^2 g/2).$$

Then, from (5.22) and (5.23), we obtain

$$\mathbf{P}_\theta\{\tilde{S}_k^i \ge x\} \le (1+\epsilon)^k \exp\left(kgt^2/2 - tx\right).$$

Taking the minimum of $kgt^2/2 - tx$ w.r.t. t, write

$$\mathbf{P}_\theta\{\tilde{S}_k^i \geq x\} \leq \begin{cases} (1+\epsilon)^k \exp(-x^2/2kg), & x \leq kgT, \\ (1+\epsilon)^k \exp(-xT/2), & x > kgT. \end{cases}$$

From the definition of ϵ, we obtain

$$\mathbf{P}_\theta\{|\tilde{S}_k^i/k| \geq x\} \leq \begin{cases} \exp(-kx^2/4g), & x \leq gT, \\ \exp(-kxT/4), & x > gT. \end{cases} \tag{5.24}$$

Now, using (5.21) and (5.24), we obtain

$$\mathbf{P}_\theta\{|\tilde{S}_n/n| \geq x\} \leq \begin{cases} m_0(x)\exp\left(-x^2 n/4gm_0(x)\right), & x \leq gT, \\ m_0(x)\exp\left(-Txn/4m_0(x)\right), & x > gT. \end{cases} \tag{5.25}$$

From (5.20) and (5.25), we get

$$\mathbf{P}_\theta\{\sup_{\beta\leq t\leq\alpha} \|\eta_N(t)\| > \epsilon\} \leq m_0(\epsilon/\mathcal{R}) \begin{cases} \exp\left(-(\epsilon/\mathcal{R})^2 N\beta/4gm_0(\epsilon/\mathcal{R})\right), \\ \epsilon \leq \mathcal{R}gT \\ \exp\left(-T(\epsilon/\mathcal{R})N\beta/4m_0(\epsilon/\mathcal{R})\right), \\ \epsilon > \mathcal{R}gT. \end{cases} \tag{5.26}$$

In particular, for the case of independent observations, $m_0(\epsilon) = 1$.

From the definition of the estimate $\hat{\theta}_N$ and (5.18), we can write

$$\begin{aligned} &\mathbf{P}_\theta\left\{|\hat{\theta}_N - \theta| > \epsilon,\, \hat{\theta}_N \in \operatorname{Arg}\max_{\beta\leq t\leq\alpha} \|\mathcal{Z}_N(t)\|\right\} = \\ &= \mathbf{P}_\theta\{\|\mathcal{Z}_N(\hat{\theta}_N)\| \geq \|\mathcal{Z}_N(t)\|,\, t \in [\beta,\alpha],\, |\hat{\theta}_N - \theta| > \epsilon\} \\ &\leq \mathbf{P}_\theta\{\|\eta_N(\hat{\theta}_N)\| - \|\eta_N(\theta)\| \geq \|m(\theta)\|^2 - \|m(\hat{\theta}_N)\|^2 + L_F/N,\, |\hat{\theta}_N - \theta| > \epsilon\} \\ &\leq \mathbf{P}_\theta\left\{\sup_{\beta\leq t\leq\alpha} \|\eta_N(t)\| \geq \left[\frac{\epsilon\lambda_F}{4\|m(\theta\|}\operatorname{tr}((\mathbf{a}-\mathbf{b})^*(\mathbf{a}-\mathbf{b})) - \frac{L_F}{N}\right]\right\} \leq \\ &\leq \mathbf{P}_\theta\left\{\sup_{\beta\leq t\leq\alpha} \|\eta_N(t)\| \geq \left[\frac{\epsilon\lambda_F}{4\mathcal{M}}\operatorname{tr}((\mathbf{a}-\mathbf{b})^*(\mathbf{a}-\mathbf{b})) - \frac{L_F}{N}\right]\right\}, \end{aligned} \tag{5.27}$$

where $\mathcal{M} = \max_{\beta\leq\theta\leq\alpha} \|m(\theta\|$.

Denote $C(\epsilon, N) = \left[\frac{\epsilon\lambda_F}{4\mathcal{M}}\|\mathbf{a}-\mathbf{b}\|^2 - \frac{L_F}{N}\right]$. Then, finally, we obtain from (5.27)

$$\sup_{\beta\leq\theta\leq\alpha} \mathbf{P}_\theta\{|\hat{\theta}_N - \theta| > \epsilon\} \leq m_0\left(C(\epsilon,N)/\mathcal{R}\right) \begin{cases} \exp\left(-\dfrac{N\beta\left(C(\epsilon,N)/\mathcal{R}\right)^2}{4gm_0\left(C(\epsilon,N)/\mathcal{R}\right)}\right), \\ \text{if } C(\epsilon,N) \leq \mathcal{R}gT \\ \exp\left(-\dfrac{TN\beta\left(C(\epsilon,N)/\mathcal{R}\right)}{4m_0\left(C(\epsilon,N)/\mathcal{R}\right)}\right), \\ \text{if } C(\epsilon,N) > \mathcal{R}gT. \end{cases}$$

Remark 5. *In case of only one regression relationship and independent noises ν_i, we obtain from here*

$$\sup_{\beta\le\theta\le\alpha} \mathbf{P}_\theta\{|\hat{\theta}_N - \theta| > \epsilon\} \le \begin{cases} \exp\left(-\frac{N\beta\epsilon^2}{4g\mathcal{R}^2}\left[\frac{\lambda_F}{4\mathcal{M}}\sum_{j=1}^{k}(a_j-b_j)^2) - \frac{L_F}{N}\right]^2\right) \\ \text{if } C(\epsilon, N) \le \mathcal{R}gT \\ \exp\left(-\frac{TN\beta\epsilon}{4\mathcal{R}}\left[\frac{\lambda_F}{4\mathcal{M}}\sum_{j=1}^{k}(a_j-b_j)^2)^2 - \frac{L_F}{N}\right]\right) \\ \text{if } C(\epsilon, N) > \mathcal{R}gT. \end{cases}.$$

Theorem 5.4.1 is proved.

From the proof, we obtain the following.

Corollary 5.4.1. *Let $C > 0$ be the decision threshold and $\mathbb{C} \stackrel{def}{=} C - \frac{L_F}{N}$. Then:*

– for type-1 error, the following inequality is satisfied:

$$\mathbf{P}_0\{\max_{[\beta N]\le n\le[\alpha N]} \|\mathcal{Z}_N(n)\|^2 > C\} \le m_0(\mathbb{C}/\mathcal{R}) \begin{cases} \exp\left(-\frac{TN\mathbb{C}\beta}{4\mathcal{R}m_0(\mathbb{C}/\mathcal{R})}\right), \\ \text{if } \mathbb{C} > \mathcal{R}gT \\ \exp\left(-\frac{N\beta\mathbb{C}^2}{4\mathcal{R}^2 g m_0(\mathbb{C}/\mathcal{R}))}\right), \\ \text{if } \mathbb{C} \le \mathcal{R}gT. \end{cases} \tag{5.28}$$

– for type-2 error, the following inequality is satisfied:

$$\mathbf{P}_\theta\{\max_{[\beta N]\le n\le[\alpha N]} \|\mathcal{Z}_N(n)\|^2 \le C\} \le m_0(d) \begin{cases} \exp\left(-\frac{TN\beta d}{4m_0(d)}\right), & d > gT \\ \exp\left(-\frac{N\beta d^2}{4gm_0(d)}\right), & d \le gT, \end{cases}$$

where $d = \mathcal{R}^{-1}\left(\|m(\theta)\| - C - \frac{L_F}{N}\right) > 0$, $\|m(\theta)\|^2 = tr(B^* A^2(\theta)B)$.

Corollary 5.4.1 can be obtained (as it follows from the proof) from the estimates of $\mathbf{P}_\theta\{\sup_{\beta\le t\le\alpha} \|\eta_N(t)\| > \epsilon\}$, $\theta = 0$ or $\theta \ne 0$.

5.4.1.2 Stochastic Predictors

In this subsection, we suppose that predictors x_{ji} in (5.2.1) are random. On the probability space $(\Omega, \mathcal{F}, \mathbf{P}_\theta)$, consider filtration $\{\mathcal{F}_n\}$, $n = 1, \dots, n$, where $\{\mathcal{F}_n\} \in \mathcal{F}$, $\mathcal{F}_n$ can be interpreted as all available information up to the instant n.

Put $\mathbf{X}(n) \stackrel{\text{def}}{=} (x_{1n}, \dots, x_{Kn})^*$.

Suppose that the following conditions are satisfied:

a) there exists a continuous symmetric matrix function $V(t), t \in [0,1]$ such that the matrix $\int_{t_1}^{t_2} V(s)ds$ is positive definite for any $0 \le t_1 < t_2 \le 1$, and $\mathbf{E}_\theta \mathbf{X}(n)\mathbf{X}^*(n) = V(n/N)$;

b) the sequence of random vectors $\{(\mathbf{X}(n), \nu_n)\}$ satisfies the uniform Cramer's and ψ-mixing conditions;

c) the random sequence $\{\nu_n\}$ is a martingale-difference sequence w.r.t. the filtration $\{\mathcal{F}_n\}$; and

d) the vector of predictors $\mathbf{X}(n) \stackrel{\text{def}}{=} (x_{1n}, \dots, x_{Kn})^*$ is $\mathcal{F}_{n-1}$-measurable.

On the segment $[0,1]$, define the $K \times K$ matrix process

$$\mathcal{T}_N(t) \stackrel{\text{def}}{=} \sum_{k=1}^{[Nt]} \mathbf{X}(k)\mathbf{X}^*(k)$$

and $K \times M$ matrix process

$$z(1,l) = \sum_{k=1}^{l} \mathbf{X}(k)\mathbf{Y}^*(k),$$

and $u_N(l/N) = z(1,l)$, $l = 1, \dots, N$.

We suppose that

$$N^{-1}\, \mathcal{T}_N(t) \to \mathbb{R}(t) \stackrel{\text{def}}{=} \int_0^t V(s)ds$$

$\mathbf{P}_\theta$- a.s. as $N \to \infty$.

This assumption usually follows from the law of large numbers.

Below, we denote $\mathbb{R}(1) \stackrel{\text{def}}{=} \mathbb{R}$.

For estimation of an unknown change-point, the following statistic is used:

$$\mathbb{Z}_N(n) = N^{-1}\Big(u_N(n/N) - \mathcal{T}_N(n/N)(\mathcal{T}_N(1))^{-1} u_N(1)\Big),\ n = 1, 2, \dots, N. \tag{5.29}$$

An arbitrary point $\hat{n}$ of the set $\operatorname{Arg} \max_{[\beta N] \le n \le N} \|\mathbb{Z}_N(n)\|^2$ is assumed to be the estimate of an unknown change-point. Again we define $\hat{\theta}_N = \hat{n}/N$ as the estimate of the change-point parameter θ.

Statistic (5.29) generalizes Statistic (5.7) to the situation of stochastic predictors. Assumptions a)–d) guarantee the analogous properties of this statistic. In particular, the limit value (as $N \to \infty$) of the mathematical expectation of the statistic $\mathbb{Z}_N([Nt])$ attains its unique global maximum on the segment $[0,1]$ at the point $t^* = \theta$.

Assumptions a)–d) guarantee convergence in probability of an arbitrary point of $\operatorname{Arg} \max_{[\beta N] \le n \le N} \|\mathbb{Z}_N(n)\|^2$ to the point θ with the exponential rate. Hence, the $\mathbf{P}_\theta$-a.s. convergence of the proposed estimate to θ follows.

Theorem 5.4.2. *Suppose that the conditions a)–d) are satisfied and* $\mathrm{rank}(\mathbb{B}) = M$ *if* $\theta \in [\beta, \alpha]$, *where* $\mathbb{B} \stackrel{def}{=} \mathbb{B}(\theta) = \Big(E - \mathbb{R}^{-1}\mathbb{R}(\theta)\Big)(\mathbf{a} - \mathbf{b})^*$.

Then, the estimate $\hat{\theta}_N$ *of the change-point parameter* θ *converges to* θ $\mathbf{P}_\theta$-*a.s. as* $N \to \infty$.

Besides, there exists the number $N_1 = N_1(\{\mathbf{X}(n)\})$, *such that for* $N > N_1$, *and any fixed* ϵ, $(\min((\alpha - \beta), \|\mathbb{R}\|/2) > \epsilon > 0)$, *the following inequality holds:*

$$\sup_{\beta \le \theta \le 1} \mathbf{P}_\theta\{|\hat{\theta}_N - \theta| > \epsilon\} \le \delta_N(\epsilon) +$$

$$m_0\left(\mathbb{C}(\epsilon, N)/\mathbf{R}\right) \begin{cases} \exp\left(-\dfrac{N\beta\Big(\mathbb{C}(\epsilon, N)/\mathbf{R}\Big)^2}{4gm_0\left(\mathbb{C}(\epsilon, N)/\mathbf{R}\right)}\right), & \text{if } \mathbb{C}(\epsilon, N) \le \mathbf{R}gT \\ \exp\left(-\dfrac{TN\beta\Big(\mathbb{C}(\epsilon, N)/\mathbf{R}\Big)}{4m_0\left(\mathbb{C}(\epsilon, N)/\mathbf{R}\right)}\right), & \text{if } \mathbb{C}(\epsilon, N) > \mathbf{R}gT, \end{cases}$$

where $\mathbb{C}(\epsilon, N) = \left[\dfrac{\epsilon\lambda_V}{4\mathbb{M}}\|\mathbf{a} - \mathbf{b}\|^2 - \dfrac{L_V}{N}\right]$, $\mathbb{M} = \max_{\beta \le t \le \alpha} \|M(t)\|$, *the constants* g, T, *and* $m_0(\cdot)$ *are taken from the uniform Cramer's and* ψ-*mixing conditions, and* $M(t)$, λ_V, L_V, δ_N, *and* $\mathbf{R}$ *are described in the proof.*

In particular, for independent observations, $m_0(\cdot) = 1$.

Comparing Theorems 5.4.2 and 5.3.1, we conclude that the order of convergence of the proposed estimate of the change-point parameter to its true value is asymptotically optimal as $N \to \infty$.

The proof of Theorem 5.4.2 is given below.

The proof is based on the same ideas as in Section C, and so we give the sketch of the proof.

Let us consider the matrix random process with continuous time $\mathbb{Z}_N(t) \stackrel{\text{def}}{=} \mathbb{Z}_N([Nt])$, $t \in [0, 1]$.

From assumption d) and Slutsky lemma, it follows that there exists

$$M(t) \stackrel{\text{def}}{=} \lim_{N \to \infty} \mathbb{Z}_N(t) \qquad P_\theta - \text{ a.s. as } N \to \infty.$$

.

After simple transformation, we have

$$M(t) = \begin{cases} \mathbb{R}(t)\mathbb{R}^{-1}\left(\mathbb{R} - \mathbb{R}(\theta)\right)(\mathbf{a} - \mathbf{b})^*, & t \le \theta \\ \left(\mathbb{R} - \mathbb{R}(t)\right)\mathbb{R}^{-1}\mathbb{R}(\theta)(\mathbf{a} - \mathbf{b})^*, & t > \theta. \end{cases} \tag{5.30}$$

It can be shown from (5.30) (by the analogous arguments as in Section C) that the function $\Phi(t) \stackrel{\text{def}}{=} \|M(t)\|^2 = \mathrm{tr}\left(M(t)M^*(t)\right)$ has unique global maximum on the segment $[0, 1]$ at the point $t = \theta$, and there exists $\lambda_V > 0$, such that the following inequality holds:

$$\Phi(\theta) - \Phi(t) \ge \lambda_V |\theta - t| \mathrm{tr}\left[(\mathbf{a} - \mathbf{b})(\mathbf{a} - \mathbf{b})^*\right] \tag{5.31}$$

for any $\beta \le t \le \alpha$. The constant λ_V depends only of $V(t)$ and can be estimated analogously the constant λ_F from Section C.

Consider the matrix sequence $N^{-1}\mathcal{T}_N(1)$. Due to the assumptions, there exists the number $N_1 = N_1(\{\mathbf{X}(n)\})$, such that as $N > N_1$, we get

$$\mathbf{P}_\theta\{\|N^{-1}\mathcal{T}_N(1) - \mathbb{R}\| > \epsilon\} \le L(\epsilon)\exp(-K(\epsilon)N), \tag{5.32}$$

where functions $L(\epsilon)$, $K(\epsilon)$ can be exactly estimated (taking into account ψ-mixing condition and Cramer's condition) by the scheme of Section C. The number N_1 can be estimated by the random sequence $\{\mathbf{X}(n)\}$.

Process $\mathbb{Z}_N(t)$ can be written as follows:

$$\mathbb{Z}_N(t) = M(t) + \Gamma_N(t) + \zeta_N(t),$$

where $\Gamma_N(t) = \mathbf{E}_\theta\mathbb{Z}_N(t) - M(t)$ and $\zeta_N = \mathbb{Z}_N(t) - \mathbf{E}_\theta\mathbb{Z}_N(t)$.

Note that $\max_{0\le t\le 1}\|\Gamma_N(t)\| \le \frac{L_V}{N}$ (because this is the difference between the sum and the integral) and constant L_V can be estimated exactly for any given function $V(t)$.

Fix ϵ, $0 < \epsilon < \min((\alpha-\beta), \|\mathbb{R}\|/2)$ and consider the events

$$\begin{aligned} D_N &= \{\|N^{-1}\mathcal{T}_N(1) - \mathbb{R}\| \le \|\mathbb{R}\|/2, \\ &\max_{0\le t\le 1}\|N^{-1}\mathcal{T}_N(t) - \mathbb{R}(t)\| < \epsilon,\ \|N(\mathcal{T}_N(1)^{-1} - \mathbb{R}^{-1}\| < \epsilon\}, \\ \bar{D}_N &= \Omega\backslash D_N. \end{aligned}$$

Note that matrix $N^{-1}\mathcal{T}_N(1)$ is nondegenerate on the set D_N. Then, due to (5.32),

$$\delta_N(\epsilon) \stackrel{\text{def}}{=} \mathbf{P}_\theta(\bar{D}_N) \le 3L(\epsilon)\exp(-K(\epsilon)N). \tag{5.33}$$

Further, in analogy with (5.20), we can write on the set D_N

$$\begin{aligned} \sup_{\beta\le t\le 1}\|\zeta_N(t)\| &\le R\left[\sqrt{K} + \|\mathbb{R}\|\cdot\|\mathbb{R}^{-1}\| + \epsilon(\|\mathbb{R}\| + \|\mathbb{R}^{-1}\| + \epsilon)\right] \times \\ &\times \left(\max_{1\le i\le K}\max_{1\le l\le M}\max_{[\beta N]\le n\le N} N^{-1}|\sum_{j=1}^{n} x_{ij}\nu_{lj}|\right) \stackrel{\text{def}}{=} \\ &= \mathbf{R}\left(\max_{1\le i\le K}\max_{1\le l\le M}\max_{[\beta N]\le n\le N} N^{-1}|\sum_{j=1}^{n} x_{ij}\nu_{lj}|\right), \end{aligned} \tag{5.34}$$

where $\mathbf{R} = \mathbf{R}(V,\epsilon)$.

Now, we can use (5.26) and get (by the analogous reasons) from (5.34) on the set D_N

$$\mathbf{P}_\theta\{\sup_{\beta\le t\le 1}\|\zeta_N(t)\| > \epsilon,\ \mathbb{I}(D_N\} \le m_0(\epsilon/\mathbf{R})\begin{cases} \exp\left(-(\epsilon/\mathbf{R})^2 N\beta/4gm_0(\epsilon/\mathbf{R})\right), \\ \epsilon \le \mathbf{R}gT \\ \exp\left(-T(\epsilon/\mathbf{R})N\beta/4m_0(\epsilon/\mathbf{R})\right), \\ \epsilon > \mathbf{R}gT. \end{cases} \tag{5.35}$$

Using (5.32), (5.33), and the analogous considerations as in (5.27), we get

$$\sup_{\beta\le\theta\le 1} \mathbf{P}_\theta\{|\hat\theta_N-\theta|>\epsilon\} \le \delta_N(\epsilon)+$$

$$m_0\left(\mathbb{C}(\epsilon,N)/\mathbf{R}\right)\begin{cases}\exp\left(-\dfrac{N\beta\left(\mathbb{C}(\epsilon,N)/\mathbf{R}\right)^2}{4gm_0\left(\mathbb{C}(\epsilon,N)/\mathbf{R}\right)}\right), & \text{if } \mathbb{C}(\epsilon,N)\le \mathbf{R}gT\\ \exp\left(-\dfrac{TN\beta\left(\mathbb{C}(\epsilon,N)/\mathbf{R}\right)}{4m_0\left(\mathbb{C}(\epsilon,N)/\mathbf{R}\right)}\right), & \text{if } \mathbb{C}(\epsilon,N)>\mathbf{R}gT,\end{cases}$$

where $\mathbb{C}(\epsilon,N)=\left[\frac{\epsilon\lambda_V}{4\mathbb{M}}\|\mathbf{a}-\mathbf{b}\|^2-\frac{L_V}{N}\right]$, $\mathbb{M}=\max\limits_{\beta\le t\le 1}\|M(t)\|$.

Theorem 5.4.2 is proved.

From the proof, we obtain the following:

Corollary 5.4.2. *Let $S>0$ be the decision threshold and $\mathbb{S}\stackrel{def}{=}S-\frac{L_V}{N}$. Then:*

— for type-1 error, the following inequality is satisfied:

$$\mathbf{P}_0\{\max_{[\beta N]\le n\le N}\|\mathbb{Z}_N(n)\|^2>S\}\le\delta_N(\mathbb{S})+m_0\left(\mathbb{S}/\mathbf{R}\right)\begin{cases}\exp\left(-\dfrac{TN\mathbb{S}\beta}{4\mathbf{R}m_0\left(\mathbb{S}/\mathbf{R}\right)}\right),\\ \mathbb{S}>\mathbf{R}gT\\ \exp\left(-\dfrac{N\beta\mathbb{S}^2}{4\mathbf{R}^2gm_0\left(\mathbb{S}/\mathbf{R}\right)}\right),\\ \mathbb{S}\le\mathbf{R}gT.\end{cases}$$

— for type-2 error, the following inequality holds:

$$\mathbf{P}_\theta\{\max_{[\beta N]\le n\le N}\|\mathbb{Z}_N(n)\|^2\le S\}\le\delta_N(\mathbb{S})+m_0(r)\begin{cases}\exp\left(-\dfrac{TN\beta r}{4Rm_0(r)}\right),\\ r>RgT\\ \exp\left(-\dfrac{N\beta r^2}{4R^2gm_0(d)}\right),\\ r\le RgT,\end{cases}$$

where $r=\mathbf{R}^{-1}\left(\|M(\theta)\|-S-L_V\right)>0$; $\|M(\theta)\|^2=\mathrm{tr}(\mathbb{B}^*\mathbb{R}^2(\theta)\mathbb{B})$.

5.4.2 Multiple Change-Points

The proposed method can be generalized to problems of detection and estimation of multiple change-points in regression models. A widespread approach to solving these problems (see, e.g., Bai, Lumsdaine, and Stock (1998)) consists in decomposition of the whole obtained sample to all possible subsamples and construction of regression estimates for each of these subsamples. The decomposition for which the minimum of the general sum of regression residuals is attained is assumed to be the estimate of a true decomposition of the whole samples of obtained observations into subsamples with different regression regimes.

These methods turn out to be rather time consuming and have a low power. For example, if there are only two regression regimes in an obtained sample but we do not know this fact and are obliged to try all possible subsamples up to the order 20, then many false structural changes will be obtained.

In this chapter, we propose a new method of detection and estimation of multiple change-points that is not based upon LSE (least square estimates) of regression parameters and computation of corresponding residuals. This method is more effective and robust for possible inaccuracies in specification of regression models.

Let us explain the idea of this method by the following example of a multiple regression model (5.1) with deterministic predictors and the row-matrix $\Pi(\vartheta, n)$. In other words, let $\vartheta = (\theta_1, \theta_2, \dots, \theta_k)$, $k \geq 1$ is an unknown vector of change-point parameters, such that $0 \equiv \theta_0 < \beta \leq \theta_1 < \cdots < \theta_k \leq \alpha < \theta_{k+1} \equiv 1$, where, as before, β, and α are known numbers, and the observations have the form

$$y_n = \Pi^*(\vartheta, n)F(n/N) + \nu_n. \tag{5.36}$$

Here,

$$\Pi(\vartheta, n) = \sum_{i=1}^{k+1} a_i \, \mathbb{I}([\theta_{i-1}N] < n \leq [\theta_i N]),$$

where $a_i \neq a_{i+1}, i = 1, 2, \dots, k$ are unknown *vectors*, $F(t)$ is a given vector-function (for all assumptions and notations, see Subsection 5.4.1).

Consider our main Statistic (5.7). The mathematical expectation of this statistic converges as $N \to \infty$ to the function

$$m(t) = \int_0^t F(s)F^*(s)\Pi(\vartheta, s)ds - A(t)I^{-1} \int_0^1 F(s)F^*(s)\Pi(\vartheta, s)ds.$$

In the situation when there is no change-points, i.e., the vector of regression coefficients is constant on $[0, 1]$, the vector function $m(t)$ equals to zero for each $t \in [0, 1]$. This property of $m(t)$ makes it possible to effectively reject the null hypothesis about the absence of change-points when they are really present in an obtained sample.

Consider the following method of detection and estimation of multiple change-points. Fix a small parameter ϵ, $\min(\beta, 1 - \alpha) > \epsilon > 0$. The proposed method consists of the following steps:

1. Compute statistic (5.7) by the data in the diapason of arguments $\mathcal{N} \stackrel{\text{def}}{=} ([\beta N], \dots, [\alpha N])$. If $\max_{n \in \mathcal{N}} \|\mathcal{Z}_N(n)\|^2 > C$, where $C = C(N)$ is the decision threshold, then compute $nmax = \text{argmax}\|\mathcal{Z}_N(n)\|^2$, otherwise the sample is assumed to be stationary (without change-points).

2. Put $N^{'} = nmax - [\epsilon N]$ and compute Statistic (5.7) by the data in the diapason of arguments $\mathcal{N}^{'} \stackrel{\text{def}}{=} \left([\beta N], \dots, N^{'}\right)$ according to step 1. This cycle is repeated until:

1) we obtain a stationary subsample in the diapason of data with arguments $\left([\beta N], \dots, N^{'}\right)$, i.e., $\max_{n \in \mathcal{N}^{'}} \|Z_{N^{'}}(n)\|^2 \leq C(N^{'})$. Then, we put $n(1) = N^{'} + [\epsilon N]$ as the estimate of the first change-point and go to step 3: or

2) we obtain a sample of the size $N^{'} \leq [2\epsilon N]$. Then we put $n(1) = N^{'} + [\epsilon N]$ as the estimate of the first change-point and go to step 3.

3. Put $n^{'} = n(1) + [\epsilon N]$ and compute Statistic (7) by the data in the diapason of arguments $\left(n^{'}, \dots, [\alpha N]\right)$ (i.e., with the relative arguments $[1, \dots, [\alpha N] - n^{'} + 1]$) and do according to steps 1 and 2. The cycle is repeated until we obtain a stationary subsample in the diapason of data with arguments $[n^{'}, \dots, nmax]$ or $nmax - n^{'} \leq [2\epsilon N]$. Then, we put $n(2) = nmax$ as the estimate of the next change-point. If $N - n(2) < [2\epsilon N]$, then stop, otherwise repeat step 3 by the data in the diapason of arguments $(n(2), \dots, [\alpha N])$.

In this way, we continue to compute the estimates $n(3), \dots$ of change-points. As a result, we obtain the series of estimates $n(1), n(2), \dots$ of the true change-points $[\theta_1 N], \dots, [\theta_k N]$. The number $\hat{k}_N$ of these estimates is determined by the quantity of stationary sub-samples

$$[1, \dots, n(1)], \dots, [n(i), \dots, n(i+1)], \dots, [n(\hat{k}_N), \dots, N].$$

The proposed method is based upon reduction to the case of only one change-point and the properties of the matrix $m(t)$. The crucial point of this method is the choice of the decision threshold $C(N)$, which depends on the sample size N. Below, we give an explicit formula for computation of $C(N)$.

Let $\hat{k}_N$ be the estimate of the number of change-points in the sample, and $\hat{\vartheta}_N = (\theta_{N1}, \dots, \theta_{N\hat{k}_N})^*$ be the vector of estimated coordinates of change-point parameters. The following theorem holds for model (5.4.30).

Theorem 5.4.3. *Suppose assumptions of Theorem 5.4.1 are satisfied. Moreover, assume that there exist $h > 0$, $B > 0$, such that for all $i = 2, \dots, k+1$:*

$$\begin{gathered} 0 < \|A(\theta_{i-1}, \theta_i) A^{-1}(\theta_{i-2}, \theta_{i-1})\| \leq h \\ \|A(\theta_{i-1}, \theta_i)(a_i - a_{i-1})\| \geq B > 0. \end{gathered}$$

Then, for sufficiently small $\delta > 0$:

$$\mathbf{P}\{(\hat{k}_N \neq k) \cup \{(\hat{k}_N = k) \cap (\max_{1 \leq i \leq k} |\hat{\theta}_{Ni} - \theta_i| > \delta)\}\} \leq C(\delta) \exp(-D(\delta) N),$$

where constants $C(\delta) > 0, D(\delta) > 0$ do not depend on N.

Analogous theorem can be proved also for stochastic predictors.

From Theorem 5.4.3, it follows that the estimated number of change-points converges almost surely to its unknown true value, as well as estimated coordinates of unknown change-points converge exponentially to their true values as the sample size tends to infinity. Moreover, comparing results of Theorems 2 and 5 we conclude that the proposed method of detection and estimation of multiple change-points is asymptotically optimal by the order of convergence of estimated change-point parameters to their true values.

The proof of Theorem 5.4.3 is given below.

The proposed method of multiple change-point detection and estimation is based upon the idea of recurrent reduction to the case of one change-point.

In order to prove Theorem 5.4.3, we need to prove the following two propositions:

i) In the case of a stationary subsample, the norm of the decision statistic does not exceed the threshold with the great probability. This fact is exactly the result of Corollary 2.

ii) In the case of a nonstationary subsample with at least two change-points, the norm of the decision statistic exceeds the decision threshold with the great probability.

In order to illustrate ii), let us consider a subsample of size N with two change-points, $0 < \theta_1 < \theta_2 < 1$.

In this case, the decision statistic can be decomposed into a deterministic and a stochastic term (see (5.19)).

We have from (5.19) for $0 \le t \le \theta_1$:

$$\begin{aligned} m(t) &= A(t)a_1 - A(t)A^{-1}(1)\left(A(\theta_1)a_1 + A(\theta_1,\theta_2)a_2 + A(\theta_2,1)a_3\right) \\ &= A(t)\left(a_1 - A^{-1}(1)u\right), \end{aligned} \tag{5.37}$$

where $u = A(\theta_1)a_1 + A(\theta_1,\theta_2)a_2 + A(\theta_2,1)a_3$.

Again, using (5.9), we get for $\theta_1 \le t \le \theta_2$:

$$\begin{aligned} m(t) &= A(\theta_1)a_1 + A(\theta_1,t)a_2 - A(t)A^{-1}(1)u = \\ &= A(\theta_1)\big(a_1 - A^{-1}(1)u\big) + A(\theta_1,t)\big(a_2 - A^{-1}(1)u\big). \end{aligned}$$

If

$$\|m(\theta_1)\| \ge \Lambda \overset{\text{def}}{=} \frac{B}{2(h+1)} > 0,$$

then $\max\limits_{\beta \le t \le \alpha} \|m(t)\| \ge \Lambda > 0$.

Otherwise, let $\|m(\theta_1)\| < \Lambda$. Then,

$$\begin{aligned} &\|m(\theta_2)\| \ge \|A(\theta_1,\theta_2)(a_2 - A^{-1}(1)(u)\| - \Lambda = \\ &= \|A(\theta_1,\theta_2)(a_2 - a_1 + a_1 - A^{-1}(1)u\| - \Lambda \\ &\ge \|A(\theta_1,\theta_2)(a_2 - a_1)\| - \|A(\theta_1,\theta_2)(a_1 - A^{-1}(1)(u)\| - \Lambda \\ &\ge B - \|A(\theta_1,\theta_2)A^{-1}(\theta_1)\|\,\Lambda - \Lambda \ge B - \Lambda(1+h) > \Lambda. \end{aligned}$$

Therefore, taking into account (5.37), there exists $\Lambda > 0$, such that

$$\max_{\beta \le t \le \alpha} \|m(t)\| \ge \Lambda. \tag{5.38}$$

From (5.38) ii) follows.

After these preliminary considerations, let us consider the probability of the event

$$(\hat{k}_N \neq k) \cup \{(\hat{k}_N = k) \cap (\max_{1 \le i \le k} |\hat{\theta}_{Ni} - \theta_i| > \delta) \tag{5.39}$$

for some fixed δ, $\epsilon > \delta > 0$. Let us consider the following cases:

a) $\{\hat{k}_N < k\}$, *b)* $\{\hat{k}_N > k\}$, *c)* $\{(\hat{k}_N = k) \cap (\max_{1 \le i \le k} |\hat{\theta}_{Ni} - \theta_i| > \delta)\}$.

Case a)

In this case, the proposed method does not detect at least one change-point, i.e., a certain subsample of size $\tilde{N} \ge [2\delta N]$ containing at least one true change-point is classified as stationary. Then,

$$\mathbf{P}_\vartheta\{\hat{k}_N < k\} \le \mathbf{P}_\vartheta\{\max_{\beta \le t \le \alpha} \|\mathcal{Z}_{\tilde{N}}(t)\| \le C(\tilde{N})\}, \tag{5.40}$$

where $C(\tilde{N})$ is the decision threshold for the sub-sample.

Choose $C(\tilde{N}) < \Lambda$. Then, due to (5.40) and (5.19), we have

$$\mathbf{P}_\vartheta\{\max_{\beta \le t \le \alpha} \|\mathcal{Z}_{\tilde{N}}\| \le C(\tilde{N})\} \le \mathbf{P}_\vartheta\{\max_{\beta \le t \le \alpha} \|\eta_{\tilde{N}}(t)\|$$

$$\ge \max_{\beta \le t \le \alpha} \|m(t)\| - \tfrac{L_F}{N} - C(\tilde{N})\} \le \mathbf{P}_\vartheta\{\max_{\beta \le t \le \alpha} \|\eta_{\tilde{N}}(t)\| \ge \Lambda - \tfrac{L_F}{N} - C(\tilde{N})\}.$$

Now, we can use (5.26), changing ϵ by $\{\Lambda - \frac{L_F}{N} - C(\tilde{N})\}$, and get the exponential estimate for the event $\{\hat{k}_N < k\}$.

Case b)

In this case, there exists a stationary subsample of the size $\hat{N} \ge [\delta N]$, such that it is classified as nonstationary. Then,

$$\mathbf{P}_0\{\hat{k}_N > k\} \le \mathbf{P}_0\{\max_{\beta \le t \le \alpha} \|\mathcal{Z}_{\hat{N}}(t)\| > C(\hat{N})\} \tag{5.41}$$

Therefore, we obtain the exponential estimate of the right-hand side of (5.41).

Case c)

In this case, there exists a subsample of the size $N^* \ge [2\delta N]$, such that the distance between a true change-point parameter θ_i and its estimate $\hat{\theta}_{Ni}$ is larger than δ. This is exactly the case of Theorem 5.4.1.

Theorem 5.4.3 is proved.

5.4.3 Choice of the Threshold

For practical applications of the proposed method, and, in particular, for the rational choice of the decision threshold $C(N)$, I need to study the limit distribution of the decision statistic under the null hypothesis.

Let us formulate a variant of the limit theorem for the simple case of unique change-point, deterministic predictors, statistically independent noises ν_n, and the one-dimensional dependent variable y_n.

Suppose there exists a continuous function $g(t)$, $0 \le t \le 1$, such that $\mathbf{E}_\theta \nu_n^2 = g^2(n/N)$.

Put

$$\sigma_i^2 = \frac{1}{t} \int_0^t f_i^2(s) g^2(s) ds, \; i = 1, \ldots, K$$
$$G(t) = (\sigma_1(t), \ldots, \sigma_K(t))^*, \; \mathbf{Z}(t) = G(t)W(t), \; U(t) = \mathbf{Z}(t) - A(t)I^{-1}\mathbf{Z}(1),$$

where $W(t)$ is the standard Wiener process, and $A(t)$, I are the above defined matrices (see Subsection 5.4.1).

Consider our main statistic, the vector process $\mathcal{Z}_N(t) = \mathcal{Z}_N([Nt])$. Then, for any $\theta \in [\beta, \alpha]$, the vector process $\sqrt{N}(\mathcal{Z}_N(t) - \mathbf{E}_\theta \mathcal{Z}_N(t))$ weakly converges to the vector process $U(t)$ in the Skorokhod space $D^K[\beta, \alpha]$ (see Brodsky and Darkhovsky, 2000). In particular, under the null hypothesis, the weak convergence is valid at $[0, 1]$.

Therefore, we have the following.

Theorem 5.4.4.

$$\lim_{N \to \infty} \mathbf{P}_0\{\sqrt{N} \max_{t \in [0,1]} \|\mathcal{Z}_N(t)\| > C\} = \mathbf{P}_0\{\max_{t \in [0,1]} \|U(t)\| > C\} \tag{5.42}$$

(here we use the Euclidean norm for vectors).

The vector $U(t)$ is Gaussian with zero mean and the following $K \times K$ correlation matrix $D(t)$:

$$D(t) = t\left[G(t)G^*(t) - G(t)G^*(1)I^{-1}A(t) - A(t)G(1)G^*(t)\right]$$
$$+ A(t)I^{-1}G(1)G^*(1)I^{-1}A(t).$$

Therefore, we have the following equality by distribution:

$$U(t) = \sqrt{D(t)}\zeta, \tag{5.43}$$

where $\zeta = (\zeta_1, \ldots, \zeta_K)^*$ is the standard Gaussian vector.

Taking (5.43) into account, we get

$$\max_{0 \le t \le 1} \|U(t)\| = \max_{0 \le t \le 1} \sqrt{\sum_{i=1}^{K} d_i^2(t) \zeta_i^2} \stackrel{\text{def}}{=} \rho(\zeta), \tag{5.44}$$

where $d_i^2(t)$ are eigenvalues of the matrix $D(t)$. The function $\rho(\zeta)$ can be explicitly calculated for any given family of functions $F(t), g(t)$.

Therefore, from (5.44), we have

$$\mathbf{P}_0\{\max_{0\le t\le 1} \|U(t)\| > C\} = \int_{\{u:\rho(u)>C\}} \varphi(u)du, \tag{5.45}$$

where $\varphi(u)$ is the density of the standard Gaussian distribution.

From (5.42) and (5.45), we can conclude that type-1 error goes to zero as $\exp(-const\, NC^2)$ for the proposed method. This fact allows us to choose the decision threshold. Note that the same asymptotical order can be obtained from Corollary 5.4.1 (see Subsection 5.4.1). For independent noises, we have

$$\mathbf{P}_0\{\max_{[\beta N]\le n\le N} \|\mathcal{Z}_N(n)\|^2 > C\} \le \begin{cases} \exp\left(-\dfrac{TN\mathbb{C}\beta}{4R}\right), & \mathbb{C} > gT \\ \exp\left(-\dfrac{N\beta\mathbb{C}^2}{4R^2 g m_0(\mathbb{C})}\right), & \mathbb{C} \le gT \end{cases}$$

(see the notations in Subsection 5.4.1).

Therefore, we conclude that type-1 error α_N goes to zero exponentially as $N\to\infty$ for the proposed method.

So, the threshold can be calculated from the relation

$$C = C(N) = \frac{1}{\sqrt{N}}\,|\ln \alpha_N|\,\lambda,$$

where λ is a certain calibration parameter that depends on variations of predictors, dispersions of noises, and characteristics of their statistical dependence.

A closer study allows us to obtain the following practical formula for the decision threshold $C = C(N)$:

$$C(N) = \frac{\left(\max_i \sigma_i^2 \cdot \max_i \max_{0\le t\le 1} f_i^2(t)\right)^{1/2}}{\sqrt{N}}\,\lambda,$$

where σ_i^2 is the dispersion of ν_i and $\lambda > 0$ is the calibration parameter.

5.5 Simulations

In this section, I present results of a simulation study of the proposed method in comparison with other well-known tests. The following methods are most often used for detection of structural changes in regression models:

— The Chow test most often used in econometric packages.

— The CUSUM (cumulative sums) test based upon recursive regression residuals (Brown, Durbin, and Evans, 1975).

— The CUSUM test based upon residuals of ordinary least squares method (OLS CUSUM test, Ploberger, and Kramer, 1992).

— Fluctuation test (Ploberger, Kramer, and Kontrus, 1989).

— Wald test (Andrews, 1993; Andrews, and Ploberger, 1994).

— LM test (Lagrange Multilpier test, Andrews, 1993).

However, it is well known (see, e.g., Maddala and Kim (1998)) that the Wald test (together with the QMLE – quasi-maximum likelihood estimation test) is the best and most often used for detection of changes in regression models because it has the best characteristics of power and accuracy of change-point estimation.

The Wald test statistic is defined as follows:

$$SupW = \max_{1 \le m \le N} N[\frac{S(N) - S_1(m) - S_2(N-m)}{S_1(m) + S_2(N-m)}],$$

where $S(N)$ is the sum of regression residuals constructed by the whole sample of the size N; $S_1(m)$ is the sum of regression residuals constructed by the subsample of the first m observations; $S_2(N-m)$ is the sum of residuals of the regression model constructed by the last $N-m$ observations.

It is natural to define the estimate of the change point as $n_0 \in arg \sup W$, and the corresponding estimate of the change-point parameter $\hat{\theta}_N = n_0/N$.

Comparison of characteristics of different methods is carried out in the following way. First, methods are "equalized" by the value of type-1 error by means of choice of the corresponding decision thresholds. In practice, for this purpose, we use experiments with stationary samples (without structural changes) in which the 95 percent quantiles of the variation series of the decision statistics are computed (see below, Table 5.1). Second, for the chosen sample sizes and decision thresholds, experiments with nonstationary samples are performed in which we compute estimates of the type 2 error probability and instants of change-points (see Tables 5.2 and 5.4). The method of change-point detection "a" is preferable w.r.t. the method "b" if for the same values of the type 1 error, it gives lower estimates of the type-2 error and the error of change-point estimation.

5.5.1 Deterministic Regression Plan

I compared characteristics of the proposed method with those of the Wald test using the following regression model with deterministic predictors:

$$y_i = c_0 + c_1 x_i + \xi_i, \quad i = 1, \dots, N, \tag{5.46}$$

where $(x_1, \dots, x_N)^*$ is the vector of deterministic predictors; $\{\xi_i\}$ is the Gaussian noise sequence with zero mean and unit variance; and c_0, and c_1 are regression coefficients that change at the instant $n_0 = [\theta N]$, $0 < \theta < 1$.

The number of independent trials of each experiment was equal to k=2000. The estimates of decision thresholds were obtained as follows. For each stationary sample, the 95 percent and 99 percent quantiles of the variation series

TABLE 5.1
Estimation of the Decision Thresholds for the Wald Test for Different Sample Sizes

N	100	200	300	400	500	700	1000	1200
$p = 0.95$	10.10	8.09	9.59	8.66	8.12	7.62	7.51	7.43
$p = 0.99$	12.60	10.88	14.14	12.10	12.20	9.97	11.68	10.02

TABLE 5.2
Estimation of the Change-Point Parameter $\theta = 0.30$ by the Wald Test

N		300	400	500	700	1000
$\delta = 0.3$	C	5.63	6.76	8.24	9.77	12.09
	$\hat{w}_N$	0.83	0.71	0.59	0.46	0.32
	$\hat{\theta}_N$	0.29	0.25	0.22	0.19	0.20
$\delta = 0.4$	C	9.65	10.20	11.88	15.27	19.32
	$\hat{w}_N$	0.56	0.47	0.34	0.23	0.18
	$\hat{\theta}_N$	0.28	0.25	0.22	0.20	0.23

of maximums of the decision statistic were computed in 2000 trials. These quantiles were then assumed to be estimates of the decision thresholds for 5 percent and 1 percent error level, respectively.

The values of the threshold C given in Table 5.1, were used as decision bounds for the confidence probability 95 percent in experiments with non-stationary regression models. The following cases were considered:

— Before the change-point: $c_0 = 0$, $c_1 = 1$.

— After the change-point: $c_0 = \delta$, $c_1 = 1$.

In experiments, the parameter δ and the sample size N were changed. The following characteristics of the proposed method were estimated:

— The empirical estimate of decision threshold C (more exactly, the empirical estimate of $\max_n \|\mathcal{Z}_N(n)\|$).

— The empirical estimate of type-2 error probability $\hat{w}_N$.

— The empirical estimate of the change-point parameter $\hat{\theta}_N$.

Results obtained for the Wald test are given in Table 5.2.

The same model was studied with the help of the method proposed in this chapter.

1) Decision thresholds

In the first series of experiments, Model (5.46) with constant coefficients $c_0 = 0$, $c_1 = 1$ was used. The results obtained are presented in Tables 5.3, and 5.4.

2) The estimates of the change-point parameter

Comparing results from Tables 5.2 and 5.4, I conclude that type-2 error estimates for the proposed method are lower than for the Wald test, and the error of estimation for this method is much lower than for the Wald test.

TABLE 5.3
Detection of Change-Points in Regression Models; Estimation of the Decision Thresholds

N	100	200	300	400	500	700	1000	1200
$p = 0.95$	0.401	0.257	0.202	0.182	0.150	0.125	0.103	0.081
$p = 0.99$	0.450	0.300	0.247	0.211	0.187	0.162	0.138	0.102

TABLE 5.4
Results of Estimation of the Change-Point Parameter $\theta = 0.30$

N		300	400	500	700	1000
$\delta = 0.3$	C	0.179	0.177	0.168	0.157	0.151
	$\hat{w}_N$	0.64	0.55	0.33	0.13	0.03
	$\hat{\theta}_N$	0.340	0.322	0.332	0.324	0.307
$\delta = 0.4$	C	0.220	0.211	0.208	0.195	0.192
	$\hat{w}_N$	0.28	0.24	0.11	0.02	0.005
	$\hat{\theta}_N$	0.315	0.312	0.308	0.305	0.304

Therefore, we conclude that the proposed method is essentially better by the main performance characteristics of change-point detection than the Wald test, and, therefore, we conclude that the proposed method is one of the most effective among all known tests for detection and estimation of structural changes in regression models.

Comparing results from Table 5.4 and 5.5, I can conclude that the quality of estimation of the change-point parameter θ depends on its location on the segment $[0, 1]$; estimation of θ, which is closer to the bounds of the segment $[0, 1]$, is more difficult.

5.5.2 Stochastic Regression Plan

In this series of experiments, the following model of observations was used:

$$y_i = c_0 + c_1 x_i + \xi_i, \quad i = 1, \dots, N,$$

where $(x_1, \dots, x_N)^*$ is a stationary random sequence of the following type:

$$x_i = \rho x_{i-1} + \eta_i, \quad i = 1, \dots, N, \ x_0 \equiv 0.$$

$\{\xi_i, \eta_i\}$ is the sequence of independent Gaussian r.v.'s with zero mean and unit dispersion; c_0, and c_1 are regression coefficients that change at the instant $n_0 = [\theta N]$, $0 < \theta < 1$; $|\rho| < 1$.

1) Estimation of decision thresholds

In the first series of tests, decision thresholds were estimated. For this

TABLE 5.5
Results of Estimation of the Change-Point Parameter $\theta = 0.50$

N		300	400	500	700	1000
$\delta = 0.3$	C	0.194	0.184	0.175	0.168	0.164
	$\hat{w}_N$	0.62	0.50	0.25	0.05	0.01
	$\hat{\theta}_N$	0.456	0.485	0.501	0.502	0.499
$\delta = 0.4$	C	0.231	0.221	0.215	0.214	0.211
	$\hat{w}_N$	0.26	0.22	0.003	0.02	0
	$\hat{\theta}_N$	0.495	0.495	0.489	0.501	0.499

TABLE 5.6
Estimation of Decision Thresholds (the Case of Stochastic Predictors)

N	100	200	300	400	500	700	1000	1200
$p = 0.95$	0.355	0.291	0.230	0.188	0.150	0.132	0.103	0.082
$p = 0.99$	0.401	0.332	0.273	0.218	0.192	0.171	0.141	0.100

purpose, stationary sequences (without change-points) were used: $c_0 = 0$, $c_1 = 1$, and $\rho = 0.3$. The results obtained are given in Table 5.6.

2) Estimation of the change-point parameter

In the following series of experiments, a model with a structural change in the regression coefficients was used:

— Before the change-point: $c_0 = 0$, $c_1 = 1$.

— After the change-point: $c_0 = 0$, $c_1 = 1.3$.

Results obtained are presented in Table 5.7.

TABLE 5.7
Estimation of Change-Point Parameters (the Case of Stochastic Predictors)

N		500	700	1000	1200
$\theta = 0.5$	C	0.167	0.157	0.152	0.152
	$\hat{w}_N$	0.32	0.21	0.02	0
	$\hat{\theta}_N$	0.481	0.495	0.498	0.499
$\theta = 0.3$	C	0.156	0.148	0.142	0.140
	$\hat{w}_N$	0.45	0.30	0.03	0
	$\hat{\theta}_N$	0.312	0.310	0.308	0.301

TABLE 5.8
Estimation of Decision Thresholds (the Case of a Multivariate System)

N	200	400	500	700	900	1000	1200	1500
$p = 0.95$	0.28	0.20	0.19	0.18	0.16	0.15	0.145	0.14
$p = 0.99$	0.36	0.33	0.28	0.24	0.23	0.21	0.19	0.17

5.5.3 Multiple Structural Changes

The following multivariate system was used:

$$\begin{aligned} y_i &= c_0 + c_1 y_{i-1} + c_2 z_{i-1} + c_3 x_i + \epsilon_i \\ z_i &= d_0 + d_1 y_i + d_2 x_i + \xi_i \\ x_i &= 0.5 x_{i-1} + \nu_i \\ \epsilon_i &= 0.3 \epsilon_{i-1} + \eta_i, \end{aligned}$$

where ξ_i, ν_i, η_i, $i = 1, 2, \ldots$ are independent standard Gaussian random variables.

Here, $(y_i, z_i)^*$ is the vector of endogenous variables, x_i is the vector of exogenous variables, and $(y_{i-1}, z_{i-1}, x_i)^*$ is the vector of predetermined variables of the considered system.

Dynamics of this system is characterized by the following vector of coefficients: $\mathbf{u} = [c_0\ c_1\ c_2\ c_3\ d_0\ d_1\ d_2]$. The initial vector of coefficients is [0.1 0.5 0.3 0.7 0.2 0.4 0.6]. The first structural change occurs at the instant $\theta_1 = 0.3$. The vector of coefficients $\mathbf{u}$ changes into [0.1 0.5 0 0.7 0.2 0.4 0.6]. The second structural change occurs at the instant $\theta_2 = 0.7$. Then, the vector $\mathbf{u}$ changes into [0.1 0.5 0 0.7 0.2 0.4 0.9].

In the first series of tests, the decision threshold C was estimated. For this purpose, the model with the initial vector of coefficients $\mathbf{u}$ and without change-points was used. In 2000 independent trials, the maximums of the decision statistic were computed and the variation series of these maximum was constructed. Then the 95 percent and the 99 percent quantiles of this series were computed. These values are presented in table 5.8.

The computed 95 percent quantiles were assumed to be the decision thresholds for the corresponding sample volumes.

In the next series of tests, nonstationary samples with multiple change-points were used. The true number of change-points was equal to $p = 2$, the coordinates of these change-points were $\theta_1 = 0.3$ and $\theta_2 = 0.7$. In Table 5.9, the following performance characteristics are given:

— w is the estimate of the probability $\mathbf{P}_\theta\{\hat{p}_N \neq p\}$ in 2000 independent trials, where $\hat{p}_N$ is the estimate of the number of change-points in the data.

— Δ is the estimation error on condition that $\hat{p}_N = p$, i.e. $\Delta = \sqrt{\sum_{i=1}^{p} (\hat{\theta}_i - \theta_i)^2}$.

TABLE 5.9
Estimation of Change-Point Parameters (the Case of a Multivariate System)

N	200	400	500	700	900	1000	1200	1500
w	0.96	0.54	0.39	0.21	0.04	0.03	0.02	0.01
Δ	0.02	0.05	0.04	0.02	0.03	0.02	0.01	0.005

5.6 Applications

In this section, I present results of the application of methods proposed in this chapter to retrospective detection of structural changes in the monthly model of the CPI (consumer price index) inflation in Russia in 1995(1)–2015(9) (225 observations).

For this model, the following price indices were used:

$pi = CPI/100 - 1$ — the rate of change of CPI index;

$eps = E/E(-1) - 1$ — the rate of change in exchange rate dollar/ruble

$piel = PEL/100 - 1$ — the rate of change of the price index for electricity, gas, and water

$piplod = Pplod/100 - 1$ — the rate of change of the price index for fruits and vegetables;

$pimilk = Pmilk/100 - 1$ — the rate of change of the price index for milk.

The obtained multifactor econometric model for monthly data in the period 1995(1)–2015(9) has the following form:

$$pi = 0.0047 + 0.2683 * eps + 0.1014 * piel + 0.0714 * piplod + 0.2290 * pimilk.$$

The graph of monthly CPI inflation index is presented in Figure 5.1.

The t-statistics for these coeficients are as follows: 6.01 for 0.0047; 28.1 for 0.2683; 3.76 for 0.1014; 7.96 for 0.0714; and 8.33 for 0.2290. The criterion $R2 = 0.89$, the Durbin-Watson statistic: $DW = 1.65$, is the factor of residual dispersion $\sigma = (0.00927)^2$ for this regression relationship.

However, we see that this model can be subject to structural changes. In order to detect these changes, we do as follows. First, we need to compute the decision threshold. For this purpose, we do as follows. First, we model homogenous samples (without change-points):

$$y(i) = pi(i) + \sigma v(i), \quad i = 1, \ldots, N,$$

where $pi(i)$ is defined according to the above formula, $v(i) \sim \mathcal{N}(0, 1)$. In 1000 independent trials of the Monte-Carlo test for the actual sample size $N = 225$, we compute 95 and 99 percent quantiles of the proposed statistic for this data set. Then we assume the decision threshold to be equal to 95 percent quantile of this statistic. In experiments, we computed $th = 0.00045$.

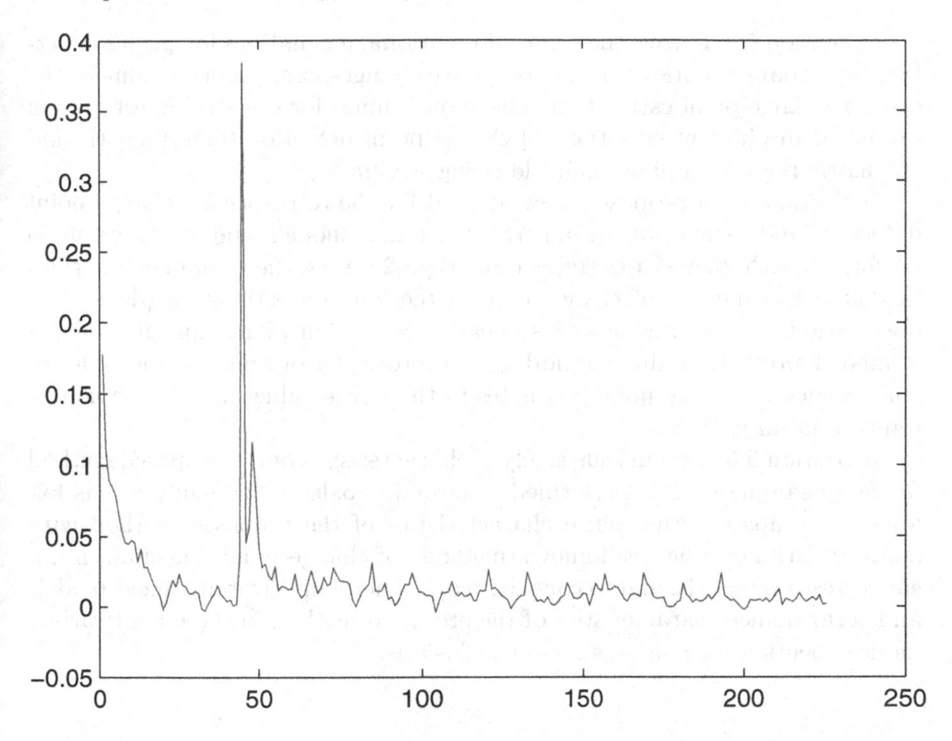

FIGURE 5.1
Graph of the monthly CPI inflation index in Russia (1995(1)–2015(9)).

Second, we use this threshold for the actual sample size to determine structural changes in the whole sample of initial data. For this purpose, the method described in this chapter can be used. Then, we compute structural changes in this sample. Finally, we detected the following points of structural changes: 12, 44, 76, 172. However, the point $n_1 = 12$ is evidently the effect of the sample bounds. Therefore, we report the following instants: 44, 76, 172.

5.7 Conclusions

In this chapter, the main problems of change-point detection in multivariate non-stationary stochastic systems are considered. In Section 5.2, the general change-point problem for multivariate linear systems is formulated and general assumptions are given.

In Section 5.3, I prove the prior informational inequalities for the main performance characteristic of the retrospective change-point problem, namely, the error of change-point estimation. The lower bounds for the error of estimation are found in different situations of change-point detection (deterministic and stochastic regression plan, multiple change-points).

In Section 5.4, I propose a new method for the retrospective change-point detection and estimation in multivariate linear models and study its main performance characteristics (type-1 and type-2 errors, the error of estimation) in different situations of change-point detection and estimation (dependent observations, deterministic and stochastic regression plan, multiple change-points). I prove that this method is asymptotically optimal by the order of convergence of change-point estimates to their true values as the sample size tends to infinity.

In Section 5.5, a simulation study of characteristics of the proposed method for finite sample sizes is performed. The main goals of this study are as follows: to compare performance characteristics of the proposed method with characteristics of other well-known methods of change-point detection in linear regression models, and to consider more general multivariate linear models and performance characteristics of the proposed method in these multivariate models. Section 5.6 contains main conclusions.

6

Retrospective Detection of Change-Points in State-Space Models

6.1 Introduction

Many applied change-point detection and estimation problems are naturally formulated in terms of state-space models, and very often in terms of the retrospective problem statement when we deal with an obtained multivariate sample of interrelated observations. To this I must add that most methods developed for solving such problems (i.e., the Kalman filter) are developed mainly for sequential problem statements when we must make decisions at every step of data acquisition (see, e.g., Fuh (2004, 3006)). However, in most applied problems, we know much more than simply a sequence of observations (see Brodsky (2011)). Most often, we know a "baseline model" (whether from a detailed experiment in the beginning of our study or on the basis of theoretical reasons). For example, an economic system can be observed via some set of key economic parameters. However, this system can be influenced by certain unobservable "shocks" taht change the vector-state of this system in an unpredictable manner. Except for the vector of observations, we know (from economic theory) the "baseline model" of this system that depicts its normative behavior. Our problem here is to detect the moments of "shocks" on the basis of all available information.

Thus, in considering applied problems, we encounter the following state-space model: Before the change-point m,

$$
\begin{aligned}
Y(t) &= JX(t) + \xi(t) \\
X(t) &= GX(t-1) + \eta(t),
\end{aligned}
$$

where

$Y(t)$ is the vector of observations $M \times 1$;
$X(t)$ is the vector of state variables in the baseline model $K \times 1$;
$\xi(t)$ is the vector of random noises in observation channels $M \times 1$;
$\eta(t)$ is the vector of random noises in the baseline model $K \times 1$;
J is the matrix of coefficients $M \times K$; and
G is the matrix of coefficients $K \times K$.

It is usually assumed that $\xi(t), \eta(t)$ are independent random sequences. Moreover, we assume that $\xi(t)$ are independent for different t, as well as $\eta(t)$.

At change-point m, there arrives an unpredictable change in observation channels or in the state model:

$$Y(t) = DZ(t) + \xi(t)$$
$$Z(t) = FZ(t-1) + \eta(t),$$

Since the instant $t = m$ on, we know only variables $Y(t)$, but we do not know variables $Z(t)$. Remark that we still can use variables X(t) of the baseline model. In usual terms, we encounter here the situation of a partly observed nonstationary system. We have:

1) the vector $Y(t)$ of observations;

2) the vector $Z(t)$ of state variables that is observable until the first change-point m; and

3) the observed vector of the "initial (baseline) model" $X(t)$ that coincides with the state vector $Z(t)$ before the change-point m.

In order to clarify this situation, let us consider the following example.

Example 6.1. Model of stochastic volatility (SV)

In applied financial analysis, the following model is often used:

$$y_t = \gamma + c\,h_t + \xi_t$$
$$h_t = \phi h_{t-1} + \eta_t,$$

where

y_t is the sequence of expected returns; and

h_t is the sequence of volatility values at different times t.

Remark that the dynamics of this system is well known in the "basis situation" that is obtained from the detailed experiment. In other words, coefficients γ, c, ϕ of this system are known to us. However, at some (a priori unknown) time m, a change in these coefficients occurs, i.e., the above system is transformed into

$$y_t = \delta + dz_t + \xi_t$$
$$z_t = \psi z_{t-1} + \eta_t.$$

The problem is to detect this change-point m using information on observations y_t and the basis model of volatility h_t only.

6.2 Problem Statement

Now, I continue with the general case. First, let us discuss the heuristic idea of this method. The difference between processes $Z(t)$ and $X(t)$ satisfies the following relationship:

$$Z(t+1) - X(t+1) = -GX(t) + FZ(t) = F(Z(t) - X(t)) + (F - G)X(t).$$

Denote $W(t) = Z(t) - X(t)$ and introduce the shift operator $LX(t+1) = X(t)$. We can write

$$W(t) = (E - FL)^{-1}(F - G)LX(t).$$

Hence, for the process $Y(t)$ and $t > m$, we obtain

$$Y(t) = D(W(t)+X(t))+\xi(t) = D(G+(E-FL)^{-1}(F-G))LX(t)+D\eta(t)+\xi(t),$$

and, therefore, for $t > m$, we can think about $Y(t)$ as the sum of the infinite series

$$\begin{aligned} Y(t) &= DFLX(t) + DF_1L^2X(t) + \cdots + \epsilon(t), \\ X(t+1) &= GX(t) + \eta(t+1), \end{aligned}$$

where $F_k = F^k(F - G)$. Remember that $\epsilon(t) = D\eta(t) + \xi(t)$.

Therefore we can consider the following system

$$\begin{aligned} Y(t+1) &= DFX(t) + DF_1X(t-1) + \cdots + DF_kX(t-k) + \Omega^k(t+1), \\ X(t+1) &= GX(t) + \eta(t+1) \end{aligned}$$

for detection of structural changes in the matrices of state-space and observations equations. Here, $\Omega^k(t) = \epsilon(t) + DF_kX^m(t-k-1) + \ldots$.

In particular, for $k = 0$ and $t > m$, we obtain

$$\begin{aligned} Y(t+1) &= DFX(t) + \Omega^0(t+1), \\ X(t+1) &= GX(t) + \eta(t+1), \end{aligned}$$

where $\Omega^0(t) = \epsilon(t) + D(F - G)X(t-1) + \ldots$.

Now, compare the initial system that is valid until $t = m$:

$$\begin{aligned} Y(t+1) &= JGX(t) + J\eta(t+1) + \xi(t+1), \\ X(t+1) &= GX(t) + \eta(t+1), \end{aligned}$$

with the previous system that is valid, since $t = m + 1$. The matrices G, J can change to F, D, respectively, at the instant m, and we must detect this change using information on observations $Y(t)$ and the model $X(t)$ only.

Finally, the considered state-space model can be written in the following form:

for $t < m$,

$$\begin{aligned} Y(t+1) &= JGX(t) + J\eta(t) + \xi(t+1), \\ X(t+1) &= GX(t) + \eta(t+1), \end{aligned}$$

and for $t \geq m$,

$$\begin{aligned} Y(t+1) &= DFX(t) + DF(F - G)X(t-1) + \cdots + DF_kX(t-k) + \Omega^k(t+1), \\ X(t+1) &= GX(t) + \eta(t+1), \end{aligned}$$

where $\Omega^k(t) = D\eta(t) + \xi(t+1) + DF_kX(t-k-1) + \ldots$, $F_k = F^k(F - G)$, and $X(t)$ is the output of the baseline model.

Remark that we purposely wrote the equation for $Y(t)$ in the form that includes variables dependent on the output of the baseline model $X(t)$ only. Our goal is to detect the change-point m (of matrices J and G into D and F, respectively) using information on the vectors $Y(t)$ and $X(t)$.

Now, let us propose the following method of m detection. First, consider the $K \times K$ matrices

$$T(1,l) = \sum_{i=1}^{l} X(i-1)X'(i-1), \qquad l = 1,\dots,N,$$

where $X(0) \equiv 0$. Second consider the $K \times M$ matrices

$$z(1,l) = \sum_{i=1}^{l} X(i-1)Y'(i), \qquad l = 1,\dots,N, \quad X(0) = 0,$$

and third, the decision statistic

$$Y_N(l) = \frac{1}{N}(z(1,l) - T(1,l)(T(1,N)^{-1}z(1,N)),$$

where $l = 1,\dots,N$, $Y_N(N) = 0$ and, by definition, $Y_N(0) = 0$. The inverse matrix here exists almost surely (see below).

In the sequel, we denote by $P_0(E_0)$ the measure (mathematical expectation) corresponding to the observed sequence without change-points and by $P_m(E_m)$ – to the sequence with the change-point m. Let H_0 denote the hypothesis of statistical homogeneity of data (no change-points); H_1 – the hypothesis about the presence of a change-point in the sample obtained.

Fix the number $0 < \beta < 1/2$. In this chapter, the following performance measures for the proposed method are used:

1) Probability of type-1 error ("false decision")

$$\alpha_N = P_0\{\max_{[\beta N] \le l \le N} \|Y_N(l)\| > C\}.$$

2) Probability of type-2 error ("missed goal"):

$$\delta_N = P_m\{\max_{[\beta N] \le l \le N} \|Y_N(l)\| \le C\}.$$

3) Probability of the error of estimation: suppose $\hat{\theta}_N = \hat{m}_N/N$ is the estimate of the change-point parameter $\theta = m/N$. Then, for every $0 < \epsilon < 1/2$, we estimate

$$P_m\{|\hat{\theta}_N - \theta| > \epsilon\}.$$

6.3 Main Results

Below, we assume that Cramer's condition is satisfied for $\xi(t)$ and $\eta(t)$; the matrices G, J are stable and

$$N^{-1}\,T_N(t) \to \mathbb{R}(t) \stackrel{\text{def}}{=} \int_0^t V(s)ds$$

$\mathbf{P}_\theta$- a.s. as $N \to \infty$.

In the following theorem, the exponential upper estimate for type-1 error probability is obtained.

Theorem 6.3.1. *Suppose the Cramer condition is satisfied for random sequences $\eta(t)$ and $\xi(t)$, and the matrices G, J are stable (i.e., the maximal eigenvalue of them is less than 1). Then,*

$$\alpha_N \le L_1 \exp(-L_2(C)N),$$

where the constants L_1, L_2 do not depend on N.

The proof of this theorem is similar to the proof of Theorem 5.4.2 from Chapter 5. Therefore, we give here the sketch of this proof.

The decision statistic $Y_N(l)$ in the considered case can be written as follows:

$$\begin{aligned}
Y_N(l) &= z(1,l) - T(1,l)(T(1,N))^{-1}z(1,N) \\
&= \sum_{k=1}^{l} X(k)X(k)'J' + \sum_{k=1}^{l} X(k)\xi'(k) - T(1,l)(T(1,N))^{-1}(\sum_{k=1}^{N} X(k)X(k)'J' \\
&+ \sum_{k=1}^{N} X(k)\xi'(k)) = \sum_{k=1}^{l} X(k)\xi'(k) - T(1,l)(T(1,N))^{-1}\sum_{k=1}^{N} X(k)\xi'(k)
\end{aligned}$$

Therefore, we can prove as in Chapter 5 that

$$\sup_{[\beta N]\le l\le N} \|Y_N(l)\| \le (1+\sqrt{K}) \sup_{[\beta N]\le l\le N} \|\sum_{k=1}^{l} X(k)\xi(k)'\|.$$

From definition of $X(k)$, we have $X(k) = \sum_{i=1}^{k} G^{k-i}\eta(i)$. Therefore,

$$\sum_{k=1}^{l} X(k)\xi'(k) = \sum_{k=1}^{l}\sum_{i=1}^{k} G^{k-i}\eta(i)\xi'(k).$$

Now, remember that $\eta(i)$ and $\xi(k)$ are independent r.v.'s and the matrix G is stable. Therefore, the Cramer condition is satisfied for each summand and

$$\alpha_N = \mathbf{P}_0\{\sup_{[\beta N]\le l\le N} \|Y_N(l)\| > C\} \le L_1 \exp(-L_2(C)N),$$

where the constants $L_1, L_2(C)$ do not depend on N.

Now, let us study the probability of type-2 error δ_N in change-point detection. For $t > m$, the linear state-space model can be written as

$$\begin{aligned} Y(t+1) &= DFX(t) + DF_1X(t-1) + \dots + DF_kX(t-k) + \Omega^k(t+1), \\ X(t+1) &= GX(t) + v(t+1), \end{aligned} \tag{6.1}$$

where $\Omega^k(t) = \epsilon(t) + DF_k X(t-1) + \dots$ and $F_k = F^k(F-G)$, $\epsilon(t) = D\xi(t) + \eta(t)$.

Remark that $E\Omega^k(t)X'(t) \neq 0$, and, therefore Cramer's condition

$$E\exp(t\Omega^k(t)X'(t)) < \infty, \qquad \forall u,$$

for $0 < t < T$ is equivalent to the following condition.

Consider the $M \times K$ matrix $a^k(u) = \Omega^k(u)X'(u)$. Then, there exist $g > 0, H > 0$, such that

$$E\exp(ta^k_{ij}(u)) < \exp(tEa^k_{ij}(u))\, e^{g\frac{t^2}{2}}, \qquad \forall u,\ i = 1,\dots,M;\ j = 1,\dots,K,$$

for $0 < t < H$.

Suppose $q(k) = \max\limits_{u,i,j} Ea^k_{ij}(u) < \infty$. Then the above condition is equivalent to the following: There exist $g > 0$, and $H > 0$, such that

$$E\exp(ta^k_{ij}(u)) \le e^{tq(k)}\, e^{g\frac{t^2}{2}}, \qquad \forall u,$$

for $0 < t < H$.

In virtue of the definition of the process $\Omega^k(t)$, the value $q(k)$ goes to zero exponentially as $k \to \infty$.

Let us demonstrate this property in detail for the following example.

Example 6.2.

Consider the following state-space model:

$$\begin{aligned} y_{t+1} &= rx_{t+1} + \xi_{t+1} \\ x_{t+1} &= ux_t + \eta_{t+1}, \end{aligned}$$

where ξ_t, η_t are sequences of independent r.v.'s with zero mean. Coefficients $|r| < 1, |u| < 1$ can change at some unknown point m for d, w: $|d| < 1, |w| < 1$, respectively. For $t > m$, we obtain the following equivalent form of this model:

$$\begin{aligned} y_{t+1} &= dwx_t + df_1x_{t-1} + \dots + df_kx_{t-k} + \Omega^k_{t+1}, \\ x_{t+1} &= ux_t + \eta_{t+1} \end{aligned} \tag{6.2}$$

where $\Omega^k_t = \epsilon_t + df_kx_{t-k-1} + df_{k+1}x_{t-k-2} + \dots$, $f_k = w^k(w-u)$, and $\epsilon_t = d\xi_t + \eta_t$.

We study the following Cramer's condition: For $0 \leq t \leq T$ and any $n > 0$,

$$E \exp(t\Omega_n^k x_n) < \infty.$$

Denote $a_n^k = \Omega_n^k x_n$. We need to study Ea_n^k. From system (6.2), we obtain:

$$\begin{aligned} Ea_n^k &= df_k u^{k+1} \frac{\sigma_n^2}{1-u^2} + df_{k+1} u^{k+2} \frac{\sigma_n^2}{1-u^2} \\ &= (w-u) d \frac{\sigma_n^2}{1-u^2} (wu)^k (u + wu^2). \end{aligned}$$

Suppose $\sigma_n^2 < \sigma^2$, $\forall n$. Then,

$$E \exp(t\Omega_n^k x_n) \leq e^{tq(k)} e^{g\frac{t^2}{2}}, \qquad \forall n,$$

for $0 < t < H$ and $q(k) = (w-u) d \dfrac{\sigma^2}{1-u^2} (wu)^k (u + wu^2)$.

We conclude that $q(k) \to 0$ exponentially as $k \to \infty$. The same ideas can be used in the general case.

In the following theorem, we study type-2 error δ_N in change-point detection.

Consider the $K \times K$ matrix

$$A(t) = \int_0^t V(\tau) d\tau, \quad 0 \leq t \leq 1. \tag{6.3}$$

Define $I = A(1)$. For any $0 < t \leq 1$, the matrix $A(t)$ is positive definite.

For any $0 \leq \theta \leq 1$, consider the function

$$h(\theta) = \|A(\theta)(E - I^{-1}A(\theta))(JG - DF)'\|, \tag{6.4}$$

where E is the unit $K \times K$ matrix.

Evidently, $h(0) = h(1) = 0$. Consider the point $\tilde{\theta}$ of the global maximum of $h(\theta)$ on the segment $[0,1]$ — the root of the equation $E = I^{-1}A(\theta) + A(\theta)I^{-1}$, i.e., $A(\theta) = I/2$. In virtue of the above assumptions, the root of this equation exists and is unique. The function $h(\theta)$ is continuously differentiable by $\theta \in (0,1)$.

First, let us choose $k^* \geq 1$, such that $q(k^*) < h(\tilde{\theta})$. Since $q(k) \to 0$ exponentially as $k \to \infty$, such k^* always exists. Second, choose the decision threshold C, such that $q(k^*) + C < h(\tilde{\theta})$. Remark that this choice is slightly more difficult than in situations of fully observed state variables.

Denote $q = q(k^*)$. The following theorem holds.

Theorem 6.3.2. *As in Theorem 6.3.1, we suppose that the Cramer condition is satisfied for random sequences $\eta(t)$, and $\xi(t)$ and the matrices G, J are stable. We also suppose that $rank(S) = M$, where $S = (E - I^{-1}A(\theta))(JG -$*

$DF)'$. Denote $d = (h(\tilde{\theta}) - C - q)/(1 + \sqrt{K})$. Then, the following exponential upper estimate holds for type-2 error:

$$\delta_N \leq L_1 \exp(-L_2(d)N), \tag{6.5}$$

where the constants L_1, and L_2 do not depend on N.

Moreover, the change-point estimator $\hat{\theta}_N$ tends a.s. to the true change-point parameter θ as $N \to \infty$.

The proof of Theorem 6.3.2 is analogous to the proof of Theorem 5.4.2 from the previous chapter. Therefore, I give here the idea of this proof. First, as in Theorem 5.4.2 , I prove that the function $M(t)$ exists. The norm of this function has a unique maximum at the point $t = \theta$ on the segment [0.1]. Then, we prove that the reversed Lipshitz condition is satisfied for $\|M(t)\|$ at the point $t = \theta$. The required property follows from the decomposition of $Y_N(t)$ into the sum of a deterministic and stochastic part:

$$Y_N(t) = M(t) + \Gamma_N(t) + \zeta_N(t),$$

where $\Gamma_N(t) = E_\theta Y_N(t) - M(t)$ and $\zeta_N(t) = Y_N(t) - E_\theta Y_N(t)$. Therefore for the probability of type-2 error, we have:

$$\mathbf{P}_m\{\max_{\beta \leq t \leq 1} \|Y_N(t)\| \leq C\} \leq \mathbf{P}_m\{\max_{\beta \leq t \leq 1} \{\zeta_N(t)\| > const{\cdot}\delta\} \leq L_1 \exp(-L_2(\delta)N).$$

The probability of the estimation error is evaluated in the analogous way. For this purpose, we use the reversed Lipshitz condition:

$$\|M(\theta)\|^2 - \|M(t)\|^2 \geq |\theta - t|\lambda_M \|a - b\|^2,$$

where the constant λ_M is estimated, as in Chapter 5. From here, we obtain

$$\mathbf{P}_m\{|\hat{\theta}_N - \theta| > \epsilon\} \leq \mathbf{P}_m\{\max_{\beta \leq t \leq 1} \|\zeta_N(t)\| \geq \epsilon \lambda_M \|a - b\|^2/2\} \leq L_1 \exp(-L_2(\epsilon)N).$$

6.4 Asymptotic Optimality

Let $(\mathbf{z}_1, \ldots, \mathbf{z}_n)$ be a random sequence of dependent vector-valued observations $\mathbf{z}_n = (z_n^1, \ldots, z_n^k)$ defined on the probability space $(\Omega, \mathcal{F}, P)$.

Suppose that for any $i = 2, \ldots, n$ there exist conditional densities $f_0(\mathbf{z_i}|\mathbf{z}_1, \ldots, \mathbf{z_{i-1}})$, and $f_1(\mathbf{z_i}|\mathbf{z}_1, \ldots, \mathbf{z_{i-1}})$. I assume that at some unknown change-point $m < n$, the density function of an observed random sequence changes from $f_0(\cdot)$ to $f_1(\cdot)$:

$$f_m(\mathbf{Z}^n) = f_m(\mathbf{z}_1, \ldots, \mathbf{z}_n) = f_0(\mathbf{z}_1) f_0(\mathbf{z}_2|\mathbf{z}_1) \ldots f_0(\mathbf{z}_m|\mathbf{z}_1, \ldots, \mathbf{z}_{m-1})$$
$$f_1(\mathbf{z}_{m+1}|\mathbf{z}_1, \ldots, \mathbf{z}_m) \ldots f_1(\mathbf{z}_n|\mathbf{z}_1, \ldots, \mathbf{z}_{n-1}), \tag{6.6}$$

i.e., we consider changes in the joint density functions of subsequent subsamples and in the conditional density functions from f_0 to f_1. The mathematical expectation corresponding to this sequence is denoted by E_m.

For example, we can think of the multiple regression model with dependent errors. Until the instant m, we have one set of regression coefficients in this model, but from the instant m, this set of coefficients abruptly changes. For such models, it is natural to assume that the d.f. $f_0(\cdot)$ and $f_1(\cdot)$ depend on the time n. Denote by H_0 the hypothesis about the d.f. $f_0(\cdot)$ and by H_1 — about $f_1(\cdot)$.

Now, consider a method of sequential detection of the change-point m. Such a method is described by its decision function $d_c(\cdot)$ depending on some large parameter c (e.g., a decision threshold for CUSUM method):

$$d_c(n) = \begin{cases} 1, & \text{stop and accept } H_1 \\ 0, & \text{continue under } H_0. \end{cases}$$

Usually, the choice of a large parameter c is conditioned by the rate of convergence to zero of the probability of a false decision:

$$\alpha_c = \sup_n P_\infty\{d_c(n) = 1\}.$$

For example, if c is equal to the decision threshold for the CUSUM method, then we obtain the exponential rate of convergence to zero for the values α_c.

Consider the stopping time

$$\tau_c = \inf\{n : d_c(n) = 1\}, \tag{6.7}$$

and the normalized (by c) delay time in change-point detection:

$$\gamma_c = (\tau_c - m)^+/c. \tag{6.8}$$

In the following theorem, the a priori information-theoretical lower bound for a performance measure depending on the value γ_c is proved. The change-point m is fixed but unknown, and the large parameter c increases to infinity.

Choose the number $M = M(c)$:

$$M = \inf\{l : \sum_{n=m+l}^{\infty} P_\infty\{\tau_c = n\} \le \alpha_c\}. \tag{6.9}$$

The following theorem holds.

Theorem 6.4.1. *Consider the model of observations (6.6) with an unknown but fixed change-point m and an arbitrary method of change-point detection depending on a large parameter $c \to \infty$. Then,*

$$E_m \int_0^{\gamma_c} j(t)dt \ge \frac{|\ln(M(c)\alpha_c|}{c} + O(\frac{1}{c}), \tag{6.10}$$

where $j(t) = J(m + [tc]), J(m) \equiv 0, t \geq 0,$ *and* $J(k), k \geq m$ *is defined as follows:*

$$J(k) = \ln \frac{f_1(\mathbf{z}_k|\mathbf{z}_1, \ldots, \mathbf{z}_{k-1})}{f_0(\mathbf{z}_k|\mathbf{z}_1, \ldots, \mathbf{z}_{k-1})}.$$

The proof of this theorem is given in the appendix.

Remark that both sides of (6.10) depend on c. However, for the method proposed in this paper, from theorem 6.4.1, it follows that $\gamma_c \to \gamma^* > 0$ P-a.s. as $c \to \infty$ and

$$\frac{|\ln(M(c)\alpha_c|}{c} \to a^* > 0$$

as $c \to \infty$.

6.5 Simulations

In this section, I consider results of Monte-Carlo experiments with the above proposed nonparametric method. Remark once more that the main advantage of this method over the Kalman filter method designed initially for sequential problems is that the estimates of change-points $\hat{\theta}_N$ are consistent. In other words, we can guarantee their proximity to the true values of these change-point parameters.

Linear model

Consider the following state-space model:

$$\begin{aligned} y_i &= d_0 + d_1 z_i + v_i \\ z_i &= c_0 + c_1 z_{i-1} + w_i, \end{aligned}$$

where $z_i, y_i,\ i = 1, 2, \ldots$ is the state variable and the observed variable, respectively; v_i, w_i are independent Gaussian random sequences $N(0, 1)$ and the baseline model is characterized by the following parameters: $d_0 = c_0 = 2,\ d_1 = 0.2,$ and $c_1 = 0.3$.

I consider possible changes in coefficients of this model: $d_1 \to d_2 \neq d_1$ or $c_1 \to c_2 \neq c_1$ at an unknown change-point m. In experiments, we estimate the false alarm probability (1st-type error) pr_1, the probability of the 2nd type error w_2, and the estimate of the change=point parameter θ. Each value was computed as the average in 5000 independent Monte-Carlo trials.

The obtained results are given in Table 6.1 (comparatively large changes in coefficients) and Table 6.2 (small changes).

Non-linear model

For this type of model, advantages of the proposed method are most evident. For example, we can think about the state equations as a certain non-linear (and sophisticated enough) model of a researched natural phenomenon

TABLE 6.1
Performance Characteristics of the Proposed Test: Linear Model; Moderate Changes in Coefficients ($\theta = 0.3$)

N		200	300	500	700	1000	1500
C		0.47	0.38	0.29	0.25	0.205	0.153
pr_1		0.05	0.04	0.04	0.04	0.04	0.04
$d_2 = 0.5$	w_2	0.27	0.062	0.02	0	0	0
	$\hat{\theta}_N$	0.327	0.321	0.314	0.308	0.302	0.300
$c_2 = 0.7$	w_2	0.37	0.148	0.018	0.01	0	0
	$\hat{\theta}_N$	0.34	0.33	0.32	0.31	0.305	0.30
$c_2 = 0.6$	w_2	0.94	0.80	0.408	0.314	0.09	0.004
	$\hat{\theta}_N$	0.55	0.46	0.34	0.33	0.33	0.31

TABLE 6.2
Performance Characteristics of the Proposed Test: Linear Model; Small Changes in Coefficients

N		1000	1500	2000
C		0.205	0.153	0.120
pr_1		0.04	0.03	0.04
$c_2 = 0.5$	w_2	0.90	0.36	0.166
	$\hat{\theta}_N$	0.43	0.36	0.33
$d_2 = 0.3$	w_2	0.20	0.132	0.04
	$\hat{\theta}_N$	0.38	0.34	0.33

(e.g., chemical, financial, or economic). At some moments, this model (we know only the baseline version of this model that follows from theory) can change and we are interested in: 1) detection of a fact of a change; and 2) estimation of change instants. Remark that sequential methods cannot give answers to both of these questions. The proposed method has the following advantages: We can control the level of type-1 and type-2 errors, and we can construct consistent estimates of change-point parameters.

Now, let us first consider models with, at maximum, one change-point. Consider the following state-space model $i = 1, \dots, N$:

$$y_i = d_4 + d_5 x_i + d_6 x_i^2 + \eta_i$$
$$x_i = \frac{d_1}{d_2 + d_3 e^{-x_{i-1}}} + \xi_i,$$

where x_i, y_i is the state variable and the observed variable, respectively; η_i, ξ_i are independent Gaussian random sequences $N(0, 1)$ and the baseline model is characterized by the following parameters: $d_1 = 1,\ d_2 = 0.3,\ d_3 = 0.1,\ d_4 = 1,$ and $d_5 = 1,\ d_6 = 1$.

TABLE 6.3
Performance Characteristics of the Proposed Test:
Nonlinear Model ($\theta = 0.70$)

N		300	500	700	1000	1500	2000
C		0.14	0.11	0.09	0.08	0.065	0.055
pr_1		0.04	0.05	0.04	0.04	0.04	0.04
$d_2 = 1.0$	w_2	0.66	0.52	0.39	0.22	0.10	0.04
	$\hat{\theta}_N$	0.60	0.65	0.67	0.68	0.68	0.69
$d_4 = 1.5$	w_2	0.16	0.03	0.02	0.01	0	0
$d_5 = 0.5$	$\hat{\theta}_N$	0.68	0.695	0.70	0.705	0.701	
$d_3 = 0.5$	w_2	0.57	0.43	0.31	0.187	0.06	0.04
	$\hat{\theta}_N$	0.59	0.67	0.67	0.68	0.69	0.695

I consider possible changes in coefficients of this model at an unknown change-point $m = [\theta N]$. In experiments, we estimate the false alarm probability (1st-type error) pr_1, the probability of the 2nd-type error w_2, and the estimate of the change-point parameter $\hat{\theta}_N$. Each value was computed as an average in 5000 independent Monte-Carlo trials.

The obtained results are given in Table 6.3.

Now, let us consider models with multiple change-points.

Consider the following state-space model $i = 1, \dots, N$:

$$y_i = d_1 + d_2 x_i + d_3 x_i^2 + \eta_i$$
$$x_i = \frac{c_1}{c_2 + c_3 e^{-x_{i-1}}} + \xi_i.$$

Suppose the following change-point parameters are observed for the above nonlinear model: $\theta(1) = 0.3; \theta(2) = 0.7$. The baseline model is characterized by the following coefficients:

$$d_1 = 1 \quad d_2 = 1 \quad d_3 = 1$$
$$c_1 = 1 \quad c_2 = 0.3 \quad c_3 = 0.1.$$

Coefficients of this model change at the change-points $m_i = [\theta(i)N]$, where N is the sample size, as follows:

at $\theta(1)$: $d_2 \to \tilde{d}_2 = 5.0 \neq d_2$

at $\theta(2)$: $c_3 \to \tilde{c}_3 = 1.0 \neq c_3$

So, in this case, the number of change-point parameters equals $k = 2$. In experiments, we estimate the false alarm probability (1st-type error) pr_1, the probability of the 2nd-type error w_2, the probability of the event $\hat{k}_N \neq k$, where $\hat{k}_N$ is the estimate of the number of change-points, and in case $\hat{k}_N = 2$ the estimates of detected change-points $\hat{\theta}_N^1, \hat{\theta}_N^2$. Each value was computed as an average in 5000 independent Monte-Carlo trials.

ABLE 6.4
rformance Characteristics of the Proposed Test: Nonlinear Model
1) = 0.30; θ(2) = 0.70)

N	300	500	700	1000	1500	2000
C	0.14	0.11	0.09	0.08	0.065	0.055
pr_1	0.04	0.05	0.04	0.04	0.04	0.04
w_2	0.23	0.15	0.07	0.04	0.01	0
$\{\hat{k}_N \neq k\}$	0.41	0.25	0.14	0.04	0.01	0.0
$\hat{\theta}_N^1, \hat{\theta}_N^2$	0.41, 0.75	0.35, 0.67	0.33, 0.72	0.32, 0.71	0.29, 0.71	0.30, 0.705

TABLE 6.5
Performance Characteristics of the Proposed Test:
Multivariate Model; Changes in Coefficients

N		300	500	700	1000	1500	2000
C		0.14	0.11	0.09	0.08	0.065	0.055
pr_1		0.04	0.05	0.04	0.04	0.04	0.04
$u = 1.0$	w_2	0.66	0.52	0.39	0.22	0.10	0.04
	$\hat{\theta}_N$	0.250	0.26	0.27	0.28	0.29	0.30

Multivariate models
Here, I consider the following multivariate model:

$$x_{t+1} = Fx_t + w_t$$
$$y_t = (1,\ 0)x_t + \epsilon_t,$$

where $t = 1, \ldots, N$, where N is the sample size, and $F = \begin{pmatrix} 0.7 & 0.1 \\ 0 & 0.7 \end{pmatrix}$ and w_t, ϵ_t are independent Gaussian with zero means, $Var(\epsilon_t) = 1$, $Cov(w_t) = \begin{pmatrix} 0.745 & -0.07 \\ -0.07 & 0.51 \end{pmatrix}$, but at an unknown change-point $1 < [\beta N] \leq [\theta N] \leq [(1-\beta)N] < N$ the matrix F changes to $G = \begin{pmatrix} u & 0.1 \\ 0 & 0.7 \end{pmatrix}$, where the parameter u varies in the diapason $[0.7; 1.2]$.

In experiments, we put $u = 1.0$, and $\theta = 0.3$ and computed the following values: the false alarm probability (1st-type error) pr_1, the probability of the 2nd-type error w_2, and the estimate of the change-point parameter $\hat{\theta}_N$. Each value was computed as an average in 5000 independent Monte-Carlo trials.

Results obtained for different sample sizes N are reported in Table 6.5.

Non-Gaussian distributions; multivariate models
The baseline model is of the following functional form

$$x_{t+1} = Fx_t + w_t$$
$$y_t = (1,\ 0)x_t + \epsilon_t,$$

TABLE 6.6
Performance Characteristics of the Proposed Test:
Non-Gaussian Noises; Changes in Coefficients

N		300	500	700	1000	1500	2000	3000
C		0.53	0.45	0.37	0.32	0.25	0.22	0.18
pr_1		0.04	0.04	0.05	0.05	0.04	0.04	0.05
$u = 1.0$	w_2	0.83	0.62	0.49	0.35	0.270	0.14	0.05
	$\hat{\theta}_N$	0.20	0.21	0.24	0.26	0.27	0.33	0.31

where $F = \begin{pmatrix} 0.7 & 0.1 \\ 0 & 0.7 \end{pmatrix}$, but w_t has the multivariate t-distribution with the correlation matrix $\sigma = \begin{pmatrix} 1.0 & 0.8 \\ 0.8 & 1.0 \end{pmatrix}$ and three degrees of freedom; ϵ_t has the standard uniform d.f. on the segment $[-0.5; 0.5]$.

At an unknown change-point $m = [\theta N]$, the matrix F changes to $G = \begin{pmatrix} 1.0 & 0.1 \\ 0 & 0.7 \end{pmatrix}$.

In experiments, we put $u = 1.0$, and $\theta = 0.3$ and computed the following values: the false alarm probability (1st-type error) pr_1, the probability of the 2nd-type error w_2, and the estimate of the change-point parameter $\hat{\theta}_N$. Each value was computed as an average in 5000 independent Monte-Carlo trials.

The results obtained are reported in Table 6.6.

6.6 Conclusions

In this chapter, I propose a new nonparametric method for retrospective detection of change-points in state-space models. I repeat here that this method is designed especially for retrospective change-point detection problems and has several essential advantages over well-known sequential methods (e.g., the Kalman filter method) applied here for retrospective change-point detection:

1) This new proposed method does not depend on the crucial assumption of the Gaussian d.f.'s of noises;

2) It provides consistent estimates of change-point parameters;

3) Using this method we can detect multiple change-points in data; and

4) This method helps to control probabilities of type-1 and type-2 errors of change-point detection.

In this chapter, I formulated Theorems 6.3.1 and 6.3.2 about the exponential rate of convergence to zero of type-1 and type-2 error probability under some general assumptions on the properties of a state-space model. Results of Monte-Carlo experiments with the proposed method for different types of state-space models are given.

7

Copula, GARCH, and Other Financial Models

7.1 Introduction

This chapter aims at research into retrospective methods for revealing structural changes in different models of financial time series. The uncertainty of financial returns plays a central role in many financial models of valuation of financial derivatives, risk management, efficient asset allocation and so on. In this short introduction, we mention the following features that most financial time series share:

1. Log prices seem to be nonstationary but without any marked systematic trend.

2. Returns (log-difference of the prices) exhibit a constant mean but with periods of clustered volatility.

3. Returns are very weakly autocorrelated or have no autocorrelation at all.

4. Squared returns are highly (positively) correlated, and in general with high persistence.

5. Returns are negatively skewed and exhibit positive excess kurtosis.

From these features, we can make the following conclusions. There is almost no room for predicting the future mean value of stock returns. There is a large amount of information coming form lagged returns to predict the volatility of stock returns. Normality does not seem an adequate distributional assumption for the log-returns because of the skewness and large kurtosis.

7.2 GARCH Models

7.2.1 What Is a GARCH Model?

The GARCH model has long been popular in modeling financial time series, and is proven to be useful in handling the data with high volatility (cf. Gourieroux (1997)). The problem of testing a parameter change hypotheses in GARCH models has attracted much attention in the last several decades. For references, see Wichern, Miller, and Hsu (1976), Inclan and Tiao (1994),

Lee and Park (2001), and the papers cited therein. In 2003 a residual based CUSUM test was proposed for detecting parameter changes in GARCH(1,1) models (see Lee et al. (2003)).

A typical GARCH(p,q) model can be written as follows (see Bollerslev (1996)):

$$\begin{aligned} y_t &= \mu_t + \sigma_t \epsilon_t \\ \sigma_t^2 &= w + \alpha_1 y_{t-1}^2 + \cdots + \alpha_p y_{t-p}^2 + \beta_1 \sigma_{t-1}^2 + \cdots + \beta_q \sigma_{t-q}^2 \end{aligned}$$

for $t = s+1, \ldots, s+n, \quad s = \max(p,q) > 0$.

Here, y_t – observations; μ_t – expected returns; σ_t – volatility; and ϵ_t – white noise process with the parameters 0 and 1; $\alpha_1, \ldots, \alpha_p$ – AR coefficients; and $\beta_1, \ldots, \beta_q$ – MA coefficients.

7.2.2 Methods for Detection of Changes in GARCH Models

For detection of structural changes in GARCH models, we can directly apply Statistic (2.6) from Chapter 2. Historically, for financial time series, it was first proposed by Kokoshka and Leipus (1999), who called this statistic (which ascends to Kolmogorov's test) the CUSUM method. However, historically, Page's CUSUM test has nothing in common with this method.

For structural changes in expected returns, Statistic (2.3.3) is translated into

$$Z_n(k) = \frac{1}{\sqrt{n\hat{\sigma}^2}} \max_{1 \le i \le n} |\sum_{i=1}^{k} y_i - \frac{k}{n} \sum_{i=1}^{n} y_i|,$$

where σ^2 is the long-run dispersion of y_t.

The decisions are based upon

$$Z = \max_{1 \le k \le n} Z_n(k).$$

If $Z \le C$, where C is a certain threshold, then we accept the hypothesis H_0, otherwise – H_A.

The CUSUM test for a volatility break is closely related to the CUSUM test for structural change in mean. To demonstrate it, let us rewrite the model $\xi_t = \sigma_t \epsilon_t$ as

$$\xi_t^2 = \sigma_t^2 + e_t,$$

where $e_t = \sigma_t^2 \left(\epsilon_t^2 - 1\right)$. Under the null hypothesis H_0, this is simply a location model for ξ_t^2 with homoscedastic and serially correlated errors. So the problem of interest is translated to testing for changing mean in ξ_t^2. For changes in volatility, the decisions are based on the following test statistic:

$$T_n = \frac{1}{\sqrt{n\hat{\sigma}^2}} \max_{1 \le k \le n} |\sum_{i=1}^{k} y_i^2 - \frac{k}{n} \sum_{i=1}^{n} y_i^2|,$$

where $(y_1^2, \ldots, y_n^2)$ is the diagnostic sequence in this problem; $\hat{\sigma}^2$ is the so-called "long-run dispersion" of y_t^2.

Under the assumption of no change in volatility:

$$T_n \to \sup_{0 \le t \le 1} |B(t)|,$$

where $\{B(t),\ 0 \le t \le 1\}$ denotes a Brownian bridge.

We are interested primarily in structural changes in expected returns ($\mu_t : \mu_1 \to \mu_2$) and volatility ($\sigma_1^2 \to \sigma_2^2$). Below, we give some results of Monte-Carlo experiments with this statistic applied to GARCH data, as well as results of detection of structural changes in NYSE (New York Stock Exchange) index with the help of this statistic.

7.2.3 Simulations

In this subsection, we consider the case of one change-point in GACH(1,1) model and the case of multiple change-points in GARCH(1,1) model with changing coefficients.

One change-point

Model

The following nonstationary GARCH(1,1) model with structural changes in volatility was considered:

$$\begin{aligned} y_i &= \mu + \sqrt{z_i} \cdot v_i \\ z_i &= w_i + \alpha y_i^2 + \beta z_{i-1}, \end{aligned}$$

where $\mu = 0.5; \alpha = 0.2; \beta = 0.7$; v_i is the sequence of independent $\mathcal{N}(0,1)$ r.v.'s, and the sequence w_i model structural changes: $w_i : 0.5 \to 0.8$ (small changes); $w_i : 0.5 \to 1.5$ (large changes) at the point $\theta = 0.3$.

For retrospective detection of change-points in this model, Statistic (2.3.3) was used. First, we compute decision thresholds for different sample sizes. For this purpose, we can either use the asymptotic relationship for this statistic without structural changes or apply direct method of computation of α-quantiles of this statistic. Below, we pursued the second path and computed 75, 95, and 99 percent quantiles of Statistic (2.3.3) applied to homogenous samples. Obtained results are given in Table 7.1.

Remark that all these percentiles were obtained as averages in 1000 independent trials of each test. Then, 75 percent quantiles were used as thresholds in computation of characteristics of the decision statistic for non-stationary samples with structural changes.

Small changes in volatility: $w_i : 0.5 \to 0.8$

I remark here that convergence to the true value of a change-point is

TABLE 7.1
75, 95, and 99 Percentiles for Statistic (2.6) in GARCH(1,1) Model

N	300	500	700	1000	1500	2000	2500	3000	3500	4000
r_{75}	1.08	0.91	0.81	0.73	0.64	0.57	0.51	0.47	0.45	0.41
r_{95}	2.23	1.66	1.37	1.20	1.11	0.98	0.83	0.75	0.74	0.64
r_{99}	3.53	3.50	2.52	2.21	1.97	1.54	1.33	1.19	1.22	1.05

TABLE 7.2
GARCH(1,1) Model. Retrospective Detection of Small Changes in Volatility: $w_i : 0.5 \to 0.8$ (C-Decision Threshold, w_2-type-2 Error; $\hat{\theta}_N$ - Estimate of the Change-Point $\theta = 0.3$)

N	1500	2000	2500	3000	3500	4000
C	0.64	0.57	0.51	0.47	0.45	0.41
w_2	0.20	0.18	0.14	0.08	0.06	0.04
$\hat{\theta}_N$	0.45	0.43	0.42	0.42	0.41	0.40

extremely low for Statistic (2.6). This conclusion is confirmed for the case of large changes in volatility (see Table 7.3).

Large changes in volatility: $w_i : 0.5 \to 1.5$

I conclude that low convergence to the true change-point θ is, probably, a characteristic feature of Statistic (2.6) applied to GARCH models with sudden and persistent switches in volatility. More effective statistics were proposed below for an SV model.

Multiple change-points

Again, the following nonstationary GARCH(1,1) model with structural changes in volatility was considered:

$$y_i = \mu + \sqrt{z_i} \cdot v_i$$
$$z_i = w_i + \alpha y_i^2 + \beta z_{i-1},$$

TABLE 7.3
GARCH(1,1) Model. Retrospective Detection of Large Changes in Volatility: $w_i : 0.5 \to 1.5$ (C-Decision Threshold, w_2-type-2 Error; $\hat{\theta}_N$ - Estimate of the Change-Point $\theta = 0.3$)

N	100	300	500	700	1000	1500
C	1.27	1.08	0.91	0.81	0.73	0.64
w_2	0.20	0.12	0.04	0.01	0.005	0
$\hat{\theta}_N$	0.49	0.47	0.46	0.42	0.41	0.40

TABLE 7.4
GARCH(1,1) Model. Multiple Change-Points

N	1500	2000	2500	3000	35000	4000
C	0.64	0.57	0.51	0.47	0.45	0.41
w_2	0.18	0.16	0.14	0.08	0.06	0.05
$P\{\hat{k}_N \neq k\}$	0.13	0.11	0.09	0.07	0.05	0.03
$\hat{\theta}_1$	0.52	0.48	0.51	0.47	0.45	0.40
$\hat{\theta}_2$	0.68	0.73	0.65	0.61	0.60	0.55
$\hat{\theta}_3$	0.88	0.87	0.86	0.84	0.82	0.80

where $\mu = 0.5; \alpha = 0.2; \beta = 0.7$; v_i is the sequence of independent $\mathcal{N}(0,1)$ r.v.' and the sequence w_i model structural changes. For the case of multiple change-points in GARCH models, we consider three change-points:

$\theta = 0.3$: $w_i : 0.3 \to 0.9$
$\theta = 0.5$: $w_i : 0.9 \to 1.5$
$\theta = 0.7$: $w_i : 1.5 \to 0.3$
and sample sizes from $N = 1500$ to $N = 4000$.

For retrospective detection of these change-points, the following criterion was used. Suppose k_N^* is the estimator of the number of change-points $k = 3$, and $\hat{\theta}_N^i$ are coordinates of estimators of change-points. Then, for $0 < \epsilon < 1/2$, we use the following value:

$$P\{\{k_N^* \neq k\} \cup (\{k_N^* = k\} \cap \{\max_{1 \le i \le k} |\hat{\theta}_N^i - \theta^i| \ge \epsilon\})\}.$$

In words, we strive to correctly estimate the number of change-points and coordinates of these change-points. For this purpose, we use the three-stage method described in Chapter 2. For the GARCH model, the following results were obtained.

7.2.4 Applications to Real Data

NASDAQ

The NASDAQ series contains daily closing values of the Nasdaq™ Composite Index. The sample period is from January 2, 1990, to December 31, 2001, for a total of 3028 daily equity index observations.

The NASDAQ Composite closing index values were downloaded directly from the Market Data section of the NASDAQ™ web page.

First, the series of daily returns was computed:

$$y_t = 100 * \log(P_{t+1}/P_t),$$

where P_t is the value of NASDAG index at time t.

The graph of this series is shown in Figure 7.1.

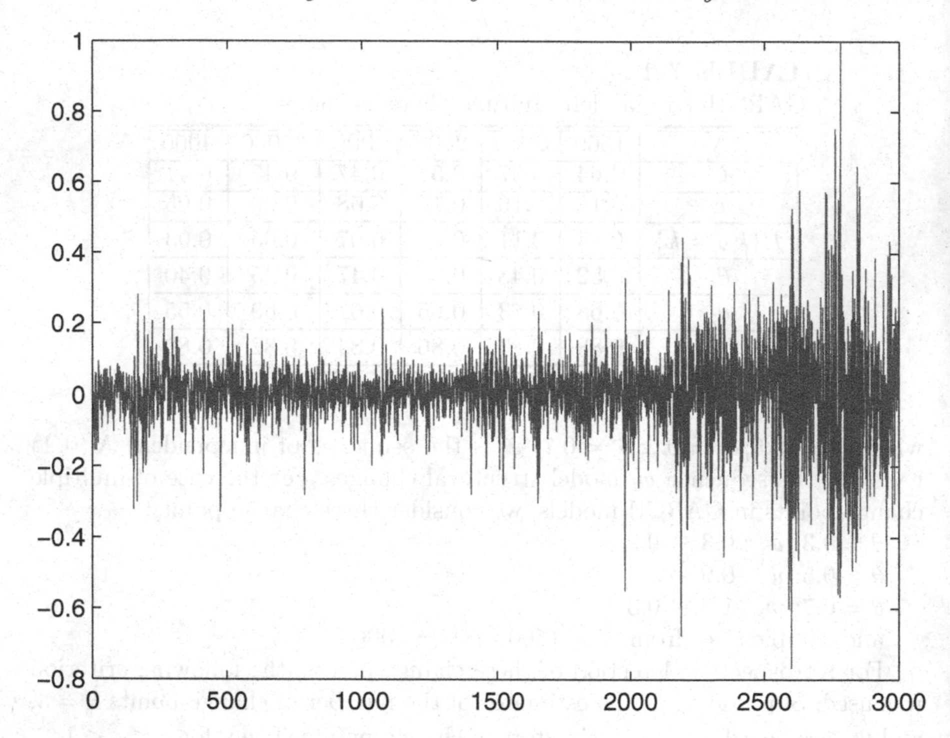

FIGURE 7.1
Graph of NASDAQ daily returns index.

The preliminary visual analysis of this series reveals tentative change-point at time 2100 − 2400. The graph of the decision statistic is given in Figure 7.2.

The above-described method detects one change-point at time $\hat{\theta} = 0.72$ ($\hat{n}_0 = 2160$).

NYSE

The NYSE series contains daily closing values of the New York Stock Exchange^{TM} Composite Index. The sample period is from January 2, 1990, to December 31, 2001, for a total of 3028 daily equity index observations of the NYSE Composite Index.

The NYSE Composite Index daily closing values were downloaded directly from the Market Information section of the NYSE^{TM} web page.

First, the series of daily returns was computed:

$$y_t = 100 * \log(P_{t+1}/P_t),$$

where P_t is the value of NASDAQ index at time t.

The graph of this series is shown in Figure 7.3.

The preliminary visual analysis of this series reveals several tentative change-points. The graph of the decision statistic is given in Figure 7.4.

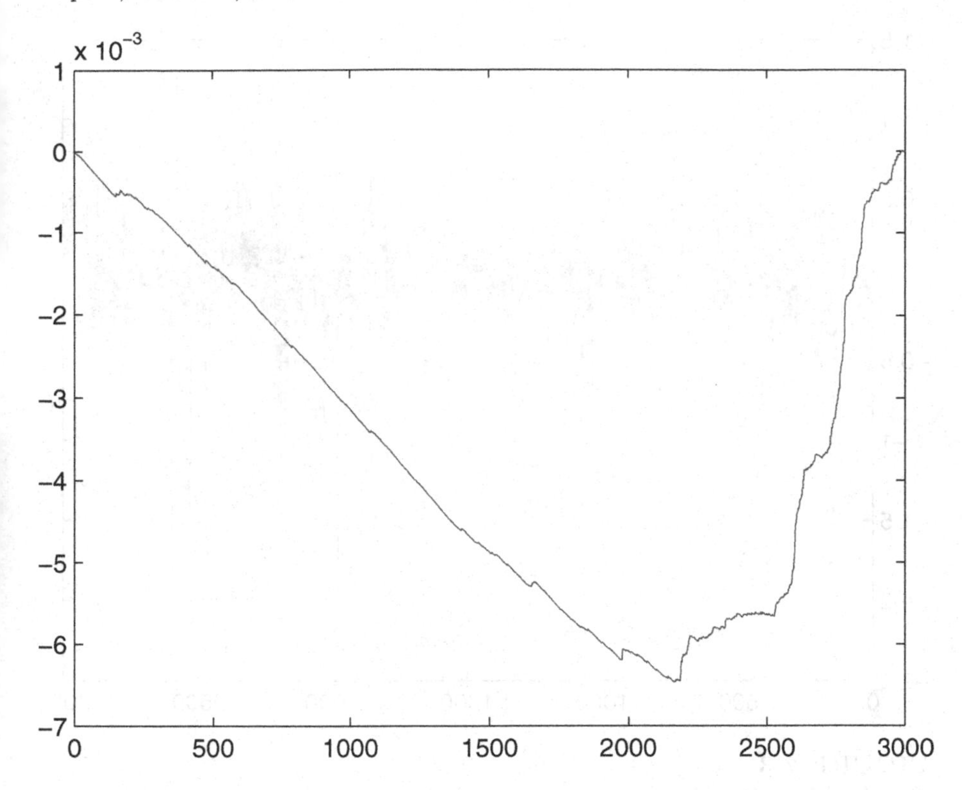

FIGURE 7.2
Graph of the decision statistic for NASDAQ.

The above-described method detects two change-points at times $\hat{\theta}_1 = 0.61$ ($\hat{n}_1 = 1830$) and $\hat{\theta}_2 = 0.82$ ($\hat{n}_2 = 2460$).

7.3 SV Models

7.3.1 What Is an SV Model?

So far, we detect changes in volatility 'in gross," not taking into account possible changes in "persistence parameter" $\alpha + \beta$ that is known as the origin for certain misspecification errors in GARCH models. Alternatively, volatility may be modelled as an unobserved component following some latent stochastic process, such as an autoregression. The resulting models are called *stochastic volatility (SV)* models created by Taylor (1986), and their interest has been increasing during the last years: see Ghysels et al. (1996), Shephard (1996), Yu and Meyer (2006), and Asai at al. (2006).

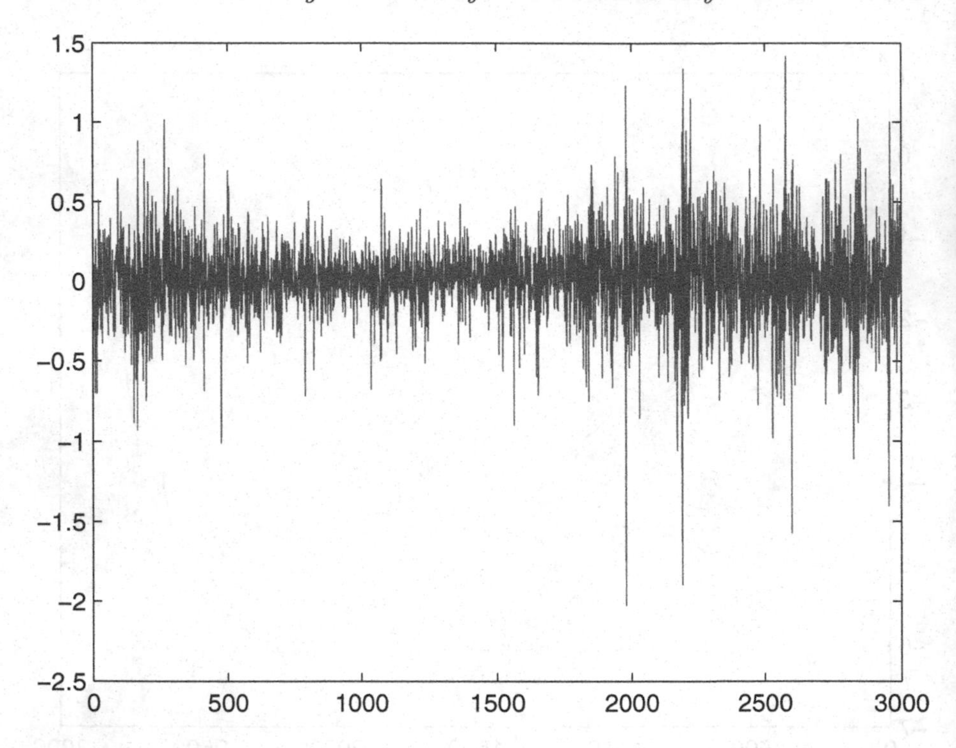

FIGURE 7.3
Graph of NYSE daily returns index.

I will follow Harvey et al. (1994) to briefly compare two types of models. For the GARCH(1,1) model we have

$$\sigma_t^2 = \gamma + \alpha y_{t-1}^2 + \beta \sigma_{t-1}^2, \quad \gamma > 0, \quad \alpha + \beta < 1. \tag{7.1}$$

Note that by letting the conditional variance be a function of squared previous returns and past variances, it is possible to capture the changes in volatility over time. Moreover, since this model is formulated in terms of the distribution of the one-step ahead prediction error, maximum likelihood estimation is straightforward. This property made GARCH models so popular and so widely used in applied research. Of course, the formulation given in (2) may be generalized by adding more lags of both, the squared returns and past variances. With respect ro its dynamics, the conditional volatility in GARCH models may be sen as ARMA(1,1) to the squared returns, where AR coefficient is the sum $p = \alpha + \beta$. Therefore, as this sum gets closer to one, the persistence of squared observations increases. This means that the autocorrelation function (ACF) of Y_t^2 will decay quite slowly, as it is commonly seen in practice.

On the other hand, SV models consider that the dynamics of the logarithm od σ_t^2, denoted h_t, is well described by a latent (unobserved) stochastic

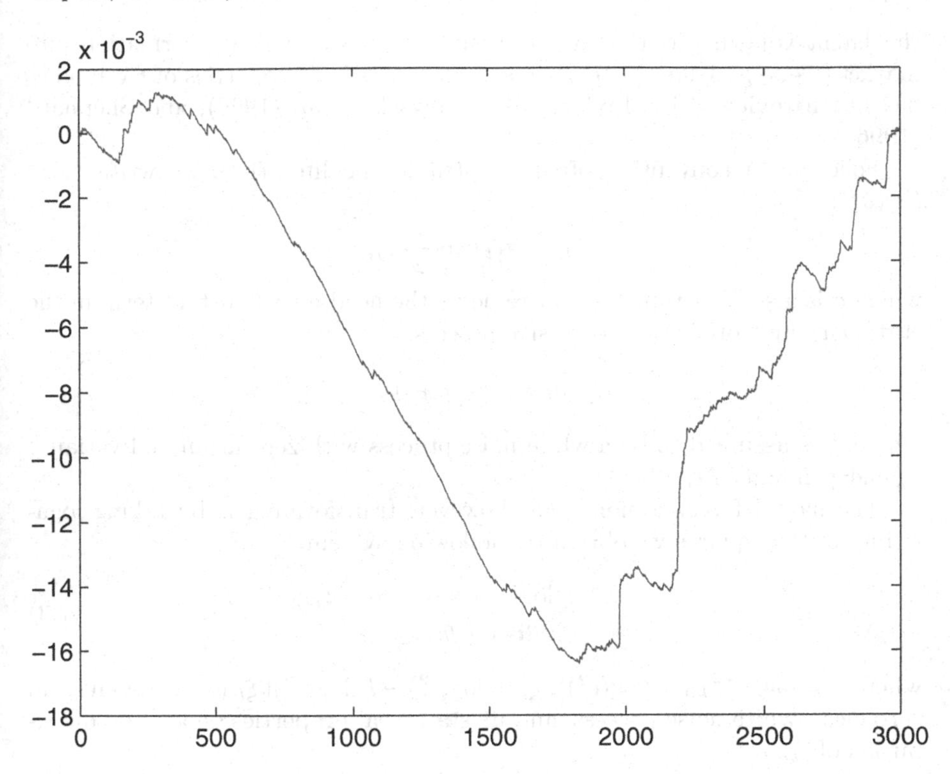

FIGURE 7.4
Graph of the decision statistic for NYSE index.

process, e.g.,

$$h_t = \psi h_{t-1} + \eta_t, \qquad 0 \le \phi < 1, \tag{7.2}$$

where η_t is the white noise process with zero mean and σ_η^2 dispersion. It is also often assumed that white noises ϵ_t and η_t in (7.1) and (7.2) are mutually independent.

SV models have several advantages over GARCH:

1) Their statistical properties are easier to find and understand.

2) They are easier to generalize to the multivariate case.

3) They are attractive because they are close to the models used in financial theory for asset pricing.

4) They capture in a more appropriate way the main empirical properties often observed in daily series of financial returns.

ARCH type of models assume that conditional volatility can be observed some steps ahead. However, a more realistic model for a conditional volatility can be based on modelling it having a predictable component that depends on past information and an unexpected noise. In this case, the conditional volatility is a latent unobserved variable (see Taylor (1986)). One interpretation of the latent volatility is that it represents an arrival of new information into market (see, e.g., Clark (1973)). The main statistical properties of SV models have been reviewed by Taylor (1994), Ghysels at al. (1996), and Shephard (1996).

Following a convention often adopted in the literature, we write $h_t = log(\sigma_t^2)$,

$$y_t = \sigma\epsilon_t \exp(\frac{1}{2}h_t),$$

where σ is a scale parameter that removes the need for a constant term in the stationary first order autoregressive process,

$$h_t = \phi h_{t-1} + \eta_t,$$

where η_t is assumed to be a white noise process with zero mean and variance σ_η^2 independent of ϵ_t.

The model for y_t is nonlinear, however, transforming it by taking logarithms of the squares we obtain the following system:

$$\begin{aligned} y_t^* &\equiv \log(y_t^2) = \gamma + h_t + \xi_t, \\ h_t &= \phi h_{t-1} + \eta_t, \end{aligned} \tag{7.3}$$

where $\gamma = \log(\sigma^{*2}) + E[\log(\epsilon_t^2)]$, $\xi_t = \log(\epsilon_t^2) - E[\log(\epsilon_t^2)]$ ξ_t is a non-Gaussian zero mean white noise process, and its statistical properties depend on distribution of ϵ_t.

7.3.2 Methods for Detection of Changes in SV Models

System (7.3) can be thought as a linear state-space model. However, the distribution of ξ_t is far from Gaussian. Therefore, all attempts to apply the Kalman filter methodology to detection of structural changes in SV models are methodologically incorrect.

The parameters γ, ϕ of this system are time variant in the general case. Therefore, prior to estimating them, we need to make sure that they are constant within the considered time interval $t = 1, \ldots, T$.

For this purpose, we can perform the nonparametric test considered in the previous chapter.

$$\begin{aligned} y_t &= y_0 + d\,h_t + \xi_t \\ h_t &= h_0 + \phi h_{t-1} + \eta_t, \end{aligned}$$

where

y_t is the sequence of expected returns;

h_t is the sequence of volatility values at different times t, and

y_0, h_0 are the initial values of y_t, h_t, respectively.

Remark that the dynamics of this system is well known in the "basis situation" that is obtained from the detailed experiment. In other words, coefficients γ, d, ϕ of this system are known to us. However, at some (a priori unknown) time m, a change in these coefficients occurs, i.e., the above system is transformed into

$$y_t = \delta + dz_t + \xi_t$$
$$z_t = \psi z_{t-1} + \eta_t.$$

The problem is to detect this change-point m using information on observations y_t and the basis model of volatility h_t only.

For this purpose, we can apply the test proposed in the previous chapter. In our case, this test can be written in the following form. Define the decision statistic: for $n = 1, \dots, N$, where N is the sample size:

$$Y_N(n) = \frac{1}{N}(\sum_{i=1}^{n} y_t h_t - (\sum_{t=1}^{n} h_t^2)(\sum_{t=1}^{N} h_t^2)^{-1}(\sum_{t=1}^{N} y_t h_t))$$

and for a certain $0 < \beta < 1/2$: the decision criterion

$$T_N = \max_{[\beta N] \le n \le [(1-\beta)N]} |Y_N(n)|.$$

For a certain threshold $C > 0$, we compare T_N with C. If $T_N \le C$, then the null hypothesis is assumed, otherwise we construct the following estimate of the change-point m:

$$\hat{m}_N \in arg \max_n |Y_N(n)|.$$

In the previous chapter, we demonstrated that

$$\frac{\hat{m}_N}{N} \to \theta, \quad P_\theta \text{ a.s. as } N \to \infty,$$

where θ is the true change-point parameter.

Moreover, probabilities of type-1 and type-2 errors go to zero exponentially as $N \to \infty$.

However, theoretical results are insufficient for a strong belief in the usage of this statistic. We might wonder that the main Statistic (2.6) proposed in Chapter 2 can also be efficiently used for retrospective detection of structural changes in SV models. In order to make the decision which statistic ought to be used, we perform here Monte-Carlo experiments with simulated SV models.

7.3.3 Simulations

I consider the following parameters of the baseline equations in SV model: $y_0 = h_0 = 0$ $d = 0.2$ $\phi = 0.3$. At the change-point $\theta = 0.7$, the parameters change as follows: $d = 0.3$; $\phi = 0.8$.

First, I tried to estimate thresholds for different sample types. Simulated α-quantiles for the proposed statistic and homogenous samples are presented in Table 7.5.

75 percentiles of the proposed statistic were then used as thresholds for

TABLE 7.5
75, 95, and 99 Percentiles for Different Sample Sizes

N	100	200	300	500	800	1000	1500	2000
r_{75}	0.0990	0.0715	0.0585	0.047	0.0403	0.0331	0.0271	0.0238
r_{95}	0.1331	0.0963	0.0777	0.0647	0.0547	0.0454	0.0363	0.0316
r_{99}	0.1662	0.1111	0.0953	0.0751	0.0674	0.0551	0.0441	0.0386

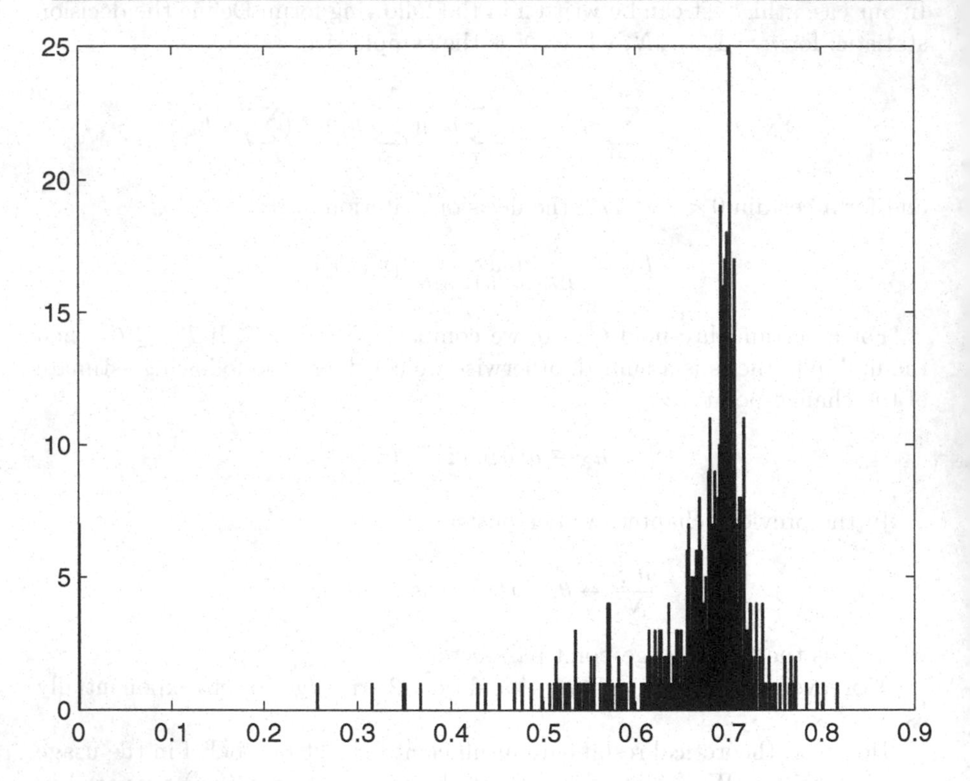

FIGURE 7.5
Histogram of change-point estimates: SV model, $\theta = 0.7, N = 1000$.

detection of structural changes in samples with structural changes. Each result was obtained as the average in 1000 independent trials of each experiment.

In particular, for the case $N = 1000$, the following estimates of the true change-point $\theta = 0.7$ were obtained (see Figure 7.5).

The same approach was used for construction of the change-point estimates with the help of Statistic (2.6). For the same sample size $N = 1000$ and parameters of the SV model, the obtained results are given in Figure 7.6.

We can see that the statistic proposed in the previous chapter for estimation of the change-point θ gives much better results: the obtained estimates

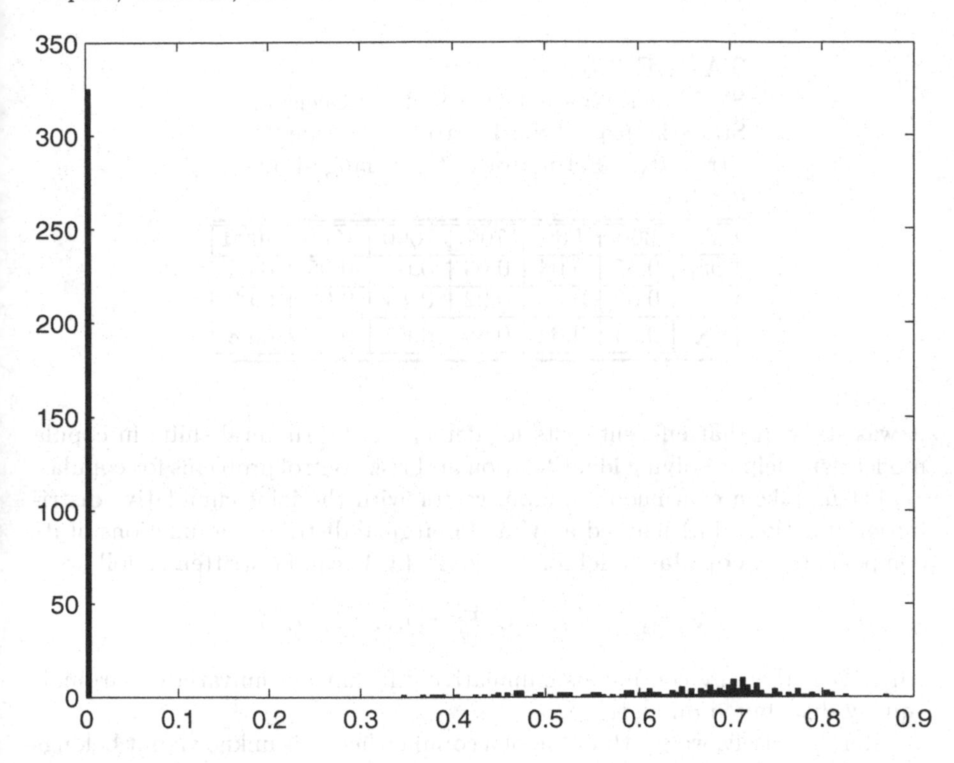

FIGURE 7.6
Histogram of change-point estimates: statistic (2.6), $\theta = 0.7, N = 1000$.

are concentrated around the true change-point, while Statistic (2.6) gives unsatisfactory results for the same sample size.

More results for the proposed statistic are reported in Table 7.6.

7.4 Copula Models

7.4.1 What is Copula Model?

In this section, I consider results concerning applications of nonstationary copula models to financial problems. Copula models are successfully applied to modeling of financial risk problems (see Frees and Valdez, 1998). In spring 2009, the Basel committee for bank supervision stressed the role of Copulas as one of the most correct methods of estimation of credit risks.

Detection of structural shifts on the Copula model is of the utmost importance for correct risk evaluation. In Brodsky, Penikas, and Safaryan (2011),

TABLE 7.6
SV Model (Characteristics of the Decision Statistic: pr_1 – type-1 Error, w_2 – type-2 Error, $\hat{\theta}_N$ – Estimator of the Change-Point $\theta = 0.7$)

N	300	500	700	1000	1500	2000
pr_1	0.05	0.04	0.06	0.05	0.05	0.05
w_2	0.39	0.33	0.22	0.13	0.03	0.02
$\hat{\theta}_N$	0.64	0.64	0.65	0.66	0.67	0.68

it was stressed that efficient tests for detection of structural shifts in copula models will help in solving identification and risk control problems for copulas.

Let us take a continuous random vector with the joint cumulative distribution function (d.f.) marked as V and marginal distribution functions of its components. A **copula** model for the joint d.f. V can be written as follows:

$$V(x_1, \ldots, x_d) = G(F_1(x_1), \ldots, F_d(x_d)),$$

where G is the only continuous cumulative d.f. that has univariate marginals equally distributed on $[0, 1]$.

More precisely, we are thinking of a copula when G is unknown but belongs to the following class:

$$\mathbf{A} = \{G_\theta : \theta \in \Theta\},$$

where Θ is an open set in $\mathbf{R}^p$ space.

Two well-known monographs containing detailed descriptions of parametric copula models are those of Joe (1997) and Nelsen (2006). Copulas are often of use in actuarian science, econometrics, and hydrology (Fries and Valdez (1998), Cui and Sun (2004), Genest and Favre (2007)). They are often applied in solving financial and risk management tasks (Cherubini et al. (2004), McNeil et al. (2005)).

Modern findings in the field of copula applications can be classified into two principal groups:

1) Parametric copula models estimation and goodness-of-fit testing (Malavergne and Sornette (2003), Shih (1998), Glidden (1999), Cui and Sun (2004));

2) Nonparametric methods of testing goodness-of-fit of copula models, including blanket tests (Genest et al. (2006), Breymann et al. (2003), Dobric and Schmid (2005), Junker and May (2005)).

Below in this chapter, I propose a nonparametric method for estimating a change-point in a copula model.

7.4.2 Methods for Detection of Changes in Copula Models

Let us start from the sample $\{X_1, \ldots, X_N\}$ of independent R^d-dimensional vectors with the cumulative d.f.'s $V_1, \ldots, V_N$.

Suppose there exist two alternatives. The null hypothesis H_0 is that the copula model is unchanged, i.e., $G_1 = G_2 = \cdots = G_N$. The alternative is that the copula changes after some instant $m = [\theta N]$. Here, we do suppose that all marginal d.f.'s $F_1, \ldots, F_d$ are unchanged. So, the joint d.f. V_k at each instant $1 \leq k \leq N$ can be written as follows:

$$V_k(x_1, \ldots, x_k) = \begin{cases} G_1(F_1(x_1), \ldots, F_d(x_d)), & 1 \leq k \leq m \\ G_2(F_1(x_1), \ldots, F_d(x_d)), & m < k \leq N. \end{cases}$$

We want to test the null hypothesis H_0 and in case it is rejected to estimate the change-point parameter θ.

In other words, we test whether there exists a structural shift in a pattern of comovement of an observed vector components. The goal is to propose a method with small type-1 (false alarm) and type-2 (false calmness) errors (which converge to zero as the sample size N tends ro infinity) and strongly consistent estimates of the change-point parameter θ.

To obtain consistent tests, let us consider empirical copulas

$$C_n(u) = \frac{1}{n}\sum_{i=1}^{n} I(U_{i,n} \leq u) = \frac{1}{n}\sum_{i=1}^{n}\prod_{l=1}^{d} I(U_{il,n} \leq u_l)$$

$$D_{N-n}(u) = \frac{1}{N-n}\sum_{i=n+1}^{N} I(V_{i,N-n} \leq u) = \frac{1}{N-n}\sum_{i=n+1}^{N}\prod_{l=1}^{d} I(V_{il,N-n} \leq u_l),$$

where $U_{i,n} = (U_{i1,n}, \ldots, U_{id,n})$, $V_{i,N-n} = (V_{i1,N-n}, \ldots, V_{id,N-n})$ and for any $i \in [1, \ldots, d]$,

$$U_{il,n} = \frac{n}{n+1} F_{l,n}(X_{il}), \quad 1 \leq i \leq n,$$
$$V_{il,N-n} = \frac{N-n}{N-n+1} G_{l,N-n}(X_{il}), \quad N \geq i > n+1,$$

with

$$F_{l,n}(x_l) = \frac{1}{n}\sum_{i=1}^{n} I(X_{il} \leq x_l), \quad G_{l,N-n}(x_l) = \frac{1}{N-n}\sum_{i=n+1}^{N} I(X_{il} \leq x_l).$$

Then, we use the following statistic as a modification of the Kolmogorov-Smirnov test:

$$\Psi_{i,N-i} = (D_l(u) - D_{N-l}(u))\sqrt{l(N-l)}/N$$

and

$$T_N = \max_{[\beta N] \leq l \leq [(1-\beta)N]} \sup_u |\Psi_{l,N-l}(u)|.$$

Therefore, we propose the following change-point estimate:

$$\hat{m}_N \in \arg \max_{[\beta N] \le l \le [(1-\beta)N]} \sup_u |\Psi_{l,N-l}(u)|.$$

Then, the estimate of the change-point parameter is $\hat{\theta}_N = \hat{m}_N/N$.

In the sequel, we use the following performance measures to evaluate the quality of change-point detection and estimation:

1) 1st-type error probability

$$\alpha_N = P_0\{T_N > C\},$$

where $C > 0$ is the decision threshold:

2) 2nd-type eror probability

$$\delta_N = P_m\{T_N \le C\}; and$$

3) estimation error probability: for every $0 < \epsilon < 1/2$,

$$\gamma_N = P_m\{|\hat{\theta}_N - \theta| > \epsilon\}.$$

Now, in Theorems 7.4.1 and 7.4.2, we formulate the main results.

Let us remember the major assumption: $(X_1, \dots, X_N)$ are independent random d-dimensional vectors with continuous marginal d.f.'s. Then, the r.v's U_{ij} defined above are independent for different $i = 1, \dots, l$. Besides, their d.f.'s are the same under the null (no change-points). U_{ij} satisfy the Cramer condition $E_0 \exp(tU_{ij}) < \infty$ given $|t| < T$ for some $T > 0$.

In Theorem 7.4.1 the exponential estimate for the 1st-type error probability is obtained.

Theorem 7.4.1.

$$\alpha_N \le L_1 \exp(-L_2 C^2 N),$$

where L_1, L_2 are positive constants not dependent on N.

The proof of Theorem 7.4.1 is sketched below. First, we remember that

$$D_l(u) = \frac{1}{l} \sum_{i=1}^{l} I(U_{i,l} \le u)$$

$$D_{N-l}(u) = \frac{1}{N-l} \sum_{i=L+1}^{N} I(U_{i,N-l} \le u).$$

So, under the null hypothesis $H_0 : G_1(\cdot) = G_2(\cdot)$, we obtain

$$\sqrt{l}(D_l(u) - G_1(u)) \to W_1(u)$$
$$\sqrt{N-l}(D_{N-l}(u) - G_2(u)) \to W_2(u),$$

and

$$(D_l(u) - D_{N-l}(u))\,\frac{\sqrt{l(N-l)}}{N} \to \frac{1}{\sqrt{N}}((1-\frac{l}{N})^{1/2}\,W_1(u) - (\frac{l}{N})^{1/2}\,W_2(u)),$$

where $W_1(\cdot), W_2(\cdot)$ are independent Wiener processes in $[0,1]^d$, and the symbol " $\to$" is used to signify the weak convergence in the Skorokhod space $D[0,1]^d$ as $N \to \infty$.

Therefore, we conclude that the process $((1-\frac{l}{N})^{1/2}\,W_1(u) - (\frac{l}{N})^{1/2}\,W_2(u)) \sim W(u)$ is distributed as a certain standard Wiener process $W(\cdot)$ in $[0,1]^d$.

Thus, for the probability of the 1st-type error, we can write

$$\alpha_N = P_0\{T_N > C\} \sim P_0\{\frac{1}{\sqrt{N}}\sup_u |W(u)| > C\} \le L_1 \exp(-L_2 N).$$

In Theorem 7.4.2, we consider the exponential upper estimates for the 2nd-type error probability and the estimation error probability.

Theorem 7.4.2. *Denote* $\eta = \sup_u |G_1(u) - G_2(u)|$ *for* $u \in [0,1]^d$ *and assume that* $0 < C < \eta\,\frac{\sqrt{m(N-m)}}{N}$. *Let* $d = \eta\,\frac{\sqrt{m(N-m)}}{N} - C$. *Then, the following exponential upper estimates hold:*

$$\delta_N \le L_1 \exp(-L_2 \min(d, d^2)N)$$
$$\gamma_N \le C_1 \exp(-C_2 \min(\epsilon, \epsilon^2)N),$$

where $L_1, L_2, C_1,$ *and* C_2 *are positive constants not dependent on* N.

Sketch of the proof.

First, we prove that the mathematical expectation of the statistic $\sup_u |\Psi_{l,N-l}(u)|$ has a unique maximum at the point $l = m$. Consider the case $[\beta N] \le l \le m$. We have $D_l(u) = \frac{1}{l}\sum_{i=1}^{l} I(U_{i,l} \le u)$, $ED_l(u) = G_1(u)$.

Further,

$$D_{N-l}(u) = \frac{1}{N-l}(\sum_{i=l+1}^{m} I(U_{i,N-l} \le u) + \sum_{i=m+1}^{N} I(U_{i,N-l} \le u)).$$

Therefore,

$$ED_{N-l}(u) = \frac{1}{N-l}((m-l)G_1(u) + (N-m)G_2(u)).$$

So,

$$E(D_l(u) - D_{N-l}(u)) = G_1(u)(1-\frac{m-l}{N-l}) - G_2(u)\frac{N-m}{N-l} = \frac{N-m}{N-l}(G_1(u) - G_2(u)).$$

The interval $m < l \leq [(1-\beta)N]$ is considered in the same way. As a result, we obtain

$$\sup_u E\Psi_{l,N-l}(u) = \begin{cases} \frac{N-m}{N}\sqrt{\frac{l}{N-l}} \sup_u |G_1(u) - G_2(u)|, & l \leq m \\ \frac{m}{N}\sqrt{\frac{N-l}{l}} \sup_u |G_1(u) - G_2(u)|, & l > m. \end{cases}$$

Therefore, the function $\sup_u E\Psi_{l,N-l}(u)$ has a unique maximum at the point $l = m$, and, in virtue of our assumptions,

$$\max_l \sup_u E\Psi_{l,N-l}(u) = \frac{\sqrt{m(N-m)}}{N} \sup_u |G_1(u) - G_2(u)| > C.$$

Moreover, since $[\beta N] \leq m \leq [(1-\beta)N]$, this function satisfies the reversed Lipshitz condition at this point.

The function $\Psi_{l,N-l}(u)$ can be decomposed into the deterministic and stochastic part:

$$\Psi_{l,N-l}(u) = E\Psi_{l,N-l}(u) + \zeta_{l,N-l}(u).$$

For the 2nd-type error probability, we obtain

$$P\{\max_l \sup_u |\Psi_{l,N-l}(u)| < C\} \leq P\{\max_l \sup_u |\zeta_{l,N-l}(u)| > \max_l \sup_u |E\Psi_{l,N-l}(u)| -$$

Since $\max_l \sup_u |E\Psi_{l,N-l}(u)| - C = d > 0$ according to assumptions of theorem, we obtain here the situation already considered in Theorem 1. Therefore,

$$\delta_N \leq L_1 \exp(-L_2 \min(d, d^2)N),$$

where the positive constants L_1 and L_2 do not depend on N.

The second exponential inequality of Theorem 2 is obtained in the same manner given the reversed Lipshitz condition for the function $\sup_u E\Psi_{l,N-l}(u)$ at the point $l = m$.

7.4.3 Simulations

The proposed method was tested using bidimensional vectors that joint d.f. was characterized by Clayton (1) and Gumbel (2) copula.

(1) Clayton copula: for any $u, v \in (0,1)$ and $\kappa > 0$:

$$C_\kappa(u,v) = (u^{-\kappa} + v^{-\kappa} - 1)^{-1/\kappa}.$$

(2) Gumbel copula: for every $u, v \in 90, 1)$ and $\kappa > 0$:

$$C_\kappa(u,v) = \exp[-\{(-\log u)_(^{1/\kappa} - \log v)^{1/\kappa}\}^\kappa].$$

We assume that the functional form of these copulas do not change within

TABLE 7.7
75, 85, 95, and 99 Percentiles for Different Sample Sizes for Clayton Copula, Homogenous set, $\kappa = 0.3$

N	50	100	200	300	500	700	1000	1500
r_{75}	0.0990	0.0715	0.0585	0.047	0.0403	0.0331	0.0271	0.0238
r_{85}	0.0990	0.0715	0.0585	0.047	0.0403	0.0331	0.0271	0.0238
r_{95}	0.1156	0.0850	0.0615	0.0492	0.0372	0.0314	0.0213	0.0197
r_{99}	0.1343	0.0945	0.0674	0.0550	0.0426	0.0318	0.0232	0.0214

TABLE 7.8
75, 85, 95, and 99 Percentiles for Different Sample Sizes for Gumbel Copula, Homogenous set, $\kappa = 0.3$

N	50	100	200	300	500	700	1000	1500
r_{75}	0.0886	0.0607	0.0426	0.0336	0.0255	0.0207	0.0186	0.0164
r_{85}	0.0958	0.0646	0.0437	0.0358	0.0273	0.0232	0.0193	0.0185
r_{95}	0.1033	0.0749	0.0500	0.0392	0.0302	0.0243	0.0206	0.0192
r_{99}	0.1187	0.0836	0.0585	0.0446	0.0332	0.0292	0.0233	0.0212

the considered period, whereas the parameter κ may change at some point $m = [\theta N]$, where N is the sample size, $0 \leq \theta < 1$.

First, we examine the critical bounds for the method proposed w.r.t. different observation sets and copula types. Initially, we deal with homogeneous samples, i.e., without change-points. For each sample of size N, the experiment was independently simulated 500 times. 95th and 99th quantiles for the decision statistic T_n were estimated as averages in 500 independent trials. 95th quantile values were then used as critical bounds for rejection of the null hypothesis for samples with change-points. Simulation results are presented in Tables 7.7–7.8.

As Tables 7.7–7.8 show, these critical bounds are not very sensitive to the concrete copula type underlying the observations. It permits us to undertake a robust parameter calibration procedure for the purpose of change-point detection and estimation. The corresponding results are provided in Tables 7.9–7.10 below.

The typical histogram of change-point estimates for the case of Clayton copula for $N = 1000$ is shown in Figure 7.7.

The typical histogram of change-point estimates for the case of Gumbel copula for $N = 500$ is shown in Figure 7.8.

Based on simulation results presented above, we can summarize the major findings:

1) The proposed method enables us to properly identify the structural shifts in copula models and to arrive at their parameter estimates.

TABLE 7.9
Clayton Copula

N	500	700	1000	1500
C	0.037	0.031	0.027	0.020
w_2	0.58	0.43	0.15	0.02
θ_N	0.337	0.335	0.303	0.30

Note: Parameter values before and after the change-point $\kappa = 0.3$, $\kappa = 1.0$, respectively; the change-point parameter $\theta = 0.3$, C – critical bound; w_2-type-2 error

TABLE 7.10
Gumbel Copula

N	100	200	300	500	700	1000	1500
C	0.07	0.05	0.04	0.03	0.02	0.017	0.015
w_2	0.69	0.60	0.44	0.33	0.04	0.01	0.0
θ_N	0.45	0.40	0.35	0.33	0.31	0.305	0.30

Note: Parameter values before and after the change-point $\kappa = 0.3$, $\kappa = 1.0$, respectively; the change-point parameter $\theta = 0.3$, C – critical bound; w_2-type-2 error, θ_N – the estimate of change-point parameter

2) The critical bounds estimated do not depend on the copula type (i.e., Clayton, Gumbel, or other) or on the copula parameters under the null hypothesis.

7.5 Long memory, Fractional Integration, and Structural Changes in Financial Time Series

7.5.1 Introduction

At the end of this chapter, we ask the following question: Is it so important to detect structural changes in financial time series? Is it the sole advantage of such knowledge (of structural changes)?

Modern findings in economic theory (see, e.g., Diebold and Inoue (2001)) tell us that there are many other advantages. First, it is known that a structural change can favor acceptance of a fractional integration hypothesis. Second, discrimination betweeen hypotheses of a structural change and a fractional integration is extremely difficult.

In 1951, H. Hurst published his paper on Niles River inflows and storage capacity. Since that time, the phenomenon of long memory, fractional inte-

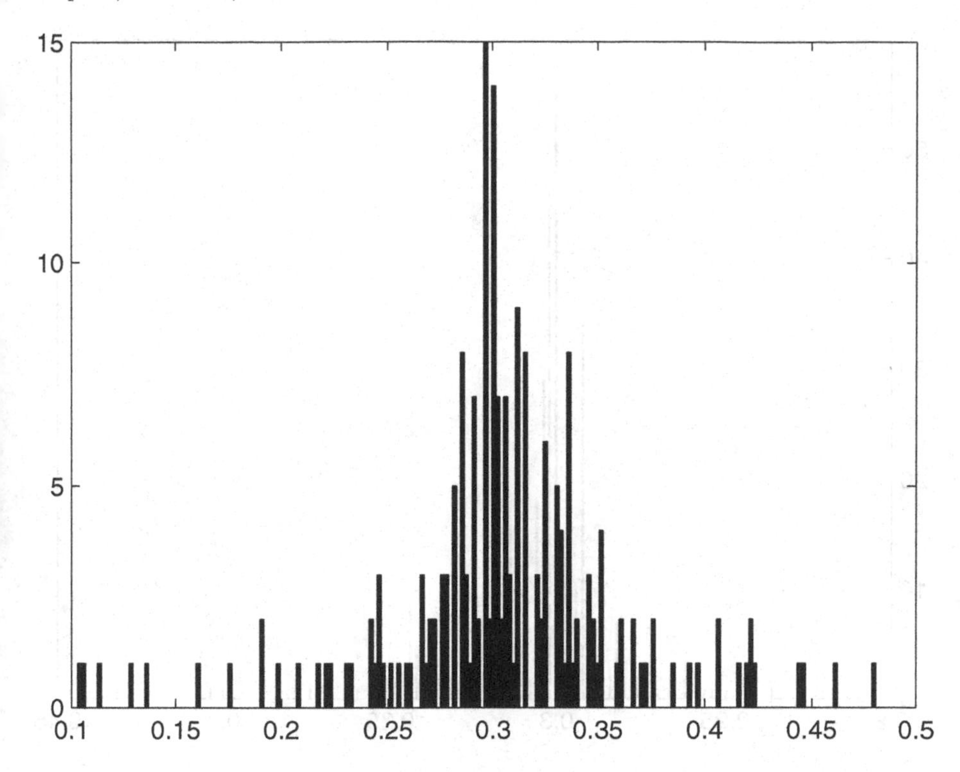

FIGURE 7.7
Histogram of change-point estimates: Clayton copula.

gration, and auto-modeling in empirical data attracted great attention from statisticians and financial analysts.

The method of R/S-analysis proposed by Hurst helps to study clustering phenomena in data, long memory, and fractal properties of time series. In the sequel, a certain modification of this method is considered that enables us to answer more sophisticated questions about discrimination between a stochastic trend and regime switching hypotheses.

We begin from a short exposition of R/S-statistic proposed by Hurst. Let $S = (S_n)_{n \geq 0}$ be a certain financial index, and h_n is a sequence of returns $h_n = \ln \dfrac{S_n}{S_{n-1}}$, $n \geq 1$. We build the values $L_n = h_1 + \cdots + h_n$, $n \geq 1$ and put

$$R_n = \max_{k \leq n}(L_k - \frac{k}{n} L_n) - \min_{k \leq n}(L_k - \frac{k}{n} L_n)$$

and

$$S_n^2 = \frac{1}{n} \sum_{k=1}^{n} h_k^2 - (\frac{1}{n} \sum_{k=1}^{n} h_k)^2.$$

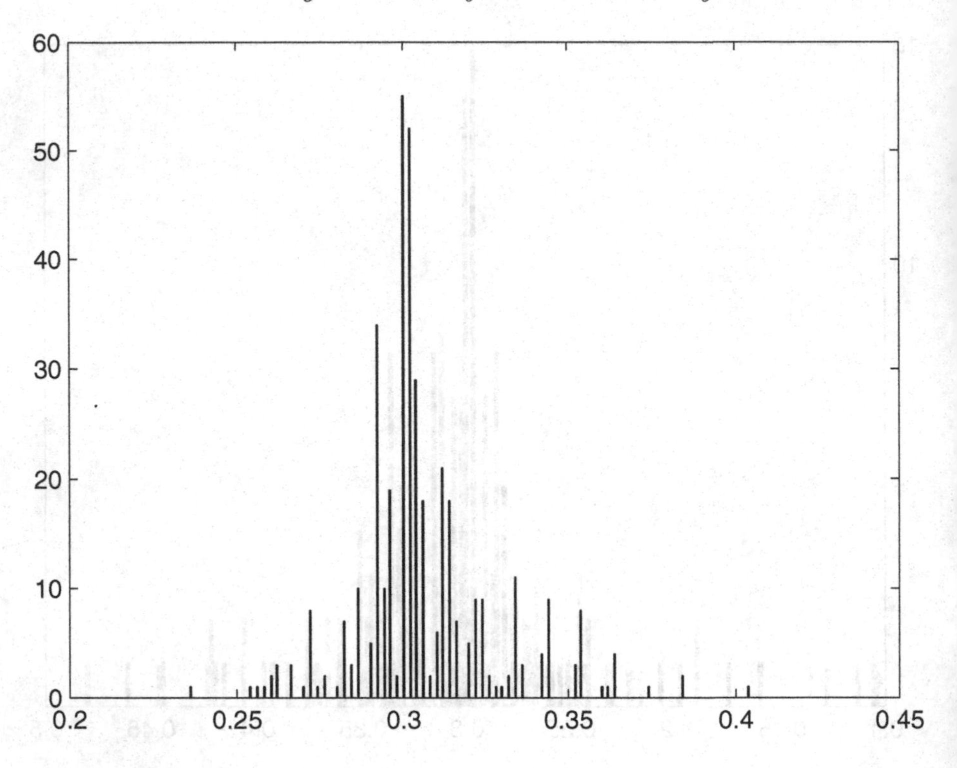

FIGURE 7.8
Histogram of change-point estimates: Gumbel copula.

We see that the value $\bar{h}_n = \dfrac{L_n}{n}$ is the empirical average constructed by the sample $(h_1, h_2, \ldots, h_n)$, and, therefore, $L_k - \dfrac{k}{n}L_n = \sum_{i=1}^{k}(h_i - \bar{h}_n)$ is the deviation of L_k from the empirical average $\dfrac{k}{n}L_n$. The value R_n characterizes the spread of these deviations $L_k - \dfrac{k}{n}L_n$.

The value R_n is too complicated to work with. Trying to simplify it, we first note that $\min_{k \le n} |L_k - \dfrac{k}{n}L_n| = 0$, and, therefore, we can simply consider $\max_{k \le n} |L_k - \dfrac{k}{n}L_n)|$ with the same qualitative conjectures. Second, we can take the value R_n (instead of R_n/S_n) and choose the decision threshold for it from the limiting behavior of R_n. After these transformations, we obtain the following statistic:

$$Y_N(n) = \max_{1 \le n \le N} |L_n - \frac{n}{N}L_N|.$$

This statistic (more precisely, $\xi(h/N) \equiv Y_N(n)$) weakly converges (in the

Skorokhod space $D(0,1)$ to the maximum of the Brownian bridge on $[0,1]$. Therefore, it is rather easy to find the decision threshold for it. Moreover, we note that the statistic $Y_N(n)$ is the variant of Kolmogorov's statistic for testing statistical homogeneity of two samples.

Now, let us discuss which hypotheses are usually tested in R/S-analysis:

H_0 — stationarity of data. Under H_0, we suppose that $h_n \sim I(0)$, i.e., the sequence of returns is stationary and with zero means.

H_1 — structural change. Under H_1, it is supposed that a structural change occurs at a certain point of the sample $h_1, \ldots, h_N$.

H_2 — integration or fractional integration of data. Under H_2, we suppose that the sequence of returns is integrated of the order $I(d)$, where $d > 0$.

In this introduction, we can qualitatively describe properties of subsequent statistics. In the work of Diebold and Inoue (2001), it was stressed that an undetected structural change usually leads to accepting H_2 instead of H_0. Hurst (1951) wrote that the real data on the Nile's inflows unexpectedly rejected H_0 and favored acceptance of H_2. In this paper, we try to say more: A method is proposed for discrimination between hypotheses H_0, H_1, and H_2 in dependence of characteristics of the decision statistic.

7.5.2 Problem Statement

Here, we give a short exposition of underlying models. We consider the following models of data $h_n = \ln \dfrac{S_n}{S_{n-1}}$ on the probability space $(\Omega, \mathcal{F}, \mathbf{P})$:

H_0: a stationary time series:

$$h_n = \rho h_{n-1} + u_n, \quad h_0 \equiv 0, \quad |\rho| < 1, \quad i = 1, \ldots, N,$$

where u_n is a sequence of i.i.d.r.v.'s, $Eu_n = 0$, and $Eu_n^2 = \sigma^2$. The process of observations $y_i = h_i$, $i = 1, 2, \ldots, N$.

H_1: the sequence of observations is:

$$y_i = hI\{i \geq [\theta N]\} + h_i,$$

where $I(A)$ is the characteristic function of the set A; the change-point parameter θ $(0 < \beta \leq \theta \leq 1 - \beta < 1$ is unknown.

H_2: fractional integration model:

$$\begin{aligned} h_i &= h_{i-1} + u_i, \quad h_0 \equiv 0, \quad i = 1, \ldots, N \\ y_i &= h_i, \end{aligned}$$

where

$$u_i = \begin{cases} o, & \text{w.p. } 1 - p_N \\ w_i, & \text{w.p. } p_N, \end{cases}$$

and w_i is a sequence of i.i.d.r.v.'s, $Ew_i \equiv 0$, $Ew_i^2 = \sigma_w^2$ and $p_N = O(N^{2d-2})$, and $0 < d < 1$.

Therefore, $var(y_N) = var(h_N) = O(N^{2(d-1)+1})$. In general, $d > 0$ and this case corresponds to the unit root and fractional integration hypotheses. Remark that the case $1/2 \le d < 1$ corresponds to the "black noise" model, and the case $0 < d < 1/2$ – to the model of "stochastic turbulence."

Below, we describe the main results for the proposed method.

The main statistic that is used for testing hypotheses H_0, H_1, and H_2 has the following form:

$$Z_N(n) = \frac{1}{N^2}(n \sum_{j=1}^{N} y_j - N \sum_{j=1}^{n} y_j), \quad n = 1, \ldots, N.$$

Remark that this statistic is exactly the above modification of the Hurst (1951) statistic and can also be considered as modification of Kolmogorov's test for detection of changes in statistical characteristics of two samples.

Let us fix the parameter $0 < \alpha < 1/2$. The decision rule based upon the statistic

$$T_N = \max_{[\alpha N] \le n \le [(1-\alpha)N]} |Z_N(n)|$$

and the threshold $0 < C$ takes the following form:

$$d_N = \begin{cases} 0, \text{ if} & T_N \le C \text{ the hypothesis } H_0 \text{ is accepted} \\ 1, \text{ if} & T_N > C \text{ the joint alternative } H_1 \& H_2 \text{ is accepted} \end{cases}$$

Here, the first stage of the proposed method is completed. At the second stage, we discriminate between the alternatives H_1 and H_2.

First, we describe the situation of only one structural break. Here, we test hypotheses H_1 and H_2 as follows:

1) Suppose at the first stage, we obtain the inequality: $T_N > C$ and the maximum of the absolute value of the statistic $Z_N(n)$ is attained at the point $\hat{n}_1$ (in the case of several such points, we take the minimum of them).

2) At the second stage, we fix the parameter $0 < \delta < \min(\theta, 1 - \theta)$ and compute two subsamples: $X_1(\hat{n}_1) = \{1, \ldots, \hat{n}_1 - [\delta N]\}$ and $X_2(\hat{n}_1) = \{\hat{n}_1 + [\delta N], \ldots, N\}$.

3) For each of these subsamples, the corresponding values of statistic $T_N(X_1)$ and $T_N(X_2)$ are computed.

4) The thresholds $C_1 = C_1(\hat{n}_1 - [\delta N])$ and $C_2 = C_2(N - \hat{n}_1 - [\delta N])$ for subsamples X_1 and X_2 are computed.

5) The decision rule at the second stage can be written as follows:

$$d_2(N) = \begin{cases} 1, & T_N(X_1) < C_1, \text{ or } T_N(X_2) < C_2 \\ 2, & \text{otherwise.} \end{cases}$$

The value $d_2(N) = 1$ corresponds to the hypothesis H_1, $d_2(N) = 2$ — to the hypothesis H_2.

In the general case of several change-points, we can do as follows. As before, we fix the numbers $0 < \delta < \theta(1-\theta)$ and $nmin \le N * \max(\alpha, 1-\alpha)$. We begin from stages 1)–5) of the above algorithm. If $T_N(X_1) \geq C_1$, or $T_N(X_2) \geq C_2$, then we repeat steps 1)–5) for the subsample, with $T_N(\cdot) \geq C1(C2)$, etc. At the end, we either obtain a homogenous subsample with $T_N(\cdot) < C1(C2)$ or still obtain $T_{nmin}(\cdot) \geq C1(C2)$. In the first case, the hypothesis H_1 is accepted, otherwise – H_2.

7.5.3 Main Results

7.5.3.1 Testing Stationarity Hypothesis

Results of this section strongly resemble Chapter 3. However, I give here detailed proofs and motivations. All subsequent considerations are based upon the following observation. For the hypothesis H_0, the decision statistic $T_N(\cdot)$ decreases as $N \to \infty$, for the hypothesis H_1, $T_N(\cdot)$ tends to a certain constant a.s. as $N \to \infty$, and for H_2, the decision statistic increases a.s. as $N \to \infty$ for any $d > 0$.

Here, I assume that the Cramer condition is satisfied for the sequence u_i. The following theorem can be formulated.

Theorem 7.5.1.
Suppose $d > 1/2$ and $0 < C < \sigma|h|\theta(1-\theta)$. Then the error probabilities $P_0\{H_1 \text{ or } H_2\}$, $P_1\{H_0\}$, and $P_2\{H_0\}$ converge to zero as the sample size $N \to \infty$.

Moreover, in the case of only one change-point, under H_1, the normalized estimator of the change-point $\hat{n}/N$ converges a.s. to the true change-point parameter, θ as $N \to \infty$.

Proof.
Consider properties of the statistic $Z_N(\cdot)$ under different alternatives.
1) First, for H_0:

$$y_i = \rho(\rho y_{i-2} + u_{i-1}) + u_i = \cdots = \sum_{k=0}^{i-1} u_{i-k}\rho^k.$$

Therefore,

$$\sum_{j=1}^{n} e_j = \sum_{j=1}^{n}\sum_{k=0}^{j-1} u_{j-k}\rho^k = \sum_{k=0}^{n-1}\rho^k \sum_{m=1}^{n-k} u_m = \sum_{m=1}^{n} u_m \sum_{k=0}^{n-m}\rho^k.$$

Under H_0, the statistic $Z_N(i)$ can be written as follows:

$$\begin{aligned} Z_N(i) &= (i\sum_{m=1}^{N} u_m \frac{1-\rho^{N-m+1}}{1-\rho} - N\sum_{m=1}^{i} u_m \frac{1-\rho^{i-m+1}}{1-\rho})/N^2 \\ &= \frac{1}{(1-\rho)N^2}(\sum_{m=1}^{i} u_m[i(1-\rho^{N-m+1}) - N(1-\rho^{i-m+1})] \\ &\quad + \sum_{m=i+1}^{N} u_m i(1-\rho^{N-m+1})). \end{aligned}$$

From independence of u_i, we obtain:

$$E_0 Z_N^2(i) \le \frac{\sigma^2}{(1-\rho)^2 N^4}\left(\sum_{m=1}^{i}(N-i)^2 + \sum_{m=i+1}^{N} i^2\right) = \sigma^2\,\frac{i(N-i)}{N^3(1-\rho)^2}.$$

Therefore,

$$\max_i E_0 Z_N^2(i) \le O\left(\frac{1}{N}\right).$$

In virtue of Cramer's condition for the sequence u_i, we can write:

$$P_0\{H_1 \text{ or } H_2\} = P_0\{\max_n |Z_N(n)| > C\} \le$$

$$\begin{cases} \exp\left(-\dfrac{(1-\rho)^2 C^2 N}{2\sigma^2 g}\right), & 0 < C < NTg\sigma^2 \\ \exp\left(-\dfrac{(1-\rho)CNT}{2}\right), & C \ge NTg\sigma^2. \end{cases}$$

2) Under the hypothesis H_2, we do as follows. First, consider the unit root case $d = 1, p_i \equiv 1$. Then,

$$Z_N(i) = \left(i\sum_{m=1}^{N} w_m(N-m+1) - N\sum_{m=1}^{i} w_m(i-m+1)\right)/N^2$$
$$= \left(\sum_{m=1}^{i} w_m(N-i)(m-1) + \sum_{m=i+1}^{N} w_m i(N-m+1)\right)/N^2.$$

Under the hypothesis H_2 for independent u_i, we obtain:

$$E_2 Z_N^2(i) = \frac{\sigma^2}{N^4}\left(\sum_{m=1}^{i}(N-i)^2(m-1)^2 + \sum_{m=i+1}^{N} i^2(N-m+1)^2\right)$$
$$= \frac{\sigma^2}{N^4}\left((N-i)^2\sum_{k=0}^{i-1} k^2 + i^2\sum_{k=1}^{N-i} k^2\right) = \sigma^2\frac{i(N-i)}{6N^3}(2i(N-i)+1).$$

Therefore, there exists the limit

$$\lim_{N\to\infty}\frac{1}{[Nt]} E_2 Z_N^2([Nt]) = \sigma^2\frac{t(1-t)^2}{3} = \sigma_t^2.$$

In virtue of Peligrad (1992), we obtain from here:

$$U_t = \frac{1}{\sigma_t\sqrt{N}} Z_N([Nt]) \to W(t),$$

where $W(t)$ is the Wiener process on $[0,1]$.

Hence, for the error probability

$$P_2\{H_0\} = P_2\{\max_{[\alpha N]\le n\le[(1-\alpha)N]} |Z_N(n)| < C\},$$
$$\le P_2\{\max_{\alpha\le t\le 1-\alpha}\frac{1}{\sigma_t\sqrt{N}}|Z_N(n)| < \frac{C}{\sigma\alpha\sqrt{N}}\left(\frac{3}{1-\alpha}\right)^{1/2}\} \to 0 \quad \text{as } N\to\infty.$$

In the general case under H_2: $d > 0, p_N = O(N^{2d-2})$,

$$h_i = h_{i-1} + u_i,$$

$$u_i = \begin{cases} o, & \text{w.p. } 1 - p_N \\ w_i, & \text{w.p. } p_N, \end{cases}$$

and w_i is a sequence of i.i.d.r.v.'s, $Ew_i \equiv 0$, and $Ew_i^2 = \sigma_w^2$.

Then,

$$E_2 Z_N^2(i) = \frac{p_N \sigma^2}{N^4}\left(\sum_{m=1}^{i} (N-i)^2(m-1)^2 + \sum_{m=i+1}^{N} i^2(N-m+1)^2\right)$$
$$= \frac{p_N\sigma^2}{N^4}\left((N-i)^2 \sum_{k=0}^{i-1} k^2 + i^2 \sum_{k=1}^{N-i} k^2\right) \sim \frac{p_N\sigma^2}{N^4}((N-i)^2 i^3 + i^2(N-i)^3)$$
$$= O(N^{2d-1}).$$

Therefore,

$$\max_{[\alpha N] \le n \le [(1-\alpha)N]} E_2 Z_N^2(n) = O(N^{2d-1}).$$

Here, we can check the limiting cases:
$d = 0$ (stationary sequence):

$$\max_{[\alpha N] \le n \le [(1-\alpha)N]} E_2 Z_N^2(n) = O(N^{-1}); \text{ and}$$

$d = 1$ (unit root case):

$$\max_{[\alpha N] \le n \le [(1-\alpha)N]} E_2 Z_N^2(n) = O(N).$$

Remark that for $d > 1/2$, the value $\max_{[\alpha N] \le n \le [(1-\alpha)N]} E_2 Z_N^2(n) = O(N^{2d-1})$ increases (the case of "black noise") and for $0 \le d < 1/2$, this value decreases (the case of "stochastic turbulence").

Now, let us define:

$$Z_N(\cdot) = N^{d-1} W_N(\cdot).$$

Then, as before, for $\alpha \le t \le 1 - \alpha$:

$$W_N([Nt])/\sqrt{N} \to W(t),$$

where $W(t)$ is the Wiener process on $[0, 1]$,

and for the error probability, we obtain

$$P_2\{\max_{[\alpha N] \le n \le [(1-\alpha)N]} |Z_N(n)| < C\} \sim P\{\max_{\alpha \le t \le 1-\alpha} |W(t)| < CN^{1/2-d}\}.$$

The last probability goes to zero as $N \to \infty$ for $d > 1/2$, i.e., for the case of the "black noise." The case of "stochastic turbulence" phenomenon is much more difficult.

The hypothesis H_1 is considered for the case of only one change-point. The general case is considered in an analogous way. Under H_1, the statistic $Z_N(i)$ is equal to:

$$Z_N(i) = a_\theta(\frac{i}{N}) + \eta_N(i),$$

where

$$a_\theta(t) = \begin{cases} t(1-\theta)h, & 0 \le t \le \theta \\ \theta(1-t)h, & \theta < t \le 1, \end{cases}$$

$$\eta_N(i) = \frac{1}{(1-\rho)N^2}(\sum_{m=1}^{i} u_m[i(1-\rho^{N-m+1}) - N(1-\rho^{i-m+1})] + \sum_{m=i+1}^{N} u_m i(1-\rho^{N-m+1})).$$

Therefore, for $C(N) < \sigma|h|\theta(1-\theta)$, we obtain for the error probability

$$P_1\{H_0\} = P_1\{\max_i |Z_N(i)| \le C\} \le P_1\{\max_i |\eta_N(i)| \ge \sigma|h|\theta(1-\theta) - C\}.$$

Denote $\varkappa = \sigma|h|\theta(1-\theta) - C$. Remark that for the last probability, the exponential upper estimate holds true analogous to those proved here for the hypothesis H_0:

$$P_1\{\max_i |\eta_N(i)| \ge \varkappa\} \le \begin{cases} \exp(-\frac{(1-\rho)^2\varkappa^2 N}{2\sigma^2 g}), & 0 < \varkappa < NTg\sigma^2 \\ \exp(-\frac{(1-\rho)\varkappa NT}{2}), & \varkappa \ge NTg\sigma^2. \end{cases}$$

Thus, all error probabilities for the proposed test converge to zero as the sample size N tends to infinity.

Now, let us prove the second statement of this theorem. Suppose that $\hat{n}$ is the estimator of the change-point obtained at the first stage of this method. Then, for any $0 < \epsilon < \alpha$, consider the event $\{|\frac{\hat{n}}{N} - \theta| > \epsilon\}$. Write

$$P_1\{|\frac{\hat{n}}{N} - \theta| > \epsilon\} \le P_1\{\max_t |\eta_N(t)| > \frac{1}{2}\epsilon|h|\min(\theta, 1-\theta)\}.$$

Let $\kappa_0 = \frac{1}{2}\epsilon|h|\min(\theta, 1-\theta)$. For the probability in the right hand, the exponential upper estimate holds analogous to those proved above:

$$P_1\{\max_t |\eta_N(t)| > \kappa_0\} \le \begin{cases} \exp(-\frac{(1-\rho)^2\kappa_0^2 N}{2\sigma^2 g}), & 0 < \kappa_0 < NTg\sigma^2 \\ \exp(-\frac{(1-\rho)\kappa_0 NT}{2}), & \kappa_0 \ge NTg\sigma^2. \end{cases}$$

From this exponential estimate, it follows that the normalized estimator of the change-point parameter $\hat{n}/N$ converges a.s. to the true change-point parameter θ as the sample size N tends to infinity.

Theorem 7.5.1 is proved.

7.5.3.2 Discrimination between the Hypotheses H_1 and H_2

At the second stage of the proposed method, we discriminate between hypotheses H_1 and H_2. Let us describe this method for the case of only one structural break:

1. Suppose at the first step the inequality $T_N > C(N)$ is satisfied (we give formula for $C(N)$ in the next section), and the maximum of $|Z_N(n)|$ is attained at the point $\tilde{n}$ (in the case of several points of maximum of $|Z_N(n)|$, we choose the minimal of these points).

2. Fix the parameter $0 < \varkappa < min(\theta, 1-\theta)/2$ and form two subsamples: $Y_1(\tilde{n}) = \{y_i : i = 1, \ldots, \tilde{n} - [\varkappa N]\}$ and $Y_2(\tilde{n}) = \{y_i : i = \tilde{n} + [\varkappa N], \ldots, N\}$. Denote by N_1 and N_2 the sizes of corresponding subsamples.

3. For these subsamples, we compute the values $T_{N_1}(Y_1)$ and $T_{N_2}(Y_2)$, i.e., the maximums of the statistic $|Z_N(\cdot)|$ designed for each of these samples, respectively.

4. The decision rule at the second step is formulated as follows:

accept hypothesis $\mathbf{H_1}$ if $T_{N_1}(Y_1) < C(N_1)$ or $T_{N_2}(Y_2) < C(N_2)$; and accept hypothesis $\mathbf{H_2}$ otherwise.

From Theorem 7.5.1, it follows that the two subsamples $Y_1(\tilde{n})$ and $Y_2(\tilde{n})$, formed at the second stage of the proposed method, are almost surely (as $N \to \infty$) classified as statistically homogenous under $\mathbf{H_1}$. However, under the hypothesis $\mathbf{H_2}$, these two subsamples $Y_1(\tilde{n})$ and $Y_2(\tilde{n})$ are almost surely classified as nonstationary. This is the raw idea of the second stage of our method: to compute the maximums $T_{N_1}(Y_1)$ and $T_{N_2}(Y_2)$ of the statistic $|Z_N(\cdot)|$ for two subsamples and to compare these values with the decision boundary $C(N_i)$, $i = 1, 2$.

The quality of hypothesis testing at the second step can be characterized by the following conditional error probabilities:

$$\delta_{12} = \frac{\mathbf{P}\{\mathbf{H}_2|\mathbf{H}_1\}}{1 - \mathbf{P}\{\mathbf{H}_0|\mathbf{H}_1\}}, \quad \delta_{21} = \frac{\mathbf{P}\{\mathbf{H}_1|\mathbf{H}_2\}}{1 - \mathbf{P}\{\mathbf{H}_0|\mathbf{H}_2\}}.$$

The sense of these conditional error probabilities is as follows: They estimate classification errors at the second step on condition that the true decision was made at the first step of the proposed method.

By definition, $\mathbf{P}\{\mathbf{H_0}|\mathbf{H_1}\} = \beta_{\mathbf{01}}(\mathbf{N})$, and by the idea of the second step of our method:

$$\mathbf{P}\{\mathbf{H_2}|\mathbf{H_1}\} \leq \max(\beta_{\mathbf{01}}([\theta\mathbf{N}] - [\varkappa\mathbf{N}]), \beta_{\mathbf{01}}(\mathbf{N} - [\theta\mathbf{N}] - [\kappa\mathbf{N}])).$$

Therefore, from Theorem 7.5.1, we conclude that δ_{12} goes to zero as N increases to infinity.

Further, $\mathbf{P}\{\mathbf{H_0}|\mathbf{H_2}\} = \beta_{\mathbf{02}}(\mathbf{N})$ and

$$\mathbf{P}\{\mathbf{H_1}|\mathbf{H_2}\} \leq \max(\beta_{\mathbf{02}}([\theta\mathbf{N}] - [\varkappa\mathbf{N}]), \beta_{\mathbf{02}}(\mathbf{N} - [\theta\mathbf{N}] - [\varkappa\mathbf{N}])).$$

Again, from Theorem 7.5.1, we conclude that δ_{21} converges to zero as $N \to \infty$. Thus, we proved the following.

Theorem 7.5.2.
Suppose $0 < C < \Delta$. Then, the conditional error probabilities at the second step, δ_{12}, and δ_{21}, tend to zero with an increasing sample size N.

7.5.4 Simulations

Experimental testing of the proposed method is carried out in the following way. First, we consider simulated examples.

1) Under H_0, the following stationary random sequence is observed:

$$e_{i+1} = \rho e_i + \sigma\sqrt{1-\rho^2}u_{i+1}, \quad i = 1, \ldots, N-1,$$

where u_i is the standard white Gaussian noise: $Eu_i = 0, Du_i = 1$.

2) Under H_1, we consider here the case of only one structural change:

$$\begin{aligned} e_{i+1} &= \rho e_i + \sigma\sqrt{1-\rho^2}u_{i+1}, \quad i = 1, \ldots, N-1, \\ y_i &= e_i + hI(i \geq [\theta N]). \end{aligned}$$

3) Under H_2, we consider the following "mean-plus-noise model" in state-space form:

$$\begin{aligned} y_i &= \mu_i + \epsilon_i, \\ \mu_i &= \mu_{i-1} + v_i, \end{aligned}$$

where

$$v_i = \begin{cases} 0, & \text{w.p. } 1 - p_N \\ e_i, & \text{w.p. } p_N \end{cases}, \text{ and}$$

where $\epsilon_i \sim N(o, \sigma_\epsilon^2)$, which will display the same behavior of variances of partial sums when $p_N \sim O(N^{2d-2})$, $0 < d < 1$ (see Diebold and Inoue, 2001).

After definition of the model of observations, the decision threshold is computed. For this purpose, the following formula from Brodsky and Darkhovsky (2000) is used:

$$C(n) = \sigma * \sqrt{\frac{1+\rho}{1-\rho}} 1.3184/\sqrt{n},$$

where n here is the sample size.

Remark that the threshold in this formula is adjusted to the value of 1st-type error $\alpha = 0.05$. Then, we begin to test hypotheses H_0, H_1, and H_2. The following output values were computed in experiments (averages in 1000 independent trials):

β — the estimate of the 2nd-type error; and

δ_{12} for H_1 and δ_{21} for H_2 — the estimates of classification errors.

TABLE 7.11
Results of Hypothesis Testing: The Hypothesis H_1 of a Structural Change

N		50	100	200	300	500	700
$h = 1.0$							
$\rho = 0.3$	β	0.511	0.207	0.034	0.03	0	0
$\sigma = 1.0$							
$\theta = 0.3$	δ_{12}	0	0	0.002	0.002	0.002	0
$h = 0.5$							
$\rho = 0.3$	β	0.8	0.636	0.415	0.270	0.108	0
$\sigma = 1.0$							
$\theta = 0.3$	δ_{12}	0	0	0.001	0.001	0.002	0
$h = 1.0$							
$\rho = 0.7$	β	0.886	0.73	0.418	0.22	0.062	0.04
$\sigma = 1.0$							
$\theta = 0.3$	δ_{12}	0	0	0	0	0	0
$h = 1.0$							
$\rho = 0.3$	β	0.054	0.005	0.014	0	0	0
$\sigma = 0.5$							
$\theta = 0.3$	δ_{12}	0.033	0.076	0.112	0.04	0	0
$h = 1.0$							
$\rho = 0.3$	β	0.037	0.001	0	0	0	0
$\sigma = 0.5$							
$\theta = 0.5$	δ_{12}	0.047	0.066	0.111	0.003	0	0

TABLE 7.12
Results of Hypothesis Testing; The Hypothesis H_2 of Fractional Integration

N		50	100	200	300	500	700
$\rho = 0.3$	β	0.027	0.001	0	0	0	0
$\sigma = 0.5$							
$d = 0.05$	δ_{21}	0.933	0.636	0.190	0.089	0.072	0.009
$\rho = 0.7$	β	0.164	0.012	0	0	0	0
$\sigma = 1.0$							
$d = 0.05$	δ_{21}	0.996	0.917	0.651	0.318	0.168	0.052
$\rho = 0$	β	0.26	0.124	0.064	0.072	0.038	0.044
$\sigma = 1.0$							
$d = 0.05$	δ_{21}	0.843	0.414	0.142	0.025	0.009	0
$\rho = 0.3$	β	0	0	0	0	0	0
$\sigma = 0.5$							
$d = 0.05$	δ_{21}	0.581	0.20	0.034	0.005	0	0
$\rho = 0.3$	β	0.083	0.001	0	0	0	0
$\sigma = 1.0$							
$d = 0.7$	δ_{21}	0.9	0.587	0.192	0.060	0.012	0

7.5.5 Applications

For practical testing the above-described method, the following daily exchange rate indices were taken: Dow Jones Index (period 01.02.1990–01.02.2015), NASDAQ index (period 01.02.1990–01.02.2015), S&P 500 index (05.05.2003–01.02.2015). The sample size of data for the first and the second index was equal to $N = 6510$, and for the third sample, $N = 5630$. Three hypotheses were tested: the stationarity hypothesis H_0, the structural change hypothesis H_1, and the fractional integration hypothesis H_2.

For testing of these hypotheses, the above-described method was used. For all three samples, the hypothesis H_1 was accepted. For the first sample, the following change-points were detected: 1255, 2544, and 4194. For the second sample these change-points were detected: 2575 and 5376. For the third sample they were: 777, 1761, and 4717.

7.6 Conclusions

In this chapter, various applications of the above proposed retrospective methods for detection of change-points in financial models were considered. First, I consider GARCH models with one or several change-points. Main Statistic (2.6) is applied for detection of structural changes in volatility of financial time series. Remark that this statistic originally considered and analyzed in our papers of 1979–1980s (see, e.g., Darkhovsky and Brodsky, 1979, 1980) is nowadays called the CUSUM statistic. Besides computer simulation of this test, I consider two practical applications of it to detect change-points in NYSE and NASDAQ time series.

In the following subsection, I consider methods of detection of structural changes in SV models. Stochastic volatility models in finance are closely related to state-space models with non-Gaussian noises. Therefore, for detection of structural changes in them, the method considered in the previous chapter can be applied. I demonstrate that the main statistic of this method substantially exceeds Statistic (2.6) by the quality of change-point detection. Results of Monte-Carlo experiments with the proposed method are given.

In Section 7.4, I consider methods for detection of structural changes in copula models. The definitions of copula models are given and actuality of detection of structural changes in them are emphasized. The problem of detection of structural changes in copula models is formulated, and the main characteristics of the proposed method of detection are analyzed in Theorems 7.4.1 and 7.4.2. Then, I consider Clayton and Gumbel copula models and study the probability of the 1st- and the 2nd-type errors for them. The main conclusion about confirmed efficiency of the proposed method for detection of structural changes in copula models is underlined.

Finally, I consider the problem of discrimination between structural change and fractional integration hypothesis. Here, I must mention the paper by H. Hurst (1951), where the problem of fractional integration in real data was first posed. The method of R/S analysis proposed by Hurst can help in detection of fractional integration in real data (in particular, Hurst applied it to Nile River inflows data). Diebold and Inoue (2001), as well as Mikosch and Starica (2004), stressed that undetected structural changes in financial data can seriously distort the factual image of a financial data series. We then are strongly inclined to accept the fractional integration hypothesis and to reject the stationarity hypothesis. This conclusion was the main motivation of this research in which the method was proposed for discrimination between structural change and fractional integration hypothesis. In Theorems 7.5.1 and 7.5.2, I study statistical characteristics of our method, which strongly resembles Hurst's R/S approach and Kolmogorov's statistic for detection of discrepances between two d.f.'s. In Section 7.6, simulated examples are considered and experiments with the proposed method are carried out for these simulated random samples (which include both stationary and fractionally integrated time series). Then, I consider real applications of the proposed method for determination of the type of financial data in NYSE and NASDAQ time series.

Part II

Sequential Problems

In Part II, sequential problems of change-point detection are considered. In sequential problems, we collect observations online, i.e., in a sequential way and are obliged to test hypotheses and to make decisions about the presence of a change-point *at every step of data collection.* Therefore, in sequential problems, the notions of a false decision about a change-point (type-1 error) and a false tranquility (type-2 error), as well as the delay time in change-point detection naturally appear.

In Chapter 8, I consider problems of sequential hypotheses testing. Here, I present new prior inequalities in sequential problems of hypotheses testing and asymptotically optimal one-sided and multisided tests. Results of computer simulation of sequential tests are given.

In Chapter 9, sequential problems of change-point detection for univariate models are considered. First, I analyze performance characteristics of well-known parametric and nonparametric sequential tests: **CUSUM** test, Girshick-Rubin-Shiryaev (**GRSh**) test, the test of exponential smoothing, the moving sample test. Prior lower bounds for performance efficiency of sequential tests are proved, including the a priori lower estimate for the delay time and the Rao-Cramer type inequality for sequential change-point detection methods. Then, I give results of the asymptotic comparative analysis and asymptotic optimality of sequential change-point detection methods.

In Chapter 10, sequential change-point detection problems in multivariate linear systems are considered. I analyze fully observed systems and consider a multivariate model in which coefficients of equations can change unexpectedly. I propose a new method for change-point detection in such a system and analyze its statistical characteristics: type-1 and type-2 error probabilities, and the normalized delay time in change-point detection. Then, I prove the prior informational lower bound for performance efficiency in change-point detection problems for multivariate systems and demonstrate that the proposed method is asymptotically optimal by the order of the normalized delay time in detection. Results of computer experiments with the proposed method for different types of multivariate systems are given.

In Chapter 11, problems of early change-point detection are considered. The main difference of early detection from sequential detection consists in the fact that in early detection problems, we consider gradual changes in statistical characteristics of observations, while in classic sequential detection problems, only abrupt changes are analyzed. I propose methods of detection and analyze their statistical characteristics, including type-1 and type-2 error probabilities and the normalized delay time in detection. Then, I prove that the proposed methods are asymptotically optimal, i.e., the lower informational bounds for performance efficiency are attained for them. Results of experimental testing of proposed methods are given.

In Chapter 12, I consider sequential detection of switches in models with changing structures. These problems include decomposition of mixtures of probabilistic distributions (sequential context), classification of multivariate observations, and classifications of observations in regression models with randomly changing coefficients of regression equations. I propose nonparametric methods for detection of switches and analyze their statistical characteristics (type-1 and type-2 error probabilities, the normalized delay time in detection). Then I prove the a priori informational lower bound for sequential problems of detection of switches and demonstrate that the proposed method is asymptotically optimal by the order of detection.

In Chapter 13, problems of sequential detection and estimation of change-points are considered. The sense of this problem is as follows: After sequential detection of a change-point, we need to estimate the true point of a change. This problem is actual, because all sequential methods have a certain delay time in change-point detection. I propose a two-stage method for solving this problem (first stage — sequential detection, second stage — retrospective estimation of this change-point using a moving window of observations) and analyze its statistical characteristics. The asymptotic optimality of this method is proved. Results of computer simulation of this method are given.

Below, I give some preliminary results from the field of sequential hypothesis testing and parameter estimation that are used in subsequent chapters.

Properties of sequential methods considered later depend on characteristics of the following processes:

$$X_N(t) = N^{-1/2} \sum_{k=1}^{[\tau_N t]} g(k/N)\xi(k),$$

where $\{\xi(k)\}$, $E\xi(k) \equiv 0$ is a random sequence, N is a "large parameter" (asymptotics are found usually on condition: $N \to \infty$), $g(t)$ is a deterministic function, and τ_N is a random sequence (usually the stopping time) defined on the same probability space as $\xi(k)$.

We now formulate two groups of conditions for analyzing asymptotic properties of sequential methods.

Strong Consistency Condition

$(\mathbf{SC})_{\mathbf{s}}$

1) The sequence $\{\xi(k)\}$ satisfies the uniform Cramer condition (see Chapter 1).

2) The sequence $\{\xi(k)\}$ satisfies ψ-mixing condition.

Functional Limit Theorems Conditions

$(\mathbf{FT})_{\mathbf{s}}$

1) There exists the limit

$$\sigma^2 = \lim_{N\to\infty} N^{-1}\mathbf{E}(\sum_{k=1}^{N} \xi(k))^2.$$

2) The sequence $\{\xi(k)\}$ satisfies the uniform Cramer and ψ-mixing conditions.

3)

$$\sum_k \sqrt{\rho(2^k)} < \infty,$$

where $\rho(k)$ is ρ-mixing coefficient for the sequence $\{\xi(k)\}$.

Now, let us formulate the following result concerning the process $X_N(t)$.

Lemma *

Suppose the sequence τ_N/N tends in probability to a positive constant θ and $(FT)_s$ holds. Then, (weakly)

$$X_N(t) \to \sigma\sqrt{\theta} \int_0^t g(s)dW(s),$$

where σ is defined in $(FT)_s$.

Wald identity

Consider the probability space $(\Omega, \mathcal{F}, \mathbf{P})$ with the fixed family $\{\mathcal{F}_n\}$ of σ-algebras $\mathcal{F}_n$, $n \geq 0$, such that $\mathcal{F}_0 \subset \mathcal{F}_1 \subset \cdots \subset \mathcal{F}$.

A random variable $\tau = \tau(\omega)$ with values in the set $\{0, 1, \ldots, +\infty\}$ is called *the Markov time* (with respect to the system $\{\mathcal{F}_n\}$) or a random variable not depending on the future, if for any $n \geq 0$

$$\{\omega : \tau(\omega) = n\} \in \mathcal{F}_n.$$

If $\mathbf{P}(\tau < \infty) = 1$, then the Markov time τ is called the *stopping time.*

Let $\xi_1, \xi_2, \ldots$ be i.i.d.r.v.'s, $\mathbf{E}|\xi_i| < \infty$, and τ a stopping time (w.r.t. $\{\mathcal{F}_n^\xi, \mathcal{F}_n^\xi = \sigma\{\omega : \xi_1, \ldots, \xi_n\})$ with $\mathbf{E}\tau < \infty$. Then, the **Wald identity** holds true

$$\mathbf{E}(\xi_1 + \cdots + \xi_\tau) = \mathbf{E}\xi_1 \cdot \mathbf{E}\tau.$$

Now, we give a **generalization of the Wald identity**. Let $\{\xi_n\}_{n=1}^{\infty}$ be a random sequence. A random variable $\tau = \tau(\omega)$, with values in the set $\{0, 1, \ldots, +\infty\}$, is called *a Wald random variable* with respect to $\{\xi_n\}_{n=1}^{\infty}$ if for any $n \geq 1$ the event $\{\tau \leq n\}$ does not depend on σ-algebra $\mathcal{F}_{n+1}$ generated by the variables $\xi_{n+1}, \xi_{n+2}, \ldots$.

Consider $S_\tau = \sum_{n=1}^{\tau} \xi_n$ and put

$$m_i = \mathbf{E}\xi_i, \quad m_i^* = \mathbf{E}|\xi_i|, \quad A_n = \sum_{k=1}^{n} m_k.$$

Let τ be the Wald r.v., such that $\sum\limits_{n=1}^{\infty} \mathbf{P}\{\tau \geq n\} m_n^* < \infty$. Then, $\mathbf{E}S_\tau = \mathbf{E}A_\tau$.

8

Sequential Hypothesis Testing

8.1 Introduction

This chapter is devoted to the problems of sequential hypotheses testing for dependent random sequences. Each hypothesis is represented by the density function of observations that depends on a parameter from one of disjoint sets. New performance measures for sequential testing of composite hypotheses are proposed and a priori lower bounds for these measures are proved. Asymptotically optimal methods for sequential multihypothesis testing are found.

More than 60 years since A. Wald's seminal ideas in sequential testing of simple hypotheses (see Wald (1947)) have passed. Many efforts were made to generalize the sequential probability ratio test (SPRT) to different cases of composite hypotheses testing.

The classic Wald's sequential test is as follows. Let $x_1, x_2, \ldots$ be i.i.d. random variables with a common distribution $\mathbf{P}$. To test the null hypothesis $\mathbf{H}_0 : \mathbf{P} = \mathbf{P}_0$ against the alternative $\mathbf{H}_1 : \mathbf{P} = \mathbf{P}_1$, the SPRT stops sampling at stage

$$T = \inf\{n \geq 1 : \ R_n \geq A \ or \ R_n \leq B\}, \tag{8.1}$$

where $A > 1 > B > 0$ are stopping boundaries, and $R_n = \prod_{i=1}^{n} (f_1(x_i)/f_0(x_i))$ is the likelihood ratio, f_i being the d.f. of $\mathbf{P}_i$ with respect to some common dominating measure ν, $i = 0, 1$. When stopping occurs, $\mathbf{H}_0$ or $\mathbf{H}_1$ is accepted according to $R_n < B$ or $R_n > A$. The choice of A and B is dictated by the error probabilities $\alpha = \mathbf{P}_0\{R_N > A\}$ and $\beta = \mathbf{P}_1\{R_N < B\}$. This test was shown by Wald and Wolfowitz (see Wald and Wolfowitz (1948)) to be optimal in the following sense: Among all tests whose sample size T has a finite expectation under both $\mathbf{H}_0$ and $\mathbf{H}_1$ and whose error probabilities satisfy

$$\mathbf{P}_0\{\mathit{Reject}\, \mathbf{H}_0\} \leq \alpha \quad and \quad \mathbf{P}_1\{\mathit{Reject}\, \mathbf{H}_1\} \leq \beta, \tag{8.2}$$

the SPRT minimizes both $\mathbf{E}_0(T)$ and $\mathbf{E}_1(T)$.

Remark that error probabilities $\mathbf{P}_0\{\mathit{Reject}\ \mathbf{H}_0\}$ and $\mathbf{P}_1\{\mathit{Reject}\ \mathbf{H}_1\}$ depend on the stopping time T. Therefore, it is often difficult to verify Conditions (8.2) (see Kiefer and Weiss (1957), "admissibility problem").

In 1959–1960s, there were many attempts to apply the SPRT to different problems of composite hypotheses testing. One of the most well-known statements of this problem was formulated by Kiefer and Weiss (1957): to test the composite hypothesis $\mathbf{H}_0 : \theta < \theta_0$ against the alternative $\mathbf{H}_1 : \theta > \theta_1$, where θ is the parameter of the d.f. of observations. It turned out that the average sample volume in the indifference zone $\theta_0 < \theta < \theta_1$ can be much larger than corresponding sample volumes at the points θ_0 and θ_1. The problem formulated by Kiefer and Weiss consisted of the following: It is required to find a sequential test that minimizes the average sample size $\mathbf{E}_{\theta^*} T$ for any given parameter $\theta = \theta^*$ of the d.f. $f_\theta(x)$ subject to error probability constraints at the fixed points $\theta = \theta_0$ and $\theta = \theta_1$. Asymptotic solutions of the Kiefer-Weiss problem were proposed by Lai (1973, 1988) and Dragalin and Novikov (1988).

Asymptotic behavior of sequential tests for composite hypotheses was studied on the basis of a priori theoretical bounds for the average sample sizes that were obtained by Wald and Hoeffding. Hoeffding (1960) derived the following lower bound on $\mathbf{E}_{\theta^*} N$ subject to error probability constraints at θ_0 and θ_1 for any stopping time N with finite expectation:

$$\mathbf{E}_{\theta^*} N \geq \frac{\{[(\tau/4)^2 - \zeta \log(\alpha_1 + \alpha_2)]^{1/2} - \tau/4\}^2}{\zeta^2},$$

where

$$\begin{aligned} \zeta &= \max(\zeta_0, \zeta_1), \quad \zeta_i = \int f_{\theta^*} \ln \frac{f_{\theta^*}}{f_{\theta_i}}\, d\nu,\ i = 0, 1, \\ \tau^2 &= \int (\ln \frac{f_{\theta_1}}{f_{\theta_0}} - \zeta_0 + \zeta_1)^2 f_{\theta^*}\, d\nu, \end{aligned} \tag{8.3}$$

and $\alpha_0 \geq \mathbf{P}_0\{reject\ \mathbf{H}_0\}$, $\alpha_1 \geq \mathbf{P}_1\{reject\ \mathbf{H}_1\}$.

Lorden (1976) showed that an asymptotic solution to the Kiefer-Weiss problem is a 2-SPRT with stopping rule of the form

$$\hat{N} = \inf\{n : \prod_{i=1}^{n} \frac{f_{\theta^*}(x_i)}{f_{\theta_0}(x_i)} \geq A_0 \ or \ \prod_{i=1}^{n} \frac{f_{\theta^*}(x_i)}{f_{\theta_1}(x_i)} \geq A_1\},$$

while $\max(A_0, A_1) \to \infty$.

Lorden also showed numerically that $\mathbf{E}_{\theta^*} \hat{N}$ is close to Hoeffding's lower bound.

In all these statements of the problem of composite hypotheses testing, two reference points θ_0 and θ_1 for error probability constraints are fixed.

Bartlett (1946) and Cox (1963) considered sequential testing of composite hypotheses parametrized by the two-dimensional vector (θ, η) for i.i.d.r.v's with the density function from the exponential family. Here, η is the nuisance parameter, and the hypotheses are of the following form: $\mathbf{H}_0 : \theta = \theta_0, \eta \in Q; \mathbf{H}_1 : \theta = \theta_1, \eta \in Q$, where Q is a certain set. The likelihood ratio statistics were proposed in which the maximum likelihood estimate (MLE) of an unknown parameter η (Batlett, 1946) or two MLEs corresponding to parameters θ_0 and θ_1 (Cox, 1963) were inserted (evidently, we obtain the classical Wald's test for simple hypotheses if there is no nuisance parameter). The asymptotic properties of these statistics and their numerical characteristics were studied by Joanes (1972). Gombay (1999) (see also Cox and Roseberry, 1966) demonstrated that optimal properties of Cox and Bartlett statistics (analogous to

Wald's test) *are not preserved* without the contiguity assumption for parameters θ_0 and θ_1.

Dragalin and Novikov (1999) considered sequential testing of multiple composite hypotheses in the presence of an indifference zone for an unknown parameter θ. Two constructions of sequential tests satisfying given upper bounds on probabilities of errors were presented. These tests generalize earlier results of Lorden (1976) and Pavlov (1988). It was shown that under certain general conditions, the proposed tests have the first-order asymptotic optimality (for the average time of observations) for all values of θ, including the indifference zone when probabilities of errors are close to zero. The results obtained by Dragalin and Novikov are based upon the assumption of the *exponential family* of the distribution function of observations both in discrete and continuous time.

Dragalin, Tartakovsky, and Veeravalli (1999, 2000) considered asymptotic optimality of multihypothesis sequential probability ratio tests for simple hypotheses and i.i.d. observations, but invariant tests with nuisanse parameters and mixture based tests for composite hypotheses were also covered. The non-i.i.d. case was considered by Tartakovsky (1998) and Tartakovsky, Li and Yaralov (2003) for the problem of sequential targets detection.

Lai (1995, 2000) proposed asymptotic lower bounds for the average sample size in sequential multihypothesis testing problems for dependent sequences that coincide with asymptotic lower bounds found earlier by Simons (1967) in independent cases when the error probabilities tend to zero. These lower bounds were then used for the analysis of the generalized likelihood ratio schemes in sequential multihypothesis testing and change-point detection problems.

In this chapter, I generalize and modify the results of Pavlov and Dragalin and Novikov and Lai. First, I propose *new performance measures* for sequential tests and the *nonasymptotical* a priori lower bounds for these measures. Second, these results hold true *without the exponential family assumption* for the distribution function of observations. Third, I consider both one-sided and multisided sequential tests for composite hypotheses.

The idea of new criteria can be clearly explained for the case of two composite hypotheses. Let us introduce necessary notations. Suppose $\mathcal{P}_0 = \{\mathbf{P}_{\theta_0}\}_{\theta_0 \in \Theta_0}$, $\mathcal{P}_1 = \{\mathbf{P}_{\theta_1}\}_{\theta_1 \in \Theta_1}$, and $\Theta_0 \bigcap \Theta_1 = \emptyset$ are two parametric families of d.f.'s defined by their densities $f(\theta_0, x)$ and $f(\theta_1, x)$ and corresponding to hypotheses $\mathbf{H}_0$ and $\mathbf{H}_1$. Let $d_c(n)$ be the decision function (i.e., a certain measurable function of observations) of a one-sided sequential procedure for testing the hypothesis $\mathbf{H}_1$ (here and below, c is the "large" parameter). The value $d_c(n) = 1$ corresponds to the decision to stop at the instant n and accept the hypothesis $\mathbf{H}_1$. The value $d_c(n) = -1$ corresponds to the decision to continue observations. For an arbitrary point $\theta_1 \in \Theta_1$, it is possible to find the "least favorable alternative" $\theta_0(\theta_1) \stackrel{\text{def}}{=} \theta_0(\cdot)$. This least favorable alternative for the value θ_1 corresponds to the point $\theta_0(\cdot)$ from the set Θ_0 for which the Kullback-Leibler distance between measures $\mathbf{P}_{\theta_1}$ and $\mathbf{P}_{\theta_0(\cdot)}$ is minimal (precise definitions are given below).

Let

$$\gamma_c = \sup_n \mathbf{P}_{\theta_0(\cdot)}\{d_c(n) = 1\}, \quad \tau_c = \inf\{n : d_c(n) = 1\}.$$

In this case, the proposed performance measure for the stopping time τ_c can be written as follows:

$$\mathcal{K}(\tau_c, \theta_1) \stackrel{\text{def}}{=} \frac{\mathbf{E}_{\theta_1}\tau_c}{|\ln \gamma_c|}. \tag{8.4}$$

It turns out that this criterion has an *a priori lower bound* that does not depend on a concrete test (and depends only on the pair $(\theta_0(\cdot), \theta_1)$). Besides, the limit of this criterion as $c \to \infty$ (if it exists) has the same lower bound. Remark that the "large parameter" c is the "threshold" of a sequential testing procedure. If this threshold is exceeded, then the hypothesis $\mathbf{H}_1$ is accepted, and if $c \to \infty$, then the probability of the "false decision" tends to zero. Taking into account that the average time before a false decision $\mathbf{E}_{\theta_0(\cdot)}\tau_c$ has the order γ_c^{-1} (as $c \to \infty$, see also Darkhovsky and Brodsky, 1980), we obtain that the limit of this new criterion is the ratio of the average time before the right and false decision (properly scaled) given an arbitrary hypothesis from the family $\mathcal{P}_1$ and its most unfavorable alternative from the family $\mathcal{P}_0$.

This ratio characterizes the quality of sequential testing procedures *no worse than conventional criteria*, i.e., the average time before the right decision given error probabilities.

Obviously, we should find such a method of sequential hypothesis testing that minimizes the proposed criterion (i.e., the method that gives the minimal average time before the right decision and the maximal average time before the false decision). However, since there exists the a priori lower bound for the proposed criterion, we can try to find the asymptotically optimal (as $c \to \infty$) method of sequential testing for which this lower bound is attained. Here we stress the analogy with the Rao-Cramer inequality in estimation problems.

Analogous considerations are valid for multisided tests. Here, we can also propose a new performance measure that has the a priori lower bound and enables us to find the asymptotically optimal method of sequential testing.

Our asymptotically optimal tests for the case of a finite number of simple hypotheses are close to tests proposed earlier by Lai (2000). However, in the case of multiple composite hypotheses, our tests differ from tests proposed by Lai (2000), since they do not use the a priori probability distribution on the parametric set. It should be noted also that the result of Lai (2000) for the case of multiple composite hypotheses is based upon some unrealizable conditions.

Comparing new tests with those proposed earlier by Dragalin and Novikov (1999) for composite multihypothesis testing, I observe that these tests are much more general, since they do not use the exponential family assumption.

In this chapter, I pursue the following main goals:

1) to propose new performance measures for one-sided and multisided tests in the problem of sequential testing of composite hypotheses; and

2) to find the *asymptotically optimal* sequential tests for which the a priori lower bounds for performance measures are attained.

The structure of this chapter is as follows. In Section 8.2 all assumptions are formulated. In Section 8.3, the asymptotically optimal one-sided tests are considered for composite hypotheses. In Section 8.4, the problem of sequential composite multihypothesis testing is considered, the a priori lower bounds for new performance measures are proved, and the asymptotically optimal test is found. All proofs are given in Section 8.5. In Section 8.6, some experimental results are given.

The results of the chapter hold true for the non-i.i.d. case (dependent random sequences with mixing conditions) and can be generalized for the nonparametric case. We do not stay at these points here.

8.2 Assumptions

Let $I \stackrel{\text{def}}{=} \{0, 1, \ldots, k-1\}$, $\theta = (\theta_0, \theta_1, \ldots, \theta_{k-1})$, $k \geq 2$, $\theta \in \Theta$, and $\theta_i \in \Theta_i$, where Θ is a certain parametric set that belongs to some open set U in the finite dimensional space, $\Theta = \bigcup_i \Theta_i, \bigcap_i \Theta_i = \emptyset$. We observe a sequence of independent random vectors $\{\xi_k\}_{k=1}^{\infty}$, with the d.f. w.r.t. some σ-finite measure μ equal to $f(\theta_i, x), \theta_i \in \Theta_i$ if hypothesis $\mathbf{H}_i$ holds true. The d.f. f is known and defined for all parameter values from U. In what follows, we denote by $\mathbf{P}_{\theta_i}(\mathbf{E}_{\theta_i})$ the measure (mathematical expectation) corresponding to a sequence $\{\xi_k\}_{k=1}^{\infty}$ with the fixed value of the parameter θ_i.

Everywhere below, we assume that the following conditions are satisfied:

8.2.1. Θ is a compact set.

8.2.2. $\mu\{x : f(a_1, x) \neq f(a_2, x)\} > 0$ if $a_1 \neq a_2$.

8.2.3. For μ-a.s. x the functions $f(\theta_i, x)$, and $i \in I$ are continuous and positive with respect to $\theta_i \in \Theta_i$.

Everywhere below for any fixed $\theta_i \in \Theta_i$, $i \in I$ denote $\mathcal{U}_i \stackrel{\text{def}}{=} \Theta \setminus \Theta_i$, $\mathcal{Z}_i \stackrel{\text{def}}{=} \Theta_i \times \mathcal{U}_i = \{z_i\}$.

8.2.4. The functions $\left(\mathbf{E}_{\theta_i} \ln \frac{f(\theta_i,\cdot)}{f(u,\cdot)}\right)^{-1}$ are continuous with respect to $u \in \mathcal{U}_i$ for any $\theta_i \in \Theta_i$, $i \in I$.

8.2.5. For any $i \in I$.

$$\infty > \sup_{\theta_i^* \in \Theta_i} \sup_{\theta_i \in \Theta_i} \sup_{u \in \mathcal{U}_i} \mathbf{E}_{\theta_i^*} \ln \frac{f(\theta_i,\cdot)}{f(u,\cdot)}, \qquad -\infty < \inf_{\theta_i^* \in \Theta_i} \inf_{\theta_i \in \Theta_i} \inf_{u \in \mathcal{U}_i} \mathbf{E}_{\theta_i^*} \ln \frac{f(u,\cdot)}{f(\theta_i,\cdot)}$$

$$\sup_{\theta_i^* \in \Theta_i} \sup_{\theta_i \in \Theta_i} \sup_{u \in \mathcal{U}_i} \mathbf{E}_{\theta_i^*} \left(\ln \frac{f(\theta_i,\cdot)}{f(u,\cdot)}\right)^2 < \infty,$$

$$\sup_{\theta_i^* \in \Theta_i} \sup_{\theta_i \in \Theta_i} \sup_{u \in \mathcal{U}_i} \mathbf{E}_{\theta_i^*} \left(\ln \frac{f(u,\cdot)}{f(\theta_i,\cdot)}\right)^2 < \infty.$$

8.2.6. For any $z_i \in \mathcal{Z}_i$, $u^* \in \mathcal{U}_i$, $\theta_i^* \in \Theta_i$, $\mathbf{P}_{\theta_i^*}$ and $\mathbf{P}_{u^*}$- distributions of the random variable $\eta(\omega, z_i) = \ln \dfrac{f(u,\cdot)}{f(\theta_i,\cdot)}$ satisfies the uniform Cramer condition:

$$\sup_{\theta_i^* \in \Theta_i} \sup_{z_i \in \mathcal{Z}_i} \mathbf{E}_{\theta_i^*} \exp\{t\eta(z_i)\} < \infty, \quad \sup_{\theta_i^* \in \Theta_i} \sup_{z_i \in \mathcal{Z}_i} \mathbf{E}_{u_i^*} \exp\{t\eta(z_i)\} < \infty \quad \text{for} \quad |t| < H$$

8.2.7. For any $z_i \in \mathcal{Z}_i$, $u^* \in \mathcal{U}_i$, the function

$$\varkappa_i(t, z_i, u^*) \stackrel{\text{def}}{=} \ln \int \left(\frac{f(\theta_i, x)}{f(u, x)} \right)^t f(u^*, x) \mu(dx)$$

has only two zeros: 0 and $t_i^*(z_i, u^*) > 0$, the function $t_i^*(\cdot, u^*)$ is continuous for any $u^* \in \mathcal{U}_i$, and $\min_{z_i \in \mathcal{Z}_i} t_i^*(z_i, u^*) > 0$.

Remark. Compactness and continuity assumptions can be relaxed, but for the sake of simplicity of exposition, we do not do it here.

8.3 One-Sided Tests: Main Results (I)

A one-sided sequential procedure for testing the hypothesis $\mathbf{H}_i$, $i \in I$ represents a choice from two possible decisions: to stop and accept the hypothesis $\mathbf{H}_i$ or to continue observations.

Let c be a "large" parameter (we study asymptotic optimality of sequential tests as $c \to \infty$). Suppose $d_c(n)$ is the decision function of a one-sided sequential procedure for testing the hypothesis $\mathbf{H}_i$, i.e., a measurable function of observations $(x_1, x_2, \cdots, x_n)$ that equals to i if the observations are stopped and hypothesis $\mathbf{H}_i$ is accepted, and equals to (-1) if we continue to collect the observations.

Let

$$\tau_c^{(i)} \stackrel{\text{def}}{=} \inf\{n : d_c(n) = i\}, \ i \in I$$

be stopping times corresponding to one-sided tests (here and below, we assume $\inf(\emptyset) = +\infty$). Let us introduce the following classes of such stopping times (and corresponding one-sided tests)

$$\mathcal{M}_i \stackrel{\text{def}}{=} \{\tau_c^{(i)} : \sup_{\theta_i \in \Theta_i} \mathbf{E}_{\theta_i} \tau_c^{(i)} < \infty\}.$$

Consider one-sided tests generated by stopping times from the set $\mathcal{M}_i$. Suppose θ_i is an arbitrary point from the set Θ_i. Starting from the point θ_i, we define the point $u_i(\theta_i) \stackrel{\text{def}}{=} u_i(\cdot)$ as the argument of maximum of the following criterion

$$\max_{u \in \Theta \setminus \Theta_i} \left(\mathbf{E}_{\theta_i} \ln \frac{f(\theta_i, x)}{f(u, x)} \right)^{-1} \stackrel{\text{def}}{=} \mathbf{I}_i^{-1}(\theta_i). \tag{8.5}$$

If the set of maximum points in this criterion consists of more than one point, then we choose an arbitrary point from this set as $u_i(\cdot)$. (We will see below that this choice does not influence the optimal rule.)

If the point θ_i is the true parameter of the d.f. under hypothesis $\mathbf{H}_i$, then the point $u_i(\cdot)$ corresponds to the least favorable alternative for this case: the Kullback-Leibler distance dist $\left(\mathbf{P}_{\theta_i}, \mathbf{P}_{u_i(\cdot)}\right)$ between distributions $\mathbf{P}_{\theta_i}$, $\mathbf{P}_{u_i(\cdot)}$ is minimal in virtue of the definition of $u_i(\cdot)$. Define

$$\gamma_i(c) = \sup_n \mathbf{P}_{u_i(\cdot)}\{d_c(n) = i\}.$$

The value $\gamma_i(c)$ is correctly defined for an *arbitrary* "true" parameter θ_i of the d.f. and can be interpreted as the *maximal probability of the false decision* under the least favorable alternative.

We call a one-sided test d_c *nondegenerate* if $\gamma_i(c) < 1$ for all $i \in I$.

8.3.1 Main Inequalities

Theorem 8.3.1. *For any stopping time $\tau_c^{(i)} \in \mathcal{M}_i, i \in I$ corresponding to a one-sided nondegenerate test d_c and any point $\theta_i \in \Theta_i$, the following inequality holds:*

$$\mathcal{K}_i(\tau_c^{(i)}, \theta_i) \stackrel{def}{=} \frac{\mathbf{E}_{\theta_i}\tau_c^{(i)}}{|\ln \gamma_i(c)|} \geq \mathbf{I}_i^{-1}(\theta_i). \tag{8.6}$$

Under $\gamma_i(c) \to 0$, the value $\mathcal{K}_i(\tau_c^{(i)}, \theta_i)$ can be interpreted as the ratio between the average time before the right decision under the distribution $\mathbf{P}_{\theta_i}$ from the family $\mathcal{P}_i$, to the logarithm of the average time before the false decision for the least favorable alternative from $\mathcal{U}_i$, to the parameter $\theta_i \in \Theta_i$ of the d.f. In fact, it can be demonstrated (see Brodsky and Darkhovshy, 2000) that $\ln \mathbf{E}_{u_i(\cdot)}\tau_c^{(i)} \sim |\ln \gamma_i(c)|$ as $\gamma_i(c) \to 0$ (the "large" parameter c is usually chosen so that $\gamma_i(c) \to 0$ as $c \to \infty$).

Therefore, $\mathcal{K}_i(\tau_c^{(i)}, \theta_i)$ can be used as *performance measures for one-sided tests.* In our opinion, these criteria are *no worse* than conventional criteria in sequential testing problems. Inequalities (8.3.2) provide the a priori lower bounds for these criteria. These lower bounds *do not depend on the concrete decision rule* but only on the pair $(\theta_i, u_i(\cdot))$, corresponding to the "true" parametric point and its least favorable alternative.

Thus, for new criteria, the situation analogous to the classic Rao-Cramer inequality holds true: We can formulate the problem of accessibility of the a priori lower bound and find asymptotically optimal methods for which this lower bound is attained.

From Theorem 8.3.1, it follows that for any stopping times $\tau_c^{(i)} \in \mathcal{M}_i$, $i \in I$ and any (above-mentioned) pairs $(\theta_i, u_i(\cdot)) \in \Theta_i \times \mathcal{U}_i$, the existence of the limits $\lim_{c\to\infty} \mathcal{K}_i(\tau_c^{(i)}, \theta_i)$ yields the following inequalities:

$$\lim_{c\to\infty} \mathcal{K}_i(\tau_c^{(i)}, \theta_i) \geq \mathbf{I}_i^{-1}(\theta_i) \tag{8.7}$$

One-sided tests $\tau_c^{(i)}$ for which the equality sign in Inequality (8.7) is attained are called *adaptive asymptotically optimal.* The term "adaptive" in this context means that the property of asymptotic optimality of a test does not depend on the true (unknown to the statistic) d.f. of observations from families $\mathcal{P}_i$.

8.3.2 Asymptotically Optimal One-Sided Tests

Suppose $a > 0$ is an arbitrary number. Fix some $i \in I$ and denote by $\Theta(a) \subset \Theta$, $\Theta(a) = (\Theta_i(a), \mathcal{U}_i(a))$ a *finite* $1/a$-network in Θ, $\#\Theta(a) = R(a)$.

For testing the hypothesis $\mathbf{H}_i$, define the stopping time

$$T_i(c_i, a) \stackrel{\text{def}}{=} \inf \left\{ n : \mathcal{L}_i(n, a) \stackrel{\text{def}}{=} \min_{u \in \mathcal{U}_i(a)} \max_{\theta_i \in \Theta_i(a)} \sum_{k=1}^{n} \ln \frac{f(\theta_i, x_k)}{f(u, x_k)} > c_i \right\}, \quad (8.8)$$

and the corresponding decision rule

$$\tilde{d}_{c_i}^{(i)}(n, a) \stackrel{\text{def}}{=} \begin{cases} i & \text{if } \mathcal{L}_i(n, a) > c_i \\ -1 & \text{if } \mathcal{L}_i(n, a) \le c_i \end{cases} . \quad (8.9)$$

Remark that in (8.8), we can interchange operations max and min.

Put

$$\gamma_i^a(c_i) \stackrel{\text{def}}{=} \sup_n \mathbf{P}_{u_i(\cdot)} \left\{ \tilde{d}_{c_i}^{(i)}(n, a) = 0 \right\}$$

(recall that $u_i(\theta_i) = u_i(\cdot)$ is the least favorable point and corresponds to point θ_i).

Theorem 8.3.2. *Let θ_i be any point in Θ_i. Then, for every $\epsilon > 0$, there exist $a(\epsilon)$, $\rho(\epsilon)$ $(a(\epsilon) \uparrow \infty$, $\rho(\epsilon) \downarrow 0$ as $\epsilon \to 0)$, such that one-sided test $\tilde{d}_{c_i}^{(i)}(n, a(\epsilon)) \stackrel{def}{=} \tilde{d}_{c_i}^{(i)}(n, \epsilon)$, corresponding to the stopping time $T_i(c_i, a(\epsilon)) \stackrel{def}{=} T_i(c_i, \epsilon)$ and the false decision probability $\gamma_i^{a(\epsilon)}(c_i) \stackrel{def}{=} \gamma_i^{\epsilon}(c_i)$, satisfies the relationships*

i)

$$\mathbf{I}_i^{-1}(\theta_i) + \rho(\epsilon) \ge \lim_{c \to \infty} \mathcal{K}_i\left(T_i(c_i, \epsilon), \theta_i\right) \ge \mathbf{I}_i^{-1}(\theta_i) \text{ and}$$

ii)

$$\liminf_{c_i \to \infty} \frac{|\ln \gamma_i^{\epsilon}(c_i)|}{c_i} \ge 1 - \epsilon. \quad (8.10)$$

Therefore, one-sided tests corresponding to the stopping times

$$T_i(c_i) \stackrel{\text{def}}{=} \inf \left\{ n : \min_{u \in \mathcal{U}_i} \max_{\theta_i \in \Theta_i} \sum_{k=1}^{n} \ln \frac{f(\theta_i, x_k)}{f(u, x_k)} > c_i \right\}$$

are *adaptive asymptotically optimal*, and the false decision probabilities tend to zero exponentially as the "large parameter" c_i tends to infinity.

8.4 Multi-Sided Tests: Main Results (II)

Suppose $c = (c_0, c_1, \ldots, c_{(k-1)})$ is the vector-valued "large" parameter. A sequential procedure for testing k hypotheses is a choice from $(k+1)$ alternatives: a) to stop observations and accept the hypothesis $\mathbf{H}_i, i \in I$ and b) to continue observations. Therefore, the decision rule $d_c(n)$ depending on the parameter c has the following form:

$$d_c(n) \stackrel{\text{def}}{=} \begin{cases} i, \; i = 0, 1, \ldots, k-1, & \text{stop and accept } \mathbf{H}_i \text{ at step } n \\ -1 & \text{continue observations.} \end{cases} \tag{8.11}$$

Suppose $\theta_i \in \Theta_i, i \in I$ is an arbitrary point. Starting from θ_i, let us define the point $u_i(\theta_i) \stackrel{\text{def}}{=} u_i(\cdot) \in \mathcal{U}_i \stackrel{\text{def}}{=} \Theta \setminus \Theta_i$ as in (8.3.1). Therefore, if $\mathbf{P}_{\theta_i}$ is the main hypothesis, then its least favorable alternative in the set $\mathcal{U}_i$ is the hypothesis $\mathbf{P}_{u_i(\cdot)}$.

For any $\mathbf{H}_i, i \in I$ it is possible to define probabilities of false decisions in analogy with the one-sided tests:

$$\alpha_{ij}(c) \stackrel{\text{def}}{=} \sup_n \mathbf{P}_{\theta_i}\{d_c(n) = j, j \neq i\}, \quad \beta_{ij}(c) \stackrel{\text{def}}{=} \inf_n \mathbf{P}_{\theta_i}\{d_c(n) = j, j \neq i\}$$
$$\gamma_i(c) \stackrel{\text{def}}{=} \sup_n \mathbf{P}_{u_i(\cdot)}\{d_c(n) = i\}, \quad \rho_i(c) \stackrel{\text{def}}{=} \inf_n \mathbf{P}_{u_i(\cdot)}\{d_c(n) = i\}.$$

We say that the test d_c is *nondegenerate* if

$$1 > \sum_{j, j \neq i} \alpha_{ij}(c) \; \forall i, \quad 1 > \sum_i \gamma_i(c).$$

Let

$$\nu_c \stackrel{\text{def}}{=} \inf\{n : d_c(n) \neq -1\}$$

be a stopping time corresponding to a sequential test. We consider such stopping times (and corresponding tests) that belong to class $\mathcal{M} \stackrel{\text{def}}{=} \{\nu_c : \sup_{\theta_i \in \Theta_i} \mathbf{E}_{\theta_i}\nu_c < \infty, \forall i \in I\}$.

For each hypothesis $\mathbf{H}_i, i \in I$, we can introduce the performance measures in analogy with the one-sided case:

$$\mathcal{K}_i(\nu_c, \theta_i) \stackrel{\text{def}}{=} \frac{\mathbf{E}_i \nu_c}{|\ln \gamma_i(c)|}.$$

Theorem 8.4.1. *For any stopping time $\nu_c \in \mathcal{M}$ corresponding to a nonde-*

generate test and any point $\theta_i \in \Theta_i, i \in I$, the following inequality holds

$$\mathcal{K}_i(\nu_c, \theta_i) \geq \mathbf{I}_i^{-1}(\theta_i) \left\{ (1 - \sum_{j,j\neq i} \alpha_{ij}(c)) + \right.$$
$$\left. (1 - \sum_{j,j\neq i} \alpha_{ij}(c)) \frac{\ln \left(1 - \sum_{j,j\neq i} \alpha_{ij}(c) \right)}{|\ln \gamma_i(c)|} + \frac{\sum_{j,j\neq i} \beta_{ij}(c)}{|\ln \gamma_i(c)|} \ln \frac{\sum_{j,j\neq i} \beta_{ij}(c)}{(1-\rho_i(c))} \right\}. \tag{8.12}$$

The interpretation of the values $\mathcal{K}_i(\nu_c, \theta_i), i \in I$ is the same as before.

Thus, the vector $\left(\mathcal{K}_0(\nu_c, \theta_0), \mathcal{K}_1(\nu_c, \theta_1), \dots, \mathcal{K}_{(k-1)}(\nu_c, \theta_{k-1})\right)$ can be used as the *vector-valued performance measure* for the stopping time ν_c (and the corresponding multisided test). In our opinion, it is no worse than commonly used performance measures in sequential testing problems.

From Theorem 8.4.1, it follows that for any points $\theta_i \in \Theta_i$ and nondegenerate multisided tests, the existence of limits $\lim_{c\to\infty} \mathcal{K}_i(\nu_c, \theta_i)$ yields (on condition that the false decision probabilities tend to zero) inequalities

$$\lim_{c\to\infty} \mathcal{K}_i(\nu_c, \theta_i) \geq \mathbf{I}_i^{-1}(\theta_i). \tag{8.13}$$

Multisided tests ν_c for which equality signs are attained in Inequalities (8.13) (i.e., the vector-valued performance measure attains its lower bound by k components, that means the ultimate high quality of composite hypotheses testing) are called *adaptive asymptotically optimal.* The term "adaptive" in this context (as before) means that the property of the asymptotic optimality of a test does not depend on the true (unknown to the statistic) d.f. of observations from the families $\mathcal{P}_{\theta_i}$.

8.4.1 Asymptotically Optimal Multisided Tests

Define the stopping time

$$T(c,a) \stackrel{\text{def}}{=} T_0(c_0,a) \bigwedge T_1(c_1,a) \bigwedge T_2(c_2,a) \bigwedge \cdots \bigwedge T_{(k-1)}(c_{(k-1)},a),$$

where $T_i(c_i, a)$ are defined in (8.8).

Let $I\left(T(c,a)\right) \stackrel{\text{def}}{=} \{i \in I : T_i(c_i,a) = T(c,a)\}$.

The corresponding decision rule can be written as follows:

$$d_c^*(n,a) = \begin{cases} i & \text{if } \; n = T(c,a) = T_i(c_i,a) < T_j(c_j,a), j \neq i, \;\; i,j \in I \\ \text{any } j & \text{if } \; n = T(c,a), \; \#I\left(T(c,a)\right) > 1, j \in I\left(T(c,a)\right) \\ -1 & \text{if } \;\; T(c,a) > n \end{cases}$$

Define also the false decision probabilities for the rule $d_c^*(n,a)$:

$$\alpha_{ij}^*(c,a) \stackrel{\text{def}}{=} \sup_n \mathbf{P}_{\theta_i}\{d_c^*(n,a)=j, j\neq i\}, \quad \beta_{ij}^*(c,a) \stackrel{\text{def}}{=} \inf_n \mathbf{P}_{\theta_i}\{d_c^*(n,a) = j, j\neq i\}$$

$$\gamma_i^*(c,a) \stackrel{\text{def}}{=} \sup_n \mathbf{P}_{u_i(\cdot)}\{d_c^*(n,a)=i\}, \quad \rho_i^*(c,a) \stackrel{\text{def}}{=} \inf_n \mathbf{P}_{u_i(\cdot)}\{d_c^*(n,a)=i\}.$$

Theorem 8.4.2. *Let θ_i be any point in Θ_i. Then, for any $\epsilon>0$, there exist $a(\epsilon)$, $r(\epsilon)$ ($a(\epsilon)\uparrow\infty$, $r(\epsilon)\downarrow 0$ as $\epsilon\to 0$), such that the multisided test $d_c^*(n,a(\epsilon)) \stackrel{def}{=} d_c^*(n,\epsilon)$ generated by the stopping time $T(c,a(\epsilon)) \stackrel{def}{=} T(c,\epsilon)$ and cross-error probabilities $\alpha_{ij}^*(c,a(\epsilon)) \stackrel{def}{=} \alpha_{ij}^*(c,\epsilon), \gamma_i^*(c,a(\epsilon)) \stackrel{def}{=} \gamma_i^*(c,\epsilon)$ satisfies the following relationships:*

a)

$$\mathbf{I}_i^{-1}(\theta_i)+r(\epsilon) \geq \lim_{c\to\infty} \mathcal{K}_i(T(c,\epsilon),\theta_i) \geq \mathbf{I}_i^{-1}(\theta_i) \text{ and} \tag{8.14}$$

b)

$$\liminf_{c_j\to\infty} \frac{|\ln\alpha_{ij}^*(c_j,\epsilon)|}{c_j} \geq 1-\epsilon, \quad 0\leq i\neq j\leq k-1, \quad \liminf_{c_i\to\infty} \frac{|\ln\gamma_i^*(c_i,\epsilon)|}{c_i} \geq 1-\epsilon$$

Thus, the multisided test generated by the stopping time

$$T(c) \stackrel{\text{def}}{=} T_0(c_0) \bigwedge T_1(c_1) \bigwedge T_2(c_2) \bigwedge \cdots \bigwedge T_{(k-1)}(c_{(k-1)})$$

is *adaptive asymptotically optimal* and the cross-error probabilities tend to zero exponentially as the "large parameter" c tends to infinity.

8.5 Proofs

8.5.1 One-Sided Tests

Proof of Theorem 8.3.1

Fix some $i\in I$ and $\theta_i\in\Theta_i$. Let N be an arbitrary positive integer. Then, in virtue of the definition of $\gamma_i(c)$ and $u_i(\cdot)\in\mathcal{U}_i$, we obtain from Jensen's inequality

$$\begin{aligned}
\gamma_i(c)N &\geq \textstyle\sum_{k=1}^N \mathbf{P}_{u_i(\cdot)}(\tau_c^{(i)}=k) \\
&= \textstyle\sum_{k=1}^N \mathbf{E}_{\theta_i}\left\{\mathbb{I}(\tau_c^{(i)}=k)\prod_{i=1}^k \frac{f(u_i(\cdot),x_i)}{f(\theta_i,x_i)}\right\} \\
&= \mathbf{E}_{\theta_i}\exp\left(-\sum_{i=1}^{\tau_c^{(i)}\wedge N} \ln\frac{f(\theta_i,x_i)}{f(u_i(\cdot),x_i)}\right) \\
&\geq \exp\left(-\mathbf{E}_{\theta_i}\left(\sum_{i=1}^{\tau_c^{(i)}\wedge N} \ln\frac{f(\theta_i,x_i)}{f(u_i(\cdot),x_i)}\right)\right)
\end{aligned} \tag{8.15}$$

(here and below, $\mathbb{I}(A)$ is an indicator of the set A).

Using Wald's identity, we obtain from (8.15)

$$\gamma_i(c)N \geq \exp\left(-\mathbf{E}_{\theta_i}(\tau_c^{(i)} \wedge N)\mathbf{I}_i(\theta_i)\right). \tag{8.16}$$

Since the stopping time $\tau_c^{(i)} \in \mathcal{M}_i$, then $\mathbf{E}_{\theta_i}(\tau_c^{(i)} \wedge N) \leq \mathbf{E}_{\theta_i}\tau_c^{(i)}$, and from (8.16), we can write

$$\frac{\mathbf{E}_{\theta_i}\tau_c^{(i)}}{|\ln \gamma_i(c)|} \geq \mathbf{I}_i^{-1}(\theta_i)\left(1 + \frac{\ln N^{-1}}{|\ln \gamma_i(c)|}\right). \tag{8.17}$$

Since (8.17) holds true for an arbitrary N, then we put $N = 1$ in (8.17) and obtain (8.6).

Proof of Theorem 8.3.2

We give here the proof of asymptotic optimality of the stopping time $T_0(c_0)$. For other $T_i(c_i)$, the proof is the same. To simplify the notations, we will use the symbol c instead of c_0.

The proof of Theorem 3.2 is split into several lemmas. For an arbitrary element $z_0 = (\theta_0, u)$, $\theta_0 \in \Theta_0$, $u \in \mathcal{U}_0 = \Theta \setminus \Theta_0$, consider the stopping time

$$\tau_c(z_0) = \inf\left\{n : \sum_{k=1}^{n} \ln \frac{f(\theta_0, x_k)}{f(u, x_k)} > c\right\}$$

and put

$$\gamma_0(c, z_0) \stackrel{\text{def}}{=} \sup_n \mathbf{P}_{u_0(\cdot)}\left\{\sum_{k=1}^{n} \ln \frac{f(\theta_0, x_k)}{f(u, x_k)} > c\right\}$$

(recall that θ_0 is an arbitrary point from Θ_0, and $u_0(\cdot)$ is defined by θ_0 as in Section 3).

Lemma 8.5.1. *For any $z_0 \in \mathcal{Z}_0$, $\theta_0^* \in \Theta_0$, the following relationships hold true:*

i)

$$\lim_{c\to\infty} \frac{\tau_c(z_0)}{c} = \left(\mathbf{E}_{\theta_0^*} \ln \frac{f(\theta_0, \cdot)}{f(u, \cdot)}\right)^{-1} \qquad \mathbf{P}_{\theta_0^*} - a.s. \text{ and} \tag{8.18}$$

ii)

$$\lim_{c\to\infty} \frac{\mathbf{E}_{\theta_0^*}\tau_c(z_0)}{c} = \left(\mathbf{E}_{\theta_0^*} \ln \frac{f(\theta_0, \cdot)}{f(u, \cdot)}\right)^{-1}.$$

In virtue of Assumptions 8.2.5, the result of this lemma follows from Woodroofe (1982, pp. 42–47).

Lemma 8.5.2. *The following relationship holds*

$$\lim_{c\to\infty} \frac{|\ln \gamma_0(c, z_0)|}{c} = t_0^*(z_0, u_0(\cdot)), \tag{8.19}$$

where $t_0^*(z_0, u^*)$ was defined in Assumption 8.2.7.

Convergence in (8.19) is uniform by $z_0 \in \mathcal{Z}_0$ for any $u \in \mathcal{U}_0$.

The proof of this lemma can be found in Brodsky and Darkhovsky, 2000 (p. 261). This proof uses Assumptions 8.2.6 and 8.2.7. Uniform convergence follows from (8.17).

Lemma 8.5.3. *For any* $\theta_0^* \in \Theta_0$*, the following relationship holds* $\mathbf{P}_{\theta_0^*}$*-a.s.:*

$$\lim_{c\to\infty} \frac{\mathcal{T}_0(c)}{c} = \max_{u\in\mathcal{U}_0} \min_{\theta_0\in\Theta_0} \left(\mathbf{E}_{\theta_0^*} \ln \frac{f(\theta_0,\cdot)}{f(u,\cdot)} \right)^{-1} = \mathbf{I}_0^{-1}(\theta_0^*). \tag{8.20}$$

Remark that the second equality on the right hand of (8.5.6) follows from inequality

$$\mathbf{E}_{\theta_0^*} \ln \frac{f(\theta_0,\cdot)}{f(u,\cdot)} \le \mathbf{E}_{\theta_0^*} \ln \frac{f(\theta_0^*,\cdot)}{f(u,\cdot)},$$

and the definition of the point $u_0^*(\cdot) \stackrel{\text{def}}{=} u_0(\theta_0^*)$.

The proof of this lemma is split into several propositions. For an arbitrary $u \in \mathcal{U}_0$, denote

$$\mathcal{T}_0(c,u) \stackrel{\text{def}}{=} \inf \left\{ n : \max_{\theta_0\in\Theta_0} \sum_{k=1}^{n} \ln \frac{f(\theta_0,x_k)}{f(u,x_k)} > c \right\}.$$

Suppose $\tilde{\Theta} \subset \Theta$, $\tilde{\Theta} = (\tilde{\Theta}_0, \tilde{\mathcal{U}}_0)$ is a countable everywhere dense set.

Proposition 8.5.1. *For any* $\theta_0^* \in \Theta_0$*, the following relationship holds* $\mathbf{P}_{\theta_0^*}$*-a.s.:*

$$\lim_{c\to\infty} \frac{\mathcal{T}_0(c,u)}{c} \le \left(\mathbf{E}_{\theta_0^*} \ln \frac{f(\theta_0^*,\cdot)}{f(u,\cdot)} \right)^{-1}. \tag{8.21}$$

Proof. In virtue of definitions of stopping times $\mathcal{T}_0(c,u)$ and $\tau_c(z_0)$, we have

$$\mathcal{T}_0(c,u,\omega) \le \tau_c(\theta_0,u,\omega)$$

for any $\theta_0 \in \Theta_0, \omega$ (and any $u \in \mathcal{U}_0$) and, therefore,

$$\mathcal{T}_0(c,u) \le \inf_{\theta_0\in\Theta_0} \tau_c(\theta_0,u). \tag{8.22}$$

From (8.22) taking into account Lemma 8.5.1, we obtain $\mathbf{P}_{\theta_0^*}$-a.s. for an arbitrary point $\theta_0^* \in \Theta_0$

$$\begin{aligned}
&\lim_{c\to\infty} \frac{\mathcal{T}_0(c,u)}{c} \le \lim_{c\to\infty} \inf_{\theta_0\in\Theta_0} \frac{\tau_c(z_0)}{c} \\
&\le \inf_{\theta_0\in\Theta_0} \lim_{c\to\infty} \frac{\tau_c(z_0)}{c} \\
&= \inf_{\theta_0\in\Theta_0} \left(\mathbf{E}_{\theta_0^*} \ln \frac{f(\theta_0,\cdot)}{f(u,\cdot)} \right)^{-1} = \left(\mathbf{E}_{\theta_0^*} \ln \frac{f(\theta_0^*,\cdot)}{f(u,\cdot)} \right)^{-1}.
\end{aligned}$$

■

Proposition 8.5.2. *For any ω, there exists an element $\theta_0(\omega) \in \tilde{\Theta}_0$, such that*

$$\tau_c\left(\theta_0(\omega), u, \omega\right) \leq \mathcal{T}_0(c, u, \omega). \tag{8.23}$$

Proof. If $\mathcal{T}_0(c, u, \omega) = +\infty$, the result is evident, and we can take an arbitrary point $\theta_0 \in \tilde{\Theta}_0$ as $\theta_0(\omega)$. So, let us consider only those ω for which $\mathcal{T}_0(c, u, \omega) < \infty$. Suppose that the result of this proposition is not true. Then, for a certain ω_0, $\mathcal{T}_0(c, u, \omega_0) = k$, we have for any $\theta_0 \in \tilde{\Theta}_0$

$$\tau_c(\theta_0, u, \omega_0) > \mathcal{T}_0(c, u, \omega_0).$$

It means that (for the sequence $\{x_s\}$ corresponding to ω_0)

$$\sum_{s=1}^{k} \ln \frac{f\left(\theta_0, x_s\right)}{f\left(u, x_s\right)} \leq c$$

for any $\theta_0 \in \tilde{\Theta}_0$, and, therefore,

$$\sup_{\theta_0 \in \tilde{\Theta}_0} \sum_{s=1}^{k} \ln \frac{f\left(\theta_0, x_s\right)}{f\left(u, x_s\right)} \leq c. \tag{8.24}$$

But for any continuous function $g(\theta_0)$, we have

$$\sup_{\theta_0 \in \tilde{\Theta}_0} g(\theta_0) = \sup_{\theta_0 \in \Theta_0} g(\theta_0).$$

Therefore, from (8.24), it follows that $\mathcal{T}_0(c, u, \omega_0) > k$, and this fact contradicts the supposition $\mathcal{T}_0(c, u, \omega_0) = k$. ■

Proposition 8.5.3. *For any $\theta_0^* \in \Theta_0$,*

$$\mathbf{P}_{\theta_0^*}\left\{\left(\mathbf{E}_{\theta_0^*} \ln \frac{f(\theta_0^*, \cdot)}{f(u, \cdot)}\right)^{-1} \leq \lim_{c \to \infty} \frac{\mathcal{T}_0(c, u)}{c}\right\} = 1. \tag{8.25}$$

Proof.
Suppose θ_0^* is an arbitrary point of Θ_0. Then, for each $\theta_0 \in \tilde{\Theta}_0$, there exists a set of $\mathbf{P}_{\theta_0^*}$ measure zero where there is no convergence in the sense of Lemma 5.1. Then we eliminate from Ω a collection w.r.t. $\theta_0 \in \tilde{\Theta}_0$ of all such null sets. In virtue of countability of $\tilde{\Theta}_0$, this collection is also of $\mathbf{P}_{\theta_0^*}$ measure zero. Denote by Ω^* a remaining set. We have $\mathbf{P}_{\theta_0^*}(\Omega^*) = 1$, and Lemma 8.5.1 holds true for all points from Ω^*.

From (8.23) and (8.18), we obtain for all points $\omega \in \Omega^*$ and $\theta_0(\omega) \in \tilde{\Theta}_0$

$$\begin{aligned}
\min_{\theta_0 \in \Theta_0}\left(\mathbf{E}_{\theta_0^*} \ln \frac{f(\theta_0, \cdot)}{f(u, \cdot)}\right)^{-1} &\leq \left(\mathbf{E}_{\theta_0^*} \ln \frac{f\left(\theta_0(\omega), \cdot\right)}{f\left(u, \cdot\right)}\right)^{-1} \\
&= \lim_{c \to \infty} \frac{\tau_c\left(\theta_0(\omega), u, \omega\right)}{c}, \\
&\leq \lim_{c \to \infty} \frac{\mathcal{T}_0(c, u, \omega)}{c}
\end{aligned}$$

i.e., (8.25) holds. ■

Proposition 8.5.4. *For any points $\theta_0^* \in \Theta_0$, $u \in \mathcal{U}_0$, the following relationship holds $\mathbf{P}_{\theta_0^*}$-a.s.:*

$$\lim_{c\to\infty} \frac{\mathcal{T}_0(c,u)}{c} = \left(\mathbf{E}_{\theta_0^*} \ln \frac{f(\theta_0^*,\cdot)}{f(u,\cdot)}\right)^{-1}.$$

The proof immediately follows from (8.21) and (8.25).

Proposition 8.5.5. *For any $\theta_0^* \in \Theta_0$, the following relationship holds $\mathbf{P}_{\theta_0^*}$-a.s.*

$$\lim_{c\to\infty} \frac{T_0(c)}{c} \geq \left(\mathbf{E}_{\theta_0^*} \ln \frac{f(\theta_0^*,\cdot)}{f(u_0^*(\cdot),\cdot)}\right)^{-1} \stackrel{def}{=} \mathbf{I}_0^{-1}(\theta_0^*) \tag{8.26}$$

Proof. In virtue of definitions of stopping times $\mathcal{T}_0(c,u)$ and $T_0(c)$, we have

$$\mathcal{T}_0(c,u,\omega) \leq T_0(c,\omega)$$

for any $u \in \mathcal{U}_0, \omega$ and, therefore,

$$\sup_{u\in\mathcal{U}_0} \mathcal{T}_0(c,u) \leq T_0(c) \tag{8.27}$$

From (8.27), taking into account Proposition 8.5.4, we obtain $\mathbf{P}_{\theta_0^*}$-a.s. for any point $\theta_0^* \in \Theta_0$

$$\begin{aligned}
\lim_{c\to\infty} \frac{T_0(c)}{c} &\geq \lim_{c\to\infty} \sup_{u\in\mathcal{U}_0} \frac{\mathcal{T}_0(c,u)}{c} \\
&\geq \sup_{u\in\mathcal{U}_0} \lim_{c\to\infty} \frac{\mathcal{T}_0(c,u)}{c} \\
&= \sup_{u\in\mathcal{U}_0} \left(\mathbf{E}_{\theta_0^*} \ln \frac{f(\theta_0^*,\cdot)}{f(u,\cdot)}\right)^{-1} \stackrel{\text{def}}{=} \mathbf{I}_0^{-1}(\theta_0^*).
\end{aligned}$$

■

Proposition 8.5.6. *For any ω, there exists $u(\omega) \in \tilde{\mathcal{U}}_0$, such that*

$$\mathcal{T}_0\left(c, u(\omega), \omega\right) \geq T_0(c,\omega). \tag{8.28}$$

The proof is analogous to the proof of Proposition 8.5.2.

Proposition 8.5.7. *For any $\theta_0^* \in \Theta_0$,*

$$\mathbf{P}_{\theta_0^*}\left\{\mathbf{I}_0^{-1}(\theta_0^*) \geq \lim_{c\to\infty} \frac{T_0(c)}{c}\right\} = 1. \tag{8.29}$$

Proof. Suppose $\theta_0^* \in \Theta_0$. Then, for each $u \in \tilde{\mathcal{U}}_0$, there exists a set of $\mathbf{P}_{\theta_0^*}$ measure zero, where there is no convergence in the sense of Proposition 8.5.4. We eliminate from Ω a collection w.r.t. $u \in \tilde{\mathcal{U}}_0$ of all such null sets. In virtue of countability of $\tilde{\Theta}_0$, such a collection is also of $\mathbf{P}_{\theta_0^*}$ measure zero. Denote by Ω^{**} a remaining set. Then $\mathbf{P}_{\theta_0^*}(\Omega^{**}) = 1$, and for all points from Ω^{**}, Proposition 8.5.4 holds.

From (8.28) and Proposition 8.5.4, it follows that for all points $\omega \in \Omega^{**}$ and $u(\omega) \in \tilde{\mathcal{U}}_0$,

$$\mathbf{I}_0^{-1}(\theta_0^*) = \max_{u \in \mathcal{U}_0} \left(\mathbf{E}_{\theta_0^*} \ln \frac{f(\theta_0^*, \cdot)}{f(u, \cdot)} \right)^{-1} \geq \left(\mathbf{E}_{\theta_0^*} \ln \frac{f(\theta_0^*, \cdot)}{f(u(\omega), \cdot)} \right)^{-1}$$
$$= \lim_{c \to \infty} \frac{\mathcal{T}_0(c, u(\omega), \omega)}{c}$$
$$\geq \lim_{c \to \infty} \frac{\mathcal{T}_0(c, \omega)}{c},$$

i.e., we obtain (8.29). ■

Lemma 8.5.3 follows from Propositions 8.5.5 and 8.5.7.

Fix the number $\epsilon > 0$. In virtue of assumptions 8.2.1, 8.2.4, and 8.2.7, there exists $a(\epsilon)$, such that for the corresponding finite sets $\Theta_0(a(\epsilon)) \stackrel{\text{def}}{=} \Theta_0(\epsilon)$, $\#\Theta_0(\epsilon) \leq R(\epsilon)$, $\mathcal{U}_0(a(\epsilon)) \stackrel{\text{def}}{=} \mathcal{U}_0(\epsilon)$, $\#\mathcal{U}_0(\epsilon) \leq R(\epsilon)$, the following relationships hold:

$$\begin{aligned} &\left| \max_{u \in \mathcal{U}_0(\epsilon)} \min_{\theta_0 \in \Theta_0(\epsilon)} \left(\mathbf{E}_{\theta_0^*} \ln \frac{f(\theta_0, \cdot)}{f(u, \cdot)} \right)^{-1} - \mathbf{I}_0^{-1}(\theta_0^*) \right| \leq \epsilon \\ &\left| \max_{u \in \mathcal{U}_0(\epsilon)} \min_{\theta_0 \in \Theta_0(\epsilon)} t_0^*(z_0, u_0^*) - \max_{u \in \mathcal{U}_0} \min_{\theta_0 \in \Theta_0} t_0^*(z_0, u_0^*) \right| \leq \epsilon. \end{aligned} \tag{8.30}$$

In the sequel, we fix ϵ and consider the stopping time

$$T_0(c, \epsilon) \stackrel{\text{def}}{=} \inf \left\{ n : \min_{u \in \mathcal{U}_0(\epsilon)} \max_{\theta_0 \in \Theta_0(\epsilon)} \sum_{k=1}^{n} \ln \frac{f(\theta_0, x_k)}{f(u, x_k)} > c \right\}. \tag{8.31}$$

Lemma 8.5.4. *For stopping time (8.31) and any $\theta_0^* \in \Theta_0$, the following relationship holds:*

$$\lim_{c \to \infty} \frac{\mathbf{E}_{\theta_0^*} T_0(c, \epsilon)}{c} = \max_{u \in \mathcal{U}_0(\epsilon)} \min_{\theta_0 \in \Theta_0(\epsilon)} \left(\mathbf{E}_{\theta_0^*} \ln \frac{f(\theta_0, \cdot)}{f(u, \cdot)} \right)^{-1} \leq \mathbf{I}_0^{-1}(\theta_0^*) + \epsilon. \tag{8.32}$$

Proof. In virtue of assumption 8.2.5, the family (w.r.t. c) of r.v.'s $\{\tau_c(z_0)/c\}$ is uniformly integrable for any z_0 (see Woodroofe, 1982 (p.47)). Since the sets $\Theta_0(\epsilon), \mathcal{U}_0(\epsilon)$ are finite, the family of random variables $\{T_0(c, \epsilon)/c\}$ is also uniformly integrable. This yields the existence of limit in (8.32), and the result of lemma follows from Lemma 8.5.3 and (8.30). ■

Recall that $u_0^*(\cdot) = u_0(\theta_0^*)$ defined by $\theta_0^* \in \Theta_0$ as in Section 8.3 and

$$\gamma_0^\epsilon(c) = \sup_n \mathbf{P}_{u_0^*(\cdot)} \left\{ \tilde{d}_c^{(0)}(n, \epsilon) = 0 \right\} =$$
$$\sup_n \mathbf{P}_{u_0^*(\cdot)} \left\{ \min_{u \in \mathcal{U}_0(\epsilon)} \max_{\theta_0 \in \Theta_0(\epsilon)} \sum_{k=1}^{n} \ln \frac{f(\theta_0, x_k)}{f(u, x_k)} > c \right\}.$$

Lemma 8.5.5. *The following inequality holds:*

$$\liminf_{c\to\infty} \frac{|\ln \gamma_0^\epsilon(c)|}{c} \geq \max_{u\in\mathcal{U}_0} \min_{\theta_0\in\Theta_0} t_0^*\left(z_0, u_0^*(\cdot)\right) - \epsilon \tag{8.33}$$

(recall that $z_0 = (\theta_0, u)$).

Proof. Suppose ξ_i, η_j, $i, j = 1, \ldots, N$ are nonnegative r.v.'s. We use the elementary inequality

$$\mathbf{P}\{\min_{1\leq j\leq N} \max_{1\leq i\leq N} \xi_i\eta_j > c\} \leq N \min_{1\leq j\leq N} \max_{1\leq i\leq N} \mathbf{P}\{\xi_i\eta_j > c\}. \tag{8.34}$$

Using (8.34) and continuity Assumptions 8.2.3, we obtain

$$\begin{aligned} &\ln R(\epsilon) + \min_{u\in\mathcal{U}_0(\epsilon)} \max_{\theta_0\in\Theta_0(\epsilon)} \ln \mathbf{P}_{u_0^*(\cdot)} \left\{ \sum_{k=1}^{n} \ln \frac{f(\theta_0, x_k)}{f(u, x_k)} > c \right\} \geq \\ &\ln \mathbf{P}_{u_0^*(\cdot)} \left\{ \min_{u\in\mathcal{U}_0(\epsilon)} \max_{\theta_0\in\Theta_0(\epsilon)} \sum_{k=1}^{n} \ln \frac{f(\theta_0, x_k)}{f(u, x_k)} > c \right\}. \end{aligned} \tag{8.35}$$

Since (8.34) is valid for any N, from Lemma 8.5.2, we obtain

$$\liminf_{c\to\infty} \frac{|\ln \gamma_0^\epsilon(c)|}{c} \geq \max_{u\in\mathcal{U}_0(\epsilon)} \min_{\theta_0\in\Theta_0(\epsilon)} t_0^*\left(z_0, u_0^*(\cdot)\right). \tag{8.36}$$

From (8.36) and (8.30), we obtain the required result.

■

Now, we can finish the proof of the first point of Theorem 8.3.2 for $i = 0$. Obviously, the test $\tilde{d}_c^{(0)}(n, \epsilon)$ is nondegenerate. Suppose the upper limit in (8.8) is realized for a sequence $\{c_k\}, k \uparrow \infty$. Then, from Theorem 8.3.1, (8.32) and (8.33), we obtain

$$\begin{aligned} &\mathbf{I}_0^{-1}(\theta_0^*) \leq \liminf_{c\to\infty} \mathcal{K}_0\left(T_0(c,\epsilon), \theta_0^*\right) = \lim_{k\to\infty} \frac{\mathbf{E}_{\theta_0^*} T_0(c_k,\epsilon)/c_k}{|\ln \gamma_0^\epsilon(c_k)|/c_k} \\ &\leq \frac{\lim_{k\to\infty} \mathbf{E}_{\theta_0^*} T_0(c_k,\epsilon)/c_k}{\liminf_{k\to\infty} |\ln \gamma_0^\epsilon(c_k)|/c_k} \leq \frac{\mathbf{I}_0^{-1}(\theta_0^*) + \epsilon}{\max_{u\in\mathcal{U}_0} \min_{\theta_0\in\Theta_0} t_0^*\left(z_0, u_0^*(\cdot)\right) - \epsilon}. \end{aligned} \tag{8.37}$$

But for $z_0 = (\theta_0, u_0^*)$, we have (see the definition in 8.2.7)

$$t_0^*\left(\theta_0, u_0^*(\cdot), u_0^*(\cdot)\right) \equiv 1,$$

and, therefore,

$$\max_{u\in\mathcal{U}_0} \min_{\theta_0\in\Theta_0} t_0^*\left(z_0, u_0^*(\cdot)\right) \geq \min_{\theta_0\in\Theta_0} t_0^*\left(\theta_0, u_0^*(\cdot), u_0^*(\cdot)\right) = 1. \tag{8.38}$$

The first point of Theorem 8.3.2 for $i = 0$ follows now from (8.37) and (8.38).

The second point of Theorem 8.3.2 was established in Lemma 8.5.5.

8.5.2 Multi-Sided Tests

Proof of Theorem 8.4.1

Without loss of generality, consider the case of 3 composite hypotheses: $\mathbf{H}_0, \mathbf{H}_1$, and $\mathbf{H}_2$. Let $\theta_0 \in \Theta_0$ be arbitrary (but fixed) point, and $u_0(\theta_0) = u_0(\cdot) \in \Theta_1 \bigcup \Theta_2$ be the point from definition (8.5).

Consider the event $R_0 = \{\text{accept}\, \mathbf{H}_0\}$, in other words, reject $\mathbf{H}_1$ and $\mathbf{H}_2$. By R_0^r, denote the event $\{\text{reject}\, \mathbf{H}_0\}$.

Suppose $d_c(n)$ is the decision rule of an arbitrary nondegenerate test, $\nu_c \stackrel{\text{def}}{=} \inf\{n : d_c(n) \neq -1\}$.

Let N be an arbitrary positive integer. Then, in virtue of the definition of $\gamma_0(c)$, we have

$$\begin{aligned} &N\gamma_0(c) \geq \sum_{k=1}^{N} \mathbf{P}_{u_0(\cdot)} \{(\nu_c = k) \bigcap R_0\} = \\ &\sum_{k=1}^{N} \mathbf{E}_{\theta_0} \left(\mathbb{I}(\nu_c = k)\mathbb{I}(R_0) \prod_{i=1}^{k} \frac{f(u_0(\cdot), x_i)}{f(\theta_0, x_i)} \right) \\ &= \mathbf{E}_{\theta_0} \left(\prod_{i=1}^{\nu_c \wedge N} \frac{f(u_0(\cdot), x_i)}{f(\theta_0, x_i)} \mathbb{I}(R_0) \right). \end{aligned} \tag{8.39}$$

Since

$$\mathbf{P}_{\theta_0} \{d_c(\nu_c \wedge N) = 0\} = 1 - [\mathbf{P}_{\theta_0}\{d_c(\nu_c \wedge N) = 1\} + \mathbf{P}_{\theta_0}\{d_c(\nu_c \wedge N) = 2\}] \geq 1 - \alpha_{01}(c) - \alpha_{02}(c),$$

and the test is nondegenerate, from (8.39) and Jensen's inequality, we obtain

$$\begin{aligned} &N\gamma_0 \geq \mathbf{E}_{\theta_0} \left\{ \exp(-\sum_{i=1}^{\nu_c \wedge N} \ln \frac{f(\theta_0, x_i)}{f(u_0(\cdot), x_i)}) | R_0 \right\} (1 - \alpha_{01}(c) - \alpha_{02}(c)) \\ &\geq \exp \left\{ -\mathbf{E}_{\theta_0} (\sum_{i=1}^{\nu_c \wedge N} \ln \frac{f(\theta_0, x_i)}{f(u_0(\cdot), x_i)} \mathbb{I}(R_0)) / (1 - \alpha_{01}(c) - \alpha_{02}(c)) \right\} \\ &(1 - \alpha_{01}(c) - \alpha_{02}(c)). \end{aligned} \tag{8.40}$$

From (8.40), we obtain

$$\begin{aligned} &(1 - \alpha_{01}(c) - \alpha_{02}(c)) [- \ln N\gamma_0 + \ln(1 - \alpha_{01}(c) - \alpha_{02}(c))] \\ &\leq \mathbf{E}_{\theta_0} \left(\sum_{i=1}^{\nu_c \wedge N} \ln \frac{f(\theta_0, x_i)}{f(u_0(\cdot), x_i)} \mathbb{I}(R_0) \right). \end{aligned} \tag{8.41}$$

Further,

$$\mathbf{P}_{u_0(\cdot)} \left\{ (\nu_c = k) \bigcap R_0^r \right\} \leq 1 - \mathbf{P}_{u_0(\cdot)} \left\{ (\nu_c = k) \bigcap R_0 \right\} \leq (1 - \rho_0(c)).$$

Therefore,

$$
\begin{aligned}
N(1-\rho_0(c)) &\geq \sum_{k=1}^{N} \mathbf{P}_{u_0(\cdot)}\left\{(\nu_c = k)\bigcap R_0^r\right\} \\
&= \sum_{k=1}^{N} \mathbf{E}_{\theta_0}\left(\mathbb{I}(\nu_c = k)\mathbb{I}(R_0^r)\prod_{i=1}^{k}\frac{f(u_0(\cdot),x_i)}{f(\theta_0,x_i)}\right) \\
&= \mathbf{E}_{\theta_0}\left(\mathbb{I}(R_0^r)\prod_{i=1}^{\nu_c\wedge N}\frac{f(u_0(\cdot),x_i)}{f(\theta_0,x_i)}\right).
\end{aligned}
\tag{8.42}
$$

Since

$$
\mathbf{P}_{\theta_0}\left(d_c(\nu_c \wedge N) = 1\right) \geq \beta_{01}(c),\ \mathbf{P}_{\theta_0}\left(d_c(\nu_c \wedge N) = 2\right) \geq \beta_{02}(c),
$$

then,

$$
\begin{aligned}
&\mathbf{E}_{\theta_0}\left(\mathbb{I}(R_0^r)\prod_{i=1}^{\nu_c\wedge N}\frac{f(u_0(\cdot),x_i)}{f(\theta_0,x_i)}\right) \geq \\
&\mathbf{E}_{\theta_0}\left\{\exp\left(-\sum_{i=1}^{\nu_c\wedge N}\ln\frac{f(\theta_0,x_i)}{f(u_0(\cdot),x_i)}\right)\Big|\mathbb{I}(R_0^r)\right\}(\beta_{01}(c)+\beta_{02}(c)) \\
&\geq \exp\left\{-\mathbf{E}_{\theta_0}\left(-\sum_{i=1}^{\nu_c\wedge N}\ln\frac{f(\theta_0,x_i)}{f(u_0(\cdot),x_i)}\mathbb{I}(R_0^r)/(\beta_{01}(c)+\beta_{02}(c))\right\}\right. \\
&(\beta_{01}(c)+\beta_{02}(c)).
\end{aligned}
\tag{8.43}
$$

From (8.42) and (8.43), we obtain

$$
\begin{aligned}
&(\beta_{01}(c)+\beta_{02}(c))\left[-\ln N(1-\rho_0(c)) + \ln(\beta_{01}(c)+\beta_{02}(c))\right] \\
&\leq \mathbf{E}_{\theta_0}\left(\mathbb{I}(R_0^r)\prod_{i=1}^{\nu_c\wedge N}\frac{f(\theta_0,x_i)}{f(u_0(\cdot),x_i)}\right).
\end{aligned}
\tag{8.44}
$$

Now, summarizing (8.41) and (8.44), using Wald's identity and the relation $\mathbf{E}_{\theta_0}(\nu_c \wedge N) \leq \mathbf{E}_{\theta_0}(\nu_c)$, we obtain

$$
\begin{aligned}
&I_0(\theta_0)\mathbf{E}_{\theta_0}(\nu_c) \geq (1-\alpha_{01}(c)-\alpha_{02}(c))\left[-\ln N\gamma_0(c) + \ln(1-\alpha_{01}(c)-\alpha_{02}(c))\right] \\
&+ (\beta_{01}(c)+\beta_{02}(c))\left[-\ln N(1-\rho_0(c)) + \ln(\beta_{01}(c)+\beta_{02}(c))\right].
\end{aligned}
\tag{8.45}
$$

Since (8.45) is valid for any N, then inserting $N = 1$, we obtain the result corresponding to Theorem 8.4.1.

Proof of Theorem 8.4.2

Without loss of generality, again consider the case of 3 composite hypotheses $\mathbf{H}_0, \mathbf{H}_1$, and $\mathbf{H}_2$. Define the stopping times $T_1(c_1,\epsilon)$, and $T_2(c_2,\epsilon)$ analogous to (8.31), and consider the stopping time $T(c,\epsilon) \stackrel{\text{def}}{=} T_0(c_0,\epsilon) \bigwedge T_1(c_1,\epsilon) \bigwedge T_2(c_2,\epsilon)$ and the corresponding decision rule $d_c^*(n,\epsilon)$.

Let $\theta_i^* \in \Theta_i, i = 0,1,2$ be arbitrary (but fixed) points.

Choose the number $\epsilon > 0$ from (8.30) and analogous conditions for the functions $\mathbf{I}_i^{-1}(\theta_i^*)$, $i = 1, 2$, and $t_i^*(z_i, u_i)$, $i = 1, 2$.

From the definition of the stopping time $T(c, \epsilon)$ and Lemma 8.5.4, it follows that for any $\theta_0^* \in \Theta_0$,

$$\lim_{c_0 \to \infty} \frac{\mathbf{E}_{\theta_0^*} T(c, \epsilon)}{c_0} \le \lim_{c_0 \to \infty} \frac{\mathbf{E}_{\theta_0^*} T_0(c_0, \epsilon)}{c_0} = \max_{u \in \mathcal{U}_0(\epsilon)} \min_{\theta_0 \in \Theta_0(\epsilon)} \left(\mathbf{E}_{\theta_0^*} \ln \frac{f(\theta_0, \cdot)}{f(u, \cdot)} \right)^{-1} \le \mathbf{I}_0^{-1}(\theta_0^*) + \epsilon. \tag{8.46}$$

Let ζ be a random variable independent of observations $\{x_k\}_{k=1}^\infty$ and taking values 0, 1, and 2 with probabilities 1/3. Then, the decision rule $d_c^*(n, a)$ can be written as follows:

$$d_c^*(n, a) = \begin{cases} i & \text{if } \; n = T(c, a) = T_i(c_i, a) < T_j(c_j, a), j \neq i, \;\; i, j \in I \\ j & \text{if } \; n = T(c, a), \; \#I\left(T(c.a)\right) > 1, \zeta = j, \; j \in I\left(T(c, a)\right). \\ -1 & \text{if } \;\; T(c, a) > n \end{cases}$$

From the definition of the false decision probability, we have

$$\begin{aligned} &\gamma_0^*(c_0, \epsilon) \stackrel{\text{def}}{=} \sup_n \mathbf{P}_{u_0(\cdot)}\{d_c^*(n, \epsilon) = 0\} = \\ &\sup_n \Big(\mathbf{P}_{u_0(\cdot)}\{T(c, \epsilon) = n, I\left(T(c, \epsilon)\right) = \{0\} \text{ or} \\ &T(c, \epsilon) = n, \, 0 \in I\left(T(c, \epsilon)\right)), \, \#I\left(T(c, \epsilon)\right) > 1 \text{ and } \zeta = 0\}) \\ &\le 4/3 \sup_n \mathbf{P}_{u_0(\cdot)}\{n = T(c, \epsilon) = T_0(c, \epsilon)\} = \\ &4/3 \sup_n \mathbf{P}_{u_0(\cdot)}\{\mathcal{L}_0(n, \epsilon) > c_0\}. \end{aligned} \tag{8.47}$$

Therefore, analogous to Lemma 8.5.5 and (8.38), we obtain from here

$$\liminf_{c_0 \to \infty} \frac{|\ln \gamma_0^*(c_0, \epsilon)|}{c_0} \ge \max_{u \in \mathcal{U}_0} \min_{\theta_0 \in \Theta_0} t_0^*\left(z_0, u_0^*(\cdot)\right) - \epsilon \ge 1 - \epsilon. \tag{8.48}$$

Returning to the proof of Lemma 8.5.5, we see that inequality of (8.5.19) type is valid also for false decision probabilities $\alpha_{01}(c_1, \epsilon)$, $\alpha_{02}(c_2, \epsilon)$ as $c_1 \to \infty$, $c_2 \to \infty$.

Therefore, from (8.48) and analogous relationships for $\alpha_{01}(c_1, \epsilon)$, $\alpha_{02}(c_2, \epsilon)$, it follows that the decision rule $d_c^*(n, \epsilon)$ is nondegenerate and the probabilities of false decisions tend to zero as $c = (c_0, c_1, c_2) \to \infty$. Therefore, from Theorem 8.4.1, (8.47) and (8.48), we obtain

$$\begin{aligned} &\mathbf{I}_0^{-1}(\theta_0^*) \le \lim_{c \to \infty} \mathcal{K}_0\left(T(c, \epsilon), \theta_0^*\right) \le \lim_{c \to \infty} \frac{\mathbf{E}_{\theta_0^*} T_0(c_0, \epsilon)/c_0}{|\ln \gamma_0^*(c_0, \epsilon)|/c_0} \\ &\le \lim_{c_0 \to \infty} \frac{\mathbf{I}_0^{-1}(\theta_0^*) + \epsilon}{\max_{u \in \mathcal{U}_0} \min_{\theta_0 \in \Theta_0} t_0^*\left(z_0, u_0^*(\cdot)\right) - \epsilon} \le \frac{\mathbf{I}_0^{-1}(\theta_0^*) + \epsilon}{1 - \epsilon}. \end{aligned} \tag{8.49}$$

The analogous inequalities for $\mathcal{K}_1\left(T(c,\epsilon),\theta_1^*\right)$, $\mathcal{K}_2\left(T(c,\epsilon),\theta_2^*\right)$ can be obtained in the same way. Therefore, the first point of the theorem is established.

The second point of the theorem follows from the proof (see (8.48), and analogous relationships for $\alpha_{01}(c_1\epsilon)$, and $\alpha_{02}(c_2,\epsilon)$).

8.6 Simulations

In this section, I report some results of a small simulation study we performed in order to assess the accuracy of the lower bounds in a priori inequalities and to test efficiency of methods in multihypothesis problems for finite values of decision thresholds.

1-sided tests

Data: The following data were analyzed. The independent Gaussian sequence was simulated with the d.f. $N(\theta,1)$. Under the null hypothesis $\mathbf{H}_0$: $\theta=\theta_0\in\Theta_0=[0,1]$, under the alternative hypothesis $\mathbf{H}_1:\ \theta=\theta_1\in\Theta_1=[1.2,5]$.

Method: the minimax test

$$\tau_c=\inf\{n:\ \inf_{\theta_0\in\Theta_0}\ \sup_{\theta_1\in\Theta_1}\sum_{i=1}^{n}\ln\frac{f(\theta_1,x_i)}{f(\theta_0,x_i)}\geq C\}.$$

Tests:

We choose an arbitrary $\theta_1^*\in\Theta_1$ and compute $\mathbf{E}_{\theta_1^*}\tau_c$ for different values of the threshold C.

For the given θ_1^*, we find "the least favorable" $\theta_0^*\in\Theta_0$ and compute $\alpha_c=\sup_n\mathbf{P}_{\theta_0^*}\{d_c(n)=1\}$ for the same values of C.

Then, we compute the ratio

$$\frac{\mathbf{E}_{\theta_1^*}\tau_c}{|\ln\alpha_c|}$$

and demonstrate that it tends to the theoretical limit $\mathbf{I}_1^{-1}(\theta_1^*)$ as $C\to\infty$.

Results.

It is easy to see that for each $\theta_1^*\in\Theta_1$ "the least favorable" point is $\theta_0^*=1$ in the considered case. The estimate of the probability α_c is obtained as follows. In $k=2000$ trials, we find the number n_k of cases in which

$$\min_{1\leq n\leq T}\ \min_{\theta_0\in\Theta_0}\ \max_{\theta_1\in\Theta_1}\sum_{i=1}^{n}\ln\frac{f(\theta_1,x_i)}{f(\theta_0,x_i)}\geq C,$$

where $T=5000$ and observations x_i have the density function $f(\theta_0^*,\cdot)$.

Then, the estimate of the probability α_c is $pr=n_k/k$.

TABLE 8.1
Characteristics of the Proposed Method: One-Sided Tests

	C	1.0	2.0	3.0	4.0	5.0	6.0	
$\theta_0^* = 1$	pr	0.45	0.23	0.09	0.04	0.015	0.004	Lower
	$\lvert\ln(pr)\rvert$	0.81	1.45	2.41	3.35	4.20	5.52	bound
$\theta_1^* = 1.2$	$\mathbf{E}_{\theta_1^*}\tau_c$	43.1	81.8	129.8	182.8	232.7	277.9	
	$\mathbf{E}_{\theta_1^*}\tau_c/\lvert\ln(pr)\rvert$	53.2	56.4	53.8	58.0	54.4	50.3	50
$\theta_1^* = 1.5$	$\mathbf{E}_{\theta_1^*}\tau_c$	8.7	15.4	23.0	30.9	38.8	47.4	
	$\mathbf{E}_{\theta_1^*}\tau_c/\lvert\ln(pr)\rvert$	10.7	10.6	9.54	9.24	9.2	8.58	8.0
$\theta_1^* = 2.0$	$\mathbf{E}_{\theta_1^*}\tau_c$	3.03	4.88	6.82	8.66	10.57	12.4	
	$\mathbf{E}_{\theta_1^*}\tau_c/\lvert\ln(pr)\rvert$	3.74	3.36	2.82	2.58	2.51	2.24	2.0

TABLE 8.2
Characteristics of the Proposed Method: Two-Sided Tests

	C	1.0	2.0	3.0	4.0	5.0	6.0	
$\theta_1^* = 1.2$	pr	0.47	0.18	0.09	0.04	0.02	0.01	Lower
	$\lvert\ln(pr)\rvert$	0.76	1.70	2.41	3.22	4.09	5.12	bound
$\theta_0^* = 1$	$\mathbf{E}_{\theta_0^*}\tau_c$	41.8	90.9	137.9	185.1	237.3	293.2	
	$\mathbf{E}_{\theta_0^*}\tau_c/\lvert\ln(pr)\rvert$	55.0	53.4	57.4	57.4	58.0	57.2	50
$\theta_0^* = 0.5$	$\mathbf{E}_{\theta_0^*}\tau_c$	5.02	8.83	12.7	16.6	20.8	24.6	
	$\mathbf{E}_{\theta_0^*}\tau_c/\lvert\ln(pr)\rvert$	6.6	5.19	5.29	5.15	5.08	4.8	4.0
$\theta_0^* = 0$	$\mathbf{E}_{\theta_0^*}\tau_c$	2.44	3.75	5.07	6.50	7.90	9.4	
	$\mathbf{E}_{\theta_0^*}\tau_c/\lvert\ln(pr)\rvert$	3.2	2.2	2.11	2.01	1.90	1.83	1.4

In the following tests, I compute the estimates pr and $\mathbf{E}_{\theta_1^*}\tau_c$ in 2000 independent trials for different values of the threshold C. Then, the performance ratio $\mathbf{E}_{\theta_1^*}\tau/|\ln(pr)|$ is computed and compared with its theoretical lower bound $\mathbf{I}_1^{-1}(\theta_1^*$.

Results are presented in Table 8.1.

2-sided tests

Analogous experiments were performed for multisided tests and the same hypotheses $\mathbf{H}_1$ and $\mathbf{H}_0$. In this situation, there are two lower bounds $\mathbf{I}_0^{-1}(\theta_0^*)$ and $\mathbf{I}_1^{-1}(\theta_1^*)$. Obviously, we need to add the "reversed" one-sided test: the null hypothesis $\mathbf{H}_1$ and the alternative $\mathbf{H}_0$, the "least favorable point" is $\theta_1^* = 1.2$. Results for this reversed test are presented in Table 8.2.

These experimental results allow us to conclude that for a wide range of finite thresholds C, the results of testing are quite satisfactory in the sense that the performance ratio is close to the theoretical lower bound.

Three-sided tests

In the last series of experiments, the case of three-sided tests was considered. Three hypotheses about the expectation θ of the Gaussian d.f. of independent observations were assumed to be as follows: $\mathbf{H}_0 : \theta \in \Theta_0 = [0, 1]$,

$\mathbf{H}_1 : \theta \in \Theta_1 = [1.2, 5]$, and $\mathbf{H}_2 : \theta \in \Theta_2 = [5.2, 9]$. The dispersion of observations was equal to 1.

For this configuration of hypotheses, if $\theta_0^* \in \Theta_0 = [0, 1]$, then the "least favorable" point is $u_0^* = 1.2$, if $\theta_1^* \in \Theta_1 = [1.2, 5]$, then $u_1^* = 1$, and if $\theta_2^* \in \Theta_2 = [5.2, 9]$, then $u_2^* = 5$. Like in the previous cases of one-sided and two-sided tests, these points were used for estimation of probabilities α_c and computing the decision thresholds c_1, c_2, and c_3 that equalize these error probabilities.

In the considered case, we can assume one threshold $C = c_1 = c_2 = c_3$ in three sequential tests,

$$\tau_0 = \inf\{n : \inf_{u \in \Theta \setminus \Theta_0} \sup_{\theta_0 \in \Theta_0} \sum_{i=1}^{n} \ln \frac{f(\theta_0, x_i)}{f(u, x_i)} \geq C\},$$

$$\tau_1 = \inf\{n : \inf_{u \in \Theta \setminus \Theta_1} \sup_{\theta_1 \in \Theta_1} \sum_{i=1}^{n} \ln \frac{f(\theta_1, x_i)}{f(u, x_i)} \geq C\}, \text{and}$$

$$\tau_2 = \inf\{n : \inf_{u \in \Theta \setminus \Theta_2} \sup_{\theta_0 \in \Theta_2} \sum_{i=1}^{n} \ln \frac{f(\theta_2, x_i)}{f(u, x_i)} \geq C\}.$$

In k=2000 trials of each experiment, the following values were computed: for $i = 0, 1, 2$, $E\tau_i$ – the average sample size of each test for some real value θ_i^* of the d.f. parameter; and for pr_i – the frequency of cases when the hypothesis H_i was accepted. So, for $i = 0$, pr_1, pr_2 are estimates of cross-error probabilities.

The obtained results are reported in Table 8.3.

From these results, we can see that cross-error probabilities rapidly converge to zero for the proposed test.

TABLE 8.3
Characteristics of the Proposed Method: Three-Sided Tests

C		0.5	1.0	2.0	3.0	4.0	5.0
$\theta_0^* = 0.5$	$\mathbf{E}_{\theta_0^*}(\tau)$	2.56	4.67	8.61	12.5	16.42	20.6
	pr_0	0.89	0.95	0.99	1	1	1
	pr_1	0.11	0.05	0.01	0	0	0
	pr_2	0	0	0	0	0	0
$\theta_1^* = 2.0$	$\mathbf{E}_{\theta_1^*}(\tau)$	1.89	2.92	4.90	6.81	8.85	10.78
	pr_0	0.058	0.016	0.002	0	0	0
	pr_1	0.942	0.984	0.998	1	1	1
	pr_2	0	0	0	0	0	0
$\theta_2^* = 6.0$	$\mathbf{E}_{\theta_2^*}(\tau)$	1.87	2.85	4.71	6.68	8.72	10.61
	pr_0	0	0	0	0	0	0
	pr_1	0.06	0.02	0.004	0	0	0
	pr_2	0.94	0.98	0.996	1	1	1

8.7 Conclusions

In this chapter, the problem of sequential testing of multiple composite hypotheses is considered. Each hypothesis is represented by the density function of observations that depends on a parameter from one of disjoint sets. New performance measures for sequential testing of composite hypotheses are proposed, and a priori lower bounds for these measures are proved. Asymptotically optimal methods for sequential multihypothesis testing are found. These methods provide the basis for construction of asymptotically optimal methods of sequential change-point detection in subsequent chapters.

9

Sequential Change-Point Detection for Univariate Models

9.1 Introduction

In this chapter, I consider sequential methods of change-point detection, i.e., detection of abrupt changes in statistical characteristics of data "online", i.e., simultaneously with the process of data collection.

Any sequential method of detection is the process of decision making about the presence of "nonhomogeneity", i.e., sequential repetition of the retrospective detection scheme for every step of data collection. Therefore, it would be quite natural, in analogy with the retrospective detection scheme, to characterize the quality of a method of sequential detection by the probability of the *1st-type error ("false alarm")*, the probability of the *2nd-type error ("false tranquillity")*, and the probability of the *error of estimation* of the change-point. However, historically, this was not the case: The average "delay time" in detection and the the average time between "false alarms" (or the average time before the first "false alarm") were assumed to be the main quality characteristics of any sequential change-point procedure.

Here, I remind about the main results in the field of sequential change-point detection.

The sequential problem of statistical diagnosis consists of the following. Suppose that observations are made sequentially, and at some moment the distribution law of observations changes. It is required to detect the change-point as soon as possible on condition that "false alarms" are seldom raised.

Sequential problems of statistical diagnosis appeared in 1930s in connection with statistical quality control. One of the first diagnostic tests was proposed by Shewhart (Shewhart chart (1931)). In the 1950s, Page (1954, 1955) and Girshik and Rubin (1952) proposed much more efficient methods of sequential detection.

The initial idea of Page consisted of the following. Suppose that a sequence $X_1, \dots, X_n$ of independent random variables is observed. For each $1 \le \nu \le n$, consider the hypothesis H_ν that r.v.'s $x_1, \dots, x_{\nu-1}$ have the same density function $f_0(\cdot)$ and r.v.'s $x_\nu, \dots, x_n$ – $f_1(\cdot)$. Denote by H_0 a hypothesis of stochastic homogeneity of the sample. Then, the likelihood ratio statistic for testing the composite hypothesis $H_\nu \quad (1 \le \nu \le n)$ against H_0 is

$$\max_{0\le k\le n} (S_n - S_k) = S_n - \min_{0\le k\le n} S_k,$$

where

$$S_0 \equiv 0,\; S_k = \sum_{j=1}^{k} \ln \frac{f_1(x_j)}{f_0(x_j)}, \quad k = 1, \dots, n.$$

Page proposed the following stopping rule:

$$\tau = \inf\{n \ge 1 : S_n - \min_{0\le j\le n} S_j \ge b\}, \tag{9.1}$$

where $b > 0$ is the alarm threshold.

Using approximations of Wald's sequential test (see Wald, 1947), Page obtained the following relations:

$$\begin{aligned} \mathbf{E}_0\,\tau &= |e^b - b - 1|/|\mu_0|, \text{ and} \\ \mathbf{E}_1\,\tau &= (e^{-b} + b - 1)/\mu_1, \end{aligned} \tag{9.2}$$

where $\mathbf{E}_0\,\tau$ is the average observation time before a "false alarm", $\mathbf{E}_1\,\tau$ is the average delay time of detection, and

$$\mu_0 = \mathbf{E}_0 \log \frac{f_1(x)}{f_0(x)} < 0, \qquad \mu_1 = \mathbf{E}_1 \log \frac{f_1(x)}{f_0(x)} > 0.$$

The Cumulative Sum (CUSUM) statistic $g_n = S_n - \min_{0\le j\le n} S_j$ can be written in the recurrent form:

$$g_n = \left(g_{n-1} + \log \frac{f_1(x_n)}{f_0(x_n)}\right)^+, \qquad g_0 = 0, \tag{9.3}$$

where $a^+ = \max(a, 0)$.

The effectiveness of this statistic is easy to explain: The mathematical expectation of $y_n = \log(f_1(x_n)/f_0(x_n))$ is negative before, and positive after, the change-point.

Girshick and Rubin (1952) proposed the following method of change-point detection. Suppose that a sequence $X = \{x_1, x_2, \dots\}$ of i.r.v.'s is observed, such that the density function of $x_i, i = 1, 2, \dots$ is

$$f(x_i) = \begin{cases} f_0(x_i) & , \quad 1 \le i \le n_0 \\ f_1(x_i) & , \quad i > n_0, \end{cases}$$

where $f_0(\cdot) \ne f_1(\cdot)$ and n_0 is the change-point. Let

$$\begin{aligned} & w_i = \frac{f_1(x_i)}{f_0(x_i)},\; i \ge 1, \\ & W_i = w_i(1 + W_{i-1}) = w_i + w_i\, w_{i-1} + \dots + w_i w_{i-1} \dots w_1, \\ & W_0 = 0. \end{aligned} \tag{9.4}$$

The decision rule proposed in the paper by Girshick and Rubin (1952) is

$$d(W_n) = \mathbb{I}(W_n > C),$$

where C is the threshold of detection.

In the works of Shiryaev (1961, 1963, 1965, 1976), the problem of sequential change-point detection for random processes with discrete (independent r.v.'s) and continuous (the Wiener process) time was considered.

Let $\theta = \theta(\omega)$ be a random change-point of an observed sequence $\xi_1, \xi_2, \ldots$. Suppose that random variables $\theta, \xi_1, \xi_2, \ldots$ are defined on a probability space $(\Omega, \mathcal{F}, \mathbf{P}^\pi)$, such that

$$\mathbf{P}^\pi\{\theta = 0\} = \pi, \quad \mathbf{P}^\pi\{\theta = n\} = (1-\pi)(1-p)^{n-1}p, \quad n \geq 1,$$

where π, p are known constants, $0 < p \leq 1$, $0 \leq \pi \leq 1$.

For $\theta = n$, the r.v.'s $\xi_1, \ldots, \xi_{n-1}, \xi_n, \ldots$ are independent, $\xi_1, \ldots, \xi_{n-1}$ are identically distributed with the density function $p_0(x)$, and $\xi_n, \xi_{n+1}, \ldots$ are identically distributed with the density function $p_1(x)$.

Let τ be a stopping moment associated with the following sequence of σ - algebras

$$\mathcal{F}^\xi = \{F_n^\xi\}, \quad n \geq 0, \text{where} \quad F_0^\xi = \{\emptyset, \Omega\}, F_n^\xi = \sigma\{\omega : \xi_1, \ldots, \xi_n\}.$$

The stopping moment τ can be interpreted as the moment of time when the "alarm" is raised about the change-point. It should be chosen as nearly as possible to the real change-point θ.

The following functional was considered in Shiryaev (1976) as the "risk" from the stopping moment τ usage:

$$\rho^\pi(\tau) = \mathbf{P}^\pi\{\tau < \theta\} + c\mathbf{E}^\pi \max\{\tau - \theta, 0\}, \tag{9.5}$$

where

$\mathbf{P}^\pi\{\tau < \theta\}$ is the "false alarm" probability, and
$\mathbf{E}^\pi \max\{\tau - \theta, 0\}$ is the average delay time of disorder detection.

In Shiryaev (1976), the following theorem is proved.

Let $c > 0, p > 0$, and

$$\pi_n^\pi = \mathbf{P}^\pi\{\theta \leq n | F_n^\xi\} \tag{9.6}$$

is the *a posteriori* probability of the change-point presence at the moment n, $\pi_0^\pi = \pi$.

Then the moment

$$\tau_\pi^* = \inf\{n \geq 0 : \pi_n^\pi \geq A^*\}, \tag{9.7}$$

where A^* is a certain constant, will be a Bayes rule, i.e., it will minimize the "risk" $\rho^\pi(\tau)$ in the class of all stopping rules $\mathcal{M}[\mathcal{F}^\xi]$.

It was proved in Shiryaev (1976) that this rule is also optimal for an extremal (or conditionally extremal) formulation of the change-point problem. More precisely, the following problem was considered. Let $\pi \in [0,1), p \in (0,1]$. Denote by $\mathcal{M}^\xi(\alpha,\pi)$ a class of stopping moments $\tau \in \mathcal{M}[\mathcal{F}^\xi]$, such that

$$\mathbf{P}^\pi\{\tau < \theta\} \le \alpha, \tag{9.8}$$

where α is a constant, $\alpha \in [0,1)$.

It is required to find the stopping rule $\tilde{\tau} \in \mathcal{M}^\xi(\alpha,\pi)$, such that

$$\mathbf{E}^\pi \max\{\tilde{\tau} - \theta, 0\} \to \inf_{\tau \in \mathcal{M}^\xi(\alpha,\pi)} . \tag{9.9}$$

Shiryaev proved that the moment

$$\tilde{\tau} = \inf\{n \ge 0 : \pi_n^{\pi_0} \ge \tilde{A}_\alpha\}, \tag{9.10}$$

where $\tilde{A}_\alpha$ is a certain specially chosen threshold, is the optimal rule in Problem (9.8)–(9.9).

Later, Pollak (1985) proved that the method of Girshick and Rubin (9.1.4) can be obtained as the limit rule for the sequence of Bayes rules (which were found by Shiryaev) as $\pi_0 \to 0, p \to 0$. Therefore, we call Method (9.4) the GRSh (Girshick-Rubin-Shiryaev) method.

In Shiryaev (1976), a continuous time change-point problem for a random process with the following stochastic differential was considered:

$$d\xi_t = r\mathbb{I}(t \ge \theta)\, dt + \sigma dW_t, \qquad \xi_0 = 0, \tag{9.11}$$

where $\sigma^2 > 0$, and $r \neq 0$, $W = (W_t, t > 0)$ is a standard Wiener process.

It was also assumed that for the change-point θ,

$$\mathbf{P}\{\theta < t\} = 1 - e^{-\lambda t}, \tag{9.12}$$

where the constant λ is known.

For a continuous analog of the extremal Problem (9.8)–(9.9), Shiryaev has proved that as $\lambda \to 0$ and simultaneously $\alpha \to 1$, so that the value $(1-\alpha)/\lambda = T$ is fixed, the average delay time of detection $\bar{\tau}(T)$ for large values of T is

$$\bar{\tau}(T) = \frac{2\sigma^2}{r^2}\left(\ln(r^2T/2\sigma^2) - 1 - C + O(2\sigma^2/Tr^2)\right),$$

where C is the Euler constant, $C = 0,577...$

In Shiryaev (1963), three continuous analogs of the following discrete methods were compared: Page's CUSUM method, the method based on the Neumann-Pearson lemma, and the optimal method (9.1.10).

It was assumed that the change-point is preceded by a quasistationary regime when observations are interrupted seldom by "false alarms" so that the average time between false alarms $\bar{T} \simeq (1-\alpha)/\lambda$ is fixed.

As the signal/noise ratio $r^2/2\sigma^2 = 1$ and $\bar{T} \to \infty$:

$$\begin{aligned} \bar{\tau}(\bar{T}) &= \ln \bar{T} - 1 - C + o(1) \text{ optimal algorithm,} \\ \bar{\tau}(\bar{T}) &= \ln \bar{T} - 3/2 + o(1) \text{ CUSUM,} \\ \bar{\tau}(\bar{T}) &\simeq \frac{3}{2} \ln \bar{T} \text{ Neumann-Pearson method.} \end{aligned} \tag{9.13}$$

It follows from (9.13) that for large enough $\bar{T}$, the CUSUM method provides a good approximation to the optimal algorithm.

In the paper by Roberts (1959), five methods of change-point detection were numerically compared: the Shewhart chart, the method based on the Neumann-Pearson lemma, the exponential smoothing test, the CUSUM method, and the optimal Method (9.10) of Girshick-Rubin-Shiryaev.

Results of statistical simulation showed that the GRSh test has the minimal delay time for "small" disorders when the average time between false alarms is fixed. For "large" disorders, on the contrary, the exponential smoothing method and the Neumann-Pearson test have an advantage over other methods.

Suppose that a sequence of i.r.v.'s, $x_1, x_2, \ldots$, is observed with the density function $f(\cdot)$, such that

$$f(x_i) = \begin{cases} f_0(x_i), & 1 \le i < \nu, \\ f_\theta(x_i), & i \ge \nu, \end{cases} \tag{9.14}$$

where $\nu \ge 1$ is a change-point.

Denote by $\mathbf{P}_{\theta,\nu}$ ($\mathbf{E}_{\theta,\nu}$) a probability measure (expectation) corresponding to a sequence with a change-point at ν, and by $\mathbf{P}_\infty(\mathbf{E}_\infty)$ a probability measure (expectation) corresponding to a sequence without a change-point.

When the change-point ν is nonrandom but unknown, Lorden (1971) proved that the CUSUM test is asymptotically optimal for the following problem:

$$\bar{\mathbf{E}}_\theta \, \tau \triangleq \sup_{\nu \ge 1} esssup \, \mathbf{E}_{\theta,\nu}[(\tau - \nu + 1)^+ | x_1, \ldots, x_{\nu-1}] \to \inf_{\tau : E_\infty \tau \ge B}, \tag{9.15}$$
$$B \to \infty.$$

In this situation, Lorden proved that, for the CUSUM procedure,

$$\bar{\mathbf{E}}_\theta \, \tau = \frac{\ln B}{I(\theta)} \left(1 + o(1)\right), \tag{9.16}$$

where $I(\theta) = \mathbf{E}_\theta \ln f_\theta(X_1)/f_0(X_1)$.

Later, Moustakides (1986) showed that by means of randomization in the beginning of the CUSUM procedure, it will minimize the criterion (9.15) and in this sense is strictly optimal.

The idea of randomization in the beginning of observations was proposed by Pollak (1985, 1987) in order to prove the asymptotic optimality of the

GRSh Method (9.4) in the situation when the change-point is nonrandom but unknown.

For this method, Pollak has proved that

$$\bar{\mathbf{E}}_\theta \tau_R = \inf_\tau \bar{\mathbf{E}}_\theta \tau + o(1), \qquad B \to \infty,$$

where

$$\begin{aligned} \bar{\mathbf{E}}_\theta \tau &= \sup_{\nu \geq 1} \mathbf{E}_{\theta,\nu}(\tau - \nu + 1 | \tau \geq \nu), \\ \tau_R &= \inf\{n : R_n(\theta) > a\}, \\ R_n(\theta) &= \frac{f_\theta(X_i)}{f_0(X_i)}(1 + R_{n-1}(\theta)), \end{aligned} \tag{9.17}$$

and the initial condition $R_0(\theta)$ is a random variable with a known distribution.

More strong results were obtained by Pollak (1985, 1987) for the exponential family of distributions

$$f_\theta(x) = \exp\{\theta x - \psi(\theta)\}, \theta \in \Theta, \tag{9.18}$$

where $\psi(\theta)$ is a twice continuously differentiable function, $\psi(0) = \psi'(0) = 0$.

Pollak has proved that for the threshold $a = \gamma(\theta)B$, the nonrandomized moment τ_R has the following properties:

$$\begin{aligned} \mathbf{E}_\infty \tau_R &= B(1 + o(1)), \\ \bar{\mathbf{E}}_\omega \tau_R &= \frac{1}{\theta\psi'(\omega) - \psi(\theta)}[\ln B + C_R^{\theta,\omega}] + o(1), \quad \omega \in \Theta, \end{aligned}$$

where

$$\begin{aligned} C_R^{\theta,\omega} &= \ln \gamma(\theta) + \rho_{\theta,\omega} - C_1^{\theta,\omega}, \\ C_1^{\theta,\omega} &= \mathbf{E}_{\omega,1} \ln[1 + \sum_{k-1}^{\infty} \exp\{-\theta S_k + k\psi(\theta)\}], \\ \gamma(\theta) &= \lim_{b\to\infty} \mathbf{E}_{\theta,1} \exp\{-\varkappa_b\}, \quad \rho_{\theta,\omega} = \mathbf{E}_{\omega,1}\varkappa_b, \\ \varkappa_b &= \theta S_\sigma - \sigma\psi(\theta) - b \qquad (S_n = \sum_{i=1}^n X_i), \\ \sigma &= \inf\{n : \theta S_n - n\psi(\theta) \geq b\}. \end{aligned} \tag{9.19}$$

Later, Dragalin (1988) has proved that the CUSUM method of detection has similar properties. But the case of unknown parameters θ is much more interesting in practice.

Lorden (1971) introduced a class of stopping moments, such that (9.1.16) is true for all $\theta \in \Theta$:

$$\tau_T^* = \inf\{n : \max_{0\leq k\leq n} \sup_{|\theta|\geq\theta_0} [\theta(S_n - S_k) - (n-k)\psi(\theta)] > a\}, \tag{9.20}$$

where $\theta_0 = \theta_0(a) \to 0$.

It was proved by Pollak (1985) that, for the stopping rule,

$$\tau_R^* = \inf\{n : \int_\Theta R_n(\theta)\, dF(\theta) > B\gamma(F)\}, \tag{9.21}$$

where F is the probability measure on Θ with $F(\{0\}) = 0, \gamma(F) = \int_\Theta \gamma(\theta)\, dF(\theta)$, the following relationships hold true:

$$\begin{aligned} \mathbf{E}_\infty \tau_R^* &= B\,(1 + o(1))\,, \\ \bar{\mathbf{E}}_\theta \tau_R^* &= \frac{1}{I(\theta)}[\ln B + \frac{1}{2}\ln\ln B + C_R^{\theta,F}] + o(1), \end{aligned} \tag{9.22}$$

where

$$C_R^{\theta,F} = \ln\gamma(F) + \rho_{\theta,\theta} + C_3^\theta - C_1^{\theta,\theta} - 1/2,\ C_3^\theta = \frac{1}{2}\ \ln\big(\frac{\psi''(\theta)}{2\pi\, I(\theta)(F'(\theta))^2}\big).$$

The relationship (9.22) characterizes the asymptotically optimal procedure of change-point detection (by the second term $\frac{1}{2}\ln\ln B$). The analog of (9.22) was proved by Dragalin (1988) for the method of Pollak and Siegmund (1985):

$$\begin{aligned} \tau_G^* &= \inf\{n : \max_{0\le k<n} G(S_n - S_k, n-k) > a\}, \\ G(x,n) &= \int_\Theta \exp\{\theta x - n\psi(\theta)\}\, dF(\theta). \end{aligned}$$

A review of the most important results in the field of sequential change-point detection, as well as some ideas concerning modification of the CUSUM procedure for observations depending on an unknown parameter and for statistically dependent observations, can be found in Lai (1995). This work also contains a discussion about the current state of sequential change-point detection.

In the work of Shyriaev (1996), the minimax optimality of the CUSUM method for the case of continuous time and the absense of *a priori* information on the change-point was established. It was assumed that the sequence of observations $X = (X_t)_{t\ge 0}$ is defined on the probability space $(\Omega, \mathcal{F}, \mathbf{P})$ by the following model:

$$dX_t = r\mathbb{I}(t \ge \theta)\, dt + \sigma W_t,\ X_0 = 0,$$

and $\tau = \tau(\omega)$ is the Markov moment with respect to the flow of σ-algebras $(\mathcal{F}_t)_{t\ge 0}$, $\mathcal{F}_t = \sigma\{\omega : X_s,\, s \le t\}$, $t \ge 0$. In the class of stopping moments with the fixed average time between false alarms $\mathcal{M}_T = \{\tau : \mathbf{E}_\infty\tau = T\}$, the following optimality criterion was considered:

$$R(T,\tau) = \sup_{\theta\ge 0} ess\sup_\omega \mathbf{E}_\theta([\tau-\theta]^+|\mathcal{F}_\theta) \to \inf_{\tau\in\mathcal{M}_T}\ .$$

In Shiryaev (1996), the following theorem was proved.

For every $T > 0$ in the class $\mathcal{M}_T$, there exists the optimal moment τ_T^*, such that

$$\tau_T^* = \inf\{t \ge 0 : \gamma_T \ge B_T^*\},$$

where

$$\gamma_t = \max_{\theta\le t}\frac{L_t}{L_\theta},\quad L_t = \exp(\frac{r}{\sigma^2}X_t - \frac{r^2}{2\sigma^2}t).$$

The threshold B_T^* is defined (on condition $\frac{r^2}{2\sigma^2} = 1$) as the root of the equation $T = B - 1 - \ln B$.

Besides the CUSUM and GRSh methods, there is another method of change-point detection widely used in practice — the exponential smoothing algorithm of detection proposed by Robinson and Ho (1978) and Roberts (1959). In the case when there is an unknown shift in the mean of an observed sequence $\{x_t\}, t = 1, 2, \ldots$, this statistic is

$$Y_{t+1} = (1-b)Y_y + bx_t, t = 0, 1, \ldots, Y_0 = 0,$$

$0 < b < 1$ is the smoothing coefficient, and the stopping rule is defined as follows:

$$T(A) = \inf\{t : |Y_t| \geq A\},$$

where $A > 0$ is the threshold of detection.

Analogous to this procedure for continuous time, one can define an autoregressive process Y_t generated by the following stochastic differential equation:

$$dY_t = -g(Y_t\, dt - dm_t), \quad Y_0 = 0,$$

where $g > 0$ is a certain constant, and m_t is a random process with independent increments (for example, the Wiener or the Poisson process) and one shift in distribution at the moment $\theta \geq 0$.

Parameters b and A of the exponential smoothing method must be chosen so as to minimize the average delay time of detection when the average time between "false alarms" is more than a certain constant L:

$$\mathbf{E}_0 T(A) \to \min,$$
$$\mathbf{E}_\infty T(A) \geq L.$$

The values of $\mathbf{E}_\infty T(A)$, and $\mathbf{E}_0 T(A)$ can be numerically computed, either from integral equations (see Box and Tiao, 1965), or by statistical simulation (Crowder, 1987), or analytically computed from martingale identities (Novikov, 1990). Novikov and Ergashev (1988) obtained the following result.

Put

$$\exp(v_\theta(z)) = \int_0^\infty \exp(zx)\, dF_\theta(x), \quad f_\theta(b, z) = \sum_{n=0}^\infty v_\theta(b^n z),$$

where $F_\theta(x)$ is the distribution function of observations corresponding to (9.14).

Then the following theorem holds true.

Suppose that $\mathbf{E}_\theta |x_t| < \infty$, $\mathbf{P}_\theta\{x_t > A\} > 0$, $t = 0, 1, \ldots$ and the case of a one-side boundary is considered, i.e.,

$$T(A) = \inf\{t : Y_t \geq A\}.$$

Then, $\mathbf{E}_\infty \exp\{qT(A)\} < \infty$ with some $q > 0$.

If, moreover, $v_\theta(z) < \infty$ for $0 \le z < \infty$, then

$$\mathbf{E}_\theta T(A) = |\log(1-b)|^{-1} \mathbf{E}_\theta \int_0^\infty z^{-1}\{\exp(A + R(A)) - 1\} \exp\{f_\theta(b,z)\}\, dz,$$

where $R(A) = Y_{T(A)} - A$ is the overshot over the boundary A.

For normally distributed r.v.'s with a shift in the mean value of the type $\mathcal{N}(0,1) \to \mathcal{N}(r,1)$, it is possible to compute the main term of the asymptotic for $\mathbf{E}_\theta T(A)$, which enables one to determine the optimal values of the parameters b and A:

$$\begin{aligned} \mathbf{E}_\infty T(A) &= \exp\{(A^2/b)(1+o(1))\}, \quad b \to 0 \\ \mathbf{E}_0 T(A) &= (1/b)|\log(1 - A/r)|(1+o(1)), \quad b \to 0, A < r. \end{aligned}$$

In all above-mentioned works, the term "change-point" was interpreted as a spontaneous change of the one-dimensional distribution function of the sequence of random variables, or a change in the drift of the Wiener process for the continuous time.

Thus, for the parametric formulation of the problem of sequential change-point detection, the optimal and asymptotically optimal solutions are found (it should be noted that optimal methods are not known for the retrospective change-point problem).

In works of Lai (1995, 2000), Tartakovsky (1998), Tartakovsky, Li, and Yaralov (2003), and Tartakovsky, Nikiforov, and Basseville (2014), these solutions were generalized to the non-i.i.d. case. In the paper by Brodsky and Darkhovsky (2008), the minimax optimality of CUSUM and GRSh methods was proved.

This book aims presumably at consideration of change-point detection and estimation methods in the multivariate context. Therefore, I give here results for sequential change-point detection methods in univariate models without proofs (see details in Brodsky and Darkhovsky (2000)).

Everywhere below in this chapter, I assume that an observed random sequence $X = \{x(n)\}_{n=1}^\infty$ is defined on the probability space $(\Omega, \mathcal{F}, \mathbf{P})$ by the following model:

$$x(n) = a + h(n)\mathbf{I}(n \ge m) + \xi(n), \tag{9.23}$$

where m is an unknown change-point, $\xi = \{xi(n)\}_{n=1}^\infty$ is a random sequence, such that $\mathbf{E}\xi(n) \equiv 0$, and $\{h(n)\}$ is the deterministic sequence representing the "profile" of nonstationarity. In particular, for the case of a change-point $h(n) = h \ne 0$ for $n \ge m$.

9.2 Performance Characteristics of Sequential Tests

In this section, $\mathbf{P}_m(\mathbf{E}_m)$ denotes the measure (mathematical expectation) corresponding to the sequence of observations X with a change in the mean value at the moment m (the symbol $\mathbf{P}_\infty(\mathbf{E}_\infty)$ corresponds to the sequence without change-points).

The problem of sequential detection consists in decision making about possible presence of change-points, trends, or some other significant changes of statistical characteristics of observations in every step of data collection. By $d_N(\cdot)$, we denote the decision function of the method depending on the "large" parameter N. Then, $d_N(n) = 1$ corresponds to the decision about the presence of a change-point at step n and $d_N(n) = 0$ – to the decision about the absence of a change-point at step n. Formally, we can say that $d_N(\cdot)$ is the measurable function with respect to the natural flow of σ -algebras generated by the sequence X.

Define the stopping moment

$$\tau_N = \inf\{n : d_N(n) = 1\} \tag{9.24}$$

and the normalized delay time of change-point detection:

$$\rho_N = \frac{(\tau_N - m)^+}{N}. \tag{9.25}$$

Define also the probability of the 1st-type error ("false alarm")

$$\beta_N = \sup_k \mathbf{P}_\infty(d_N(k) = 1) \tag{9.26}$$

and the corresponding normalized value

$$\delta_N = \frac{|\ln \beta_N|}{N}. \tag{9.27}$$

In the sequel, we analyze asymptotics of these values as $N \to \infty$ for different methods of sequential change-point detection.

We also use the following notation: For any sequence $\{u(k)\}$, $k \geq 1$ let

$$S_j(u) = \sum_{k=1}^{j} u(k), \; j = 1, \dots, n, \; S_0(u) \equiv 0.$$

Cumulative sums method (CUSUM)

In order to understand how to construct the nonparametric version of the **CUSUM** method, let us recall the main idea of this procedure. In early works of Page (1954, 1955), it was assumed that a sequence of independent random variables $x_1, x_2, \dots$ is observed, such that the density function of observations is equal to $f_0(\cdot)$ before the moment ν and $f_1(\cdot)$ after ν. The main idea of Page consisted in construction of statistics based on the likelihood ratio:

$$g_n = \left(g_{n-1} + \ln\frac{f_1(x_n)}{f_0(x_n)}\right)^+, \; g_0 \equiv 0.$$

Here, we note that the mathematical expectation of the random variable

$y_n = \ln(f_1(x_n)/f_0(x_n))$ is negative before the change-point and positive after it:

$$\mu_0 = \mathbf{E}_0 \ln \frac{f_1(x)}{f_0(x)} < 0, \ \mu_1 = \mathbf{E}_1 \ln \frac{f_1(x)}{f_0(x)} > 0.$$

This property of the model of observations is used in the classic **CUSUM** test.

Taking this into account, consider the following nonparametric variant of the **CUSUM** statistic:

$$y(n) = (y(n-1) + x(n))^+, \ y(0) \equiv 0, n = 1, 2, \dots \tag{9.28}$$

and the corresponding decision rule:

$$d_N(\cdot) = d_N(y(n)) = \mathbb{I}(y(n) > C_{cusum}). \tag{9.29}$$

For the **CUSUM** method, it is assumed that $a < 0$, $h + a > 0$ ($a > 0$, $h + a < 0$) in (9.1.23). In this section, for the sake of definiteness, we assume that $a < 0$, $h + a > 0$.

First, we choose such a "large" parameter N for the **CUSUM** method that the order of convergence of the "false alarm" probability to zero is exponential as $N \to \infty$. In the following theorem, we prove that for this purpose, it is necessary to choose $N = C_{cusum}$.

Theorem 9.2.1. *Suppose the sequence $\{\xi(n)\}$ fulfills the conditions* $(\mathbf{SC})_\mathbf{s}$. *Then, the following asymptotic relationship holds as $N \to \infty$*

$$\sup_n \mathbf{P}_\infty (d_N(n) = 1) \le C_1 \exp(-C_2 N), \tag{9.30}$$

where $C_1 > 0, C_2 > 0$ do not depend on N.

The detailed proof of this theorem is given in Brodsky and Darkhovsky (2000).

Now, consider the normalized delay time $\rho_N = ((\tau_N - m)^+/N)$ for the **CUSUM** method. Let $(h - |a|)^{-1} \triangleq \gamma_{cusum}$.

The following theorem holds true.

Theorem 9.2.2. *i) If the sequence $\{\xi(n)\}$ fulfills* $(\mathbf{SC})_\mathbf{s}$ *condition, then for any $m \ge 1$,*

$$\rho_N \to \gamma_{cusum} \quad \mathbf{P}_m\text{-}a.s. \ as\, N \to \infty. \tag{9.31}$$

ii) If the sequence $\{\xi(n)\}$ fulfils $(\mathbf{FT})_\mathbf{s}$ *condition, then,*

$$\limsup_{N\to\infty} \mathbf{E}_m([\sqrt{N}(\rho_N - \gamma_{cusum})]^2) \le \frac{\sigma^2}{(h-|a|)^3}, \tag{9.32}$$

where σ is the parameter of the condition $(\mathbf{FT})_\mathbf{s}$.

Again, the detailed proof of this theorem is given in Brodsky and Darkhovsky (2000).

Girshick-Rubin-Shyriaev method (GRSh)

In order to understand how to construct the nonparametric version of the GRSh method, we recall its initial idea. Girshick and Rubin (1952) supposed that a sequence of independent random variables $x_1, x_2, \dots$ is observed, such that the density function of observations before the change-point ν is equal to $f_0(\cdot)$, and after the change-point, $\nu - f_1(\cdot)$.

The main idea of Girshick and Rubin consisted of construction of statistics based on the likelihood ratio,

$$W_n = (1 + W_{n-1}) \frac{f_1(x_n)}{f_0(x_n)}, \; W_0 \equiv 0, \tag{9.33}$$

and decision making according to the rule

$$d(W_n) = \mathbb{I}(W_n > C),$$

where C is the threshold of detection.

I have already mentioned that the mathematical expectation of the random variable $y_n = \ln(f_1(x_n)/f_0(x_n))$ is negative before the change-point and positive after it. This is precisely the feature of the observation model used in **GRSh**-test as well as in **CUSUM** test.

Statistic (9.33) can be written in the following way:

$$W_n = \sum_{k=1}^{n} \exp S_k^n(\xi), \qquad S_k^n(\xi) = \sum_{i=k}^{n} \xi(i), \tag{9.34}$$

where $\xi(i) = \ln\left(f_1(x(i))/f_0(x(i))\right), \mathbf{E}_0\xi(i) = \mathbf{E}_0 \ln\left(f_1(x(i))/f_0(x(i))\right) < 0$, and $\mathbf{E}_1\xi(i) = \mathbf{E}_1 \ln\left(f_1(x(i))/f_0(x(i))\right) > 0$.

Hence, we can see how to construct the nonparametric analog of the **GRSh** method for Model (9.1.23) with (for definiteness) $a < 0$, $h + a > 0$. This analog takes the form:

$$d_N(\cdot) = \mathbb{I}(R_n > C_{grsh}), \tag{9.35}$$

where

$$R_n = \sum_{k=1}^{n} \exp S_k^n(x), \quad S_k^n(x) = \sum_{i=k}^{n} x(i),$$
$$\mathbf{E}x(i) = \begin{cases} -|a| < 0, & \text{if } 1 \le i \le m \\ h - |a| > 0, & \text{if } 1 > m. \end{cases} \tag{9.36}$$

The characteristics of the Method (9.35)–(9.36) are investigated further in the same way as for the nonparametric **CUSUM** method: First, we introduce a "large parameter" N, such that the "false alarm" probability exponentially converges to zero as $N \to \infty$. Then, we prove that the normalized delay time ρ_N converges almost surely to a certain deterministic limit as $N \to \infty$.

Let us choose the "large parameter" for the **GRSh** method as follows: $N = \ln C_{grsh}$. Then, the following theorem holds true.

Theorem 9.2.3. *Suppose the sequence $\{\xi(n)\}$ fulfills $(\mathbf{SC})_\mathbf{s}$ conditions. Then, the following asymptotic relationship holds as $N \to \infty$:*

$$\sup_n |\ln \mathbf{P}_\infty \ (d_N(n) = 1)\,| = O(N).$$

Now, consider the normalized delay time $\rho_N = ((\tau_N - m)^+/N)$ for the **GRSh** method. Let $(h - |a|)^{-1} \triangleq \gamma_{grsh}$.

The following theorem holds true.

Theorem 9.2.4. i) *If the sequence $\{\xi(n)\}$ fulfills the condition $(\mathbf{SC})_\mathbf{s}$, then, for every $m \geq 1$,*

$$\rho_N \to \gamma_{grsh} \quad \mathbf{P}_m\text{-}a.s. \ as\, N \to \infty. \tag{9.37}$$

ii) If the sequence $\{\xi(n)\}$ fulfills $(\mathbf{FT})_\mathbf{s}$ condition, then,

$$\limsup_{N\to\infty} E_m([\sqrt{N}(\rho_N - \gamma_{grsh})]^2) \leq \frac{\sigma^2}{(h - |a|)^3}, \tag{9.38}$$

where σ is the parameter of the condition $(\mathbf{FT})_\mathbf{s}$.

The detailed proof of Theorems 9.2.3 and 9.2.4 is given in Brodsky and Darkhovsky (2000).

Method of exponential smoothing

The method of exponential smoothing for sequential change-point detection problems was first proposed by Roberts (1959), and Robinson and Ho (1978). Later, it was investigated both by numerical Box and Tiao (1965) and analytical methods Novikov (1990) for sequences of independent random variables with known distribution functions before and after the change-point.

The method of exponential smoothing will be considered here for dependent random sequences in the context of the nonparametric approach to sequential change-point detection.

In model (9.1.23), we assume $a = 0, h \neq 0$. Consider the following statistic:

$$Y(n) = (1 - \nu)Y(n - 1) + \nu x(n), \quad Y(0) = 0, \tag{9.39}$$

where $0 < \nu < 1$ is the smoothing coefficient.

Let C_{exp} be the threshold of detection, such that $0 < C_{exp} < |h|$. Consider the following decision rule:

$$d_N(n) = \mathbb{I}\left(|Y(n)| > C_{exp}\right),$$

where N is the "large parameter".

The "large parameter" N for the method of exponential smoothing is equal to $N \triangleq 1/\nu$.

Theorem 9.2.5. *Let the conditions* $(\mathbf{SC})_\mathbf{s}$ *hold. Then,*

$$\lim_{N\to\infty} N^{-1}|\ln \max_{1\le n\le N} \mathbf{P}_\infty(d_N(n)=1)| = \frac{C^2_{exp}}{\sigma^2(1-e^{-2})},$$

where σ *is the parameter from the conditions* $(\mathbf{SC})_\mathbf{s}$.

Now, we analyze the asymptotics of the normalized delay time ρ_N.

Theorem 9.2.6. *Let the conditions* $(\mathbf{FT})_\mathbf{s}$ *hold. Then, as* $N\to\infty$:

i) $\rho_N \to \gamma_{exp} \overset{\triangle}{=} |\ln(1-C_{exp}/|h|)$ $\mathbf{P}_m$*-a.s. for any* $m\ge 1$*; and*

ii) $\sqrt{N}(\rho_N-\gamma_{exp}) \xrightarrow{d} \eta$, *where* η *is the Gaussian variable with zero mean and dispersion*

$$Q = \frac{\sigma^2(1-(1-C_{exp}/h)^2)}{2(h-C_{exp})^2},$$

and σ *is the parameter from the conditions* $(\mathbf{FT})_\mathbf{s}$.

The detailed proof of Theorems 9.2.5 and 9.2.6 is given in Brodsky and Darkhovsky (2000).

"Moving sample" methods

Let $X=\{x(k)\}, k\ge 1$ be an observed random sequence. Let us fix the volume M of the "moving sample" and consider sequential "windows" of observations

$$X_M(n) = \{x(k)\}_{k=n-M+1}^{n}, \qquad n = M, M+1, \dots$$

For each window $X_M(n)$, the hypothesis about a change-point presence in this window must be checked. The number of the first sample $X_M(\cdot)$ in which a change-point is detected is assumed to be the estimate of an unknown change-point m.

Later, we consider two methods based upon this idea: the nonparametric method proposed in our paper Darkhovsky and Brodsky (1980) — we will call it **BD method** — and the general linear "moving sample" method. The "large parameter" N for any "moving sample" method can be chosen as $N=M$. It is natural to suppose here that the change-point is equal to $m=N+k$, $k\ge 0$.

For the "moving sample" methods, a is arbitrary, and $h\ne 0$ in the main Model (9.1.23). Later, without the loss of generality, we assume $h>0$.

A. BD method

Let us fix the parameters $0<\alpha<0,5, C_{bd}>0$ and consider the following statistic:

$$Y_N(k,n) = \frac{1}{k}\sum_{i=1}^{k} x(n-N+i) - \frac{1}{N-k}\sum_{i=k+1}^{N} x(n-N+i), \tag{9.40}$$

where $k = 1, 2, \ldots, N-1; n = N, N+1, \ldots, N > [1/\alpha]$, and the sequence $\{x(i)\}$ is defined by (9.23).

Let

$$Z_N(n) = \max_{[\alpha N] \le k \le N-[\alpha N]} |Y_N(k,n)|$$

and define the decision rule as follows:

$$d_N(n) = \mathbb{I}(Z_N(n) > C_{bd}).$$

It is easy to see that (9.40) is the main retrospective Statistic (2.6) (with $\delta = 0$) modified for the sequential case.

Theorem 9.2.7. *Let the conditions* $(\mathbf{SC})_\mathbf{s}$ *hold. Then,*

$$\lim_{N\to\infty} N^{-1} |\ln \sup_{n \ge N} \mathbf{P}_\infty\{d_N(n) = 1\}| = \frac{C_{bd}^2 \alpha(1-\alpha)}{2\sigma^2}, \tag{9.41}$$

where σ *is the parameter from the conditions* $(\mathbf{SC})_\mathbf{s}$.

Theorem 9.2.8. *Let the conditions* $(\mathbf{FT})_\mathbf{s}$ *hold,* $h > C_{bd}$ *and* $m = N + k$, $k \ge 0$. *Then, for any fixed* k:

i)

$$\rho_N \xrightarrow{a.s.} \gamma_{bd} \triangleq \alpha(C_{bd}/h); \text{ and} \tag{9.42}$$

ii)

$$\zeta_N \le \sqrt{N}(h/\alpha) \min(\rho_N - \gamma_{bd}, \alpha - \gamma_{bd}) \le \eta_N,$$

where

$$\zeta_N \xrightarrow{d} \frac{\sigma W^0(1-\alpha)}{\alpha(1-\alpha)}$$

$$\eta_N \xrightarrow{d} \max_{\alpha \le t \le 1-\alpha} \frac{\sigma W^0(t)}{t(1-t)}$$

and σ *is the parameter from the conditions* $(\mathbf{FT})_\mathbf{s}$.

The detailed proof of Theorems 9.2.7 and 9.2.8 is given in Brodsky and Darkhovsky (2000).

B. General linear "moving-sample" method

Let $g(t), t \in [0,1]$ be bounded and almost everywhere (with respect to Lebesgue measure) continuous function. The following main statistic is considered:

$$Y_N(n) = N^{-1} \sum_{k=0}^{N-1} g(k/N) x(n-k), \quad n = N, N+1, \ldots,$$

and the decision rule

$$d_N(n) = \mathbb{I}(|Y_N(n)| > C_{gm}),$$

where $C_{gm} > 0$ is the threshold of detection (the parameter of the method).

In virtue of our assumptions, $g(\cdot)$ is square integrable. The value $\int_0^1 g^2(t)dt$ is the scale parameter.

It is easy to see that the general linear "moving sample" method is the direct generalization of the exponential smoothing method (in that method we used the function g_N, $\|g_N - e^{-t}\|_2 \to 0$). Therefore, analogs of **Theorems 9.2.5 and 9.2.6** hold true. The final results are formulated as follows.

Theorem 9.2.9. *i) Let the conditions* $(\mathbf{SC})_\mathbf{s}$ *hold and* $C_{gm} < h$. *Then,*

$$\lim_{N\to\infty} N^{-1}|\ln \max_{1\le n\le N} \mathbf{P}_\infty (d_N(n) = 1)| = \frac{C_{gm}^2}{2\sigma^2 \int_0^1 g^2(t)dt}, \tag{9.43}$$

where σ *is the parameter from the conditions* $(\mathbf{SC})_\mathbf{s}$.

ii) Let the conditions $\mathbf{FT})_\mathbf{s}$ *hold,* $m = N + k, k \ge 0$. *Then, for any fixed* k:

$$\rho_N \xrightarrow{a.s.} \gamma_{gm}$$

$$\sqrt{N}(\rho_N - \gamma_{gm}) \xrightarrow{d} \eta, \tag{9.44}$$

where γ_{gm} *is the minimal root of the equation* $\int_0^\gamma g(t)dt = C_{gm}/|h|$, *and* η *is the Gaussian mean zero random variable with the variance*

$$Q = (|h|)^{-2} g^{-2}(\gamma_{gm})\sigma^2 \int_0^1 g^2(t)dt,$$

where σ *is the parameter of the conditions* $(\mathbf{FT})_\mathbf{s}$.

The proof of this theorem is analogous to **Theorems 9.2.5 and 9.2.6**.

9.3 A Priori Estimates of Quality of Sequential Change-Point Detection Methods

In the last section, it was shown that for all nonparametric analogs of sequential change-point detection methods, a "large parameter" N exists, such that the probability of the error decision about the presence of a change-point ("false alarm") converges exponentially to zero as $N \to \infty$ and the normalized delay time in change-point detection converges to a certain deterministic limit as $N \to \infty$.

The goal of this section is the proof of *a priori* informational estimates that combine the characteristics the normalized "false alarm" probability and the normalized delay time in change-point detection, as well as the analog of the Rao-Cramer inequality for sequential problems that furnishes the *a priori* estimate of the rate of convergence of the normalized delay time to its limit as $N \to \infty$.

It is important to emphasize the **principal difference** of this approach to the analysis of the qualitative characteristics of sequential change-point detection methods from the traditional approach based upon the Wald sequential analysis and the nonlinear renewal theory. This traditional approach is oriented to the analysis of the mathematical expectation of the delay time and the average time between "false alarms" for sequences of independent random variables (or processes with independent increments). The **nonparametric approach** allows us to analyze different situations with spontaneous statistical nonhomogeneities (change-points, trends, outliers) for **dependent sequences** without the use of *a priori* data on distributions of observations, as well as robust properties of sequential diagnostic procedures.

I begin with the central *a priori* inequality, which combines the normalized characteristics of the delay time and the "false alarm" probability.

9.3.1 A Priori Informational Estimate for the Delay Time

Suppose $\mathbf{Z} = (\mathbf{z}_1, \mathbf{z}_2, \dots)$ is a sequence of dependent vector-valued observations $\mathbf{z}_n = (z_n^1, \dots, z_n^k)$ defined on the probability space $(\Omega, \mathcal{F}, \mathbf{P})$, and m is a change-point.

Let $\mathbf{P}_\infty$ $(\mathbf{E}_\infty)$ be the measure (mathematical expectation) corresponding to the sequence without change-points and $\mathbf{P}_m$ $(\mathbf{E}_m)$ be the measure (expectation) corresponding to the sequence with the change-point m.

Everywhere below, I assume that measures $\mathbf{P}_\infty$ and $\mathbf{P}_m$ are mutually absolutely continuous for any fixed $m \geq 1$.

Let μ be σ-finite measure on $\mathbb{R}^1$ and μ^s denote the product measure on $\mathbb{R}^s$ corresponding to μ.

I assume that the corresponding constraints of measures $\mathbf{P}_\infty$ and $\mathbf{P}_m$ on σ-algebra $\sigma(\mathbf{z} : \mathbf{z}_1, \dots, \mathbf{z}_n)$ have densities with respect to (w.r.t.) measure μ^{kn}. Denote these densities, correspondingly, as $f_{\infty,n}(\mathbf{z}_1, \dots, \mathbf{z}_n) \stackrel{\text{def}}{=} f_{\infty,n}(\mathbf{z}_1^n)$ and

$$f_{m,n}(n, \mathbf{z}_1^n) = \begin{cases} f_{\infty,n}(\mathbf{z}_1^n), & n < m \\ g_n(\frac{n-m}{N}, \mathbf{z}_1^n), & n \geq m \end{cases}, \tag{9.44}$$

where for any $n \geq m$, the map $g_n(t, \cdot) : [0, \infty) \to L_1(\mathbb{R}^{kn}, \mu^{kn})$.

In this section, I consider the a priori lower bound for a certain performance measure connected with the delay time in change-point detection. For this purpose, consider the following decision rule $d_N(n)$ depending on a large parameter N:

$$d_N(n) = \begin{cases} 1, \text{stop at time } n \text{ and accept decision about the change-point}, \\ 0, \text{continue to collect observations} \end{cases}$$

Define the following performance characteristics of a change-point detection method:

— The supremal probability of the false decision:

$$\alpha_N = \sup_n \mathbf{P}_\infty\{d_N(n) = 1\}.$$

— The stopping time:

$$\tau_N = \min\{n : \ d_N(n) = 1\}.$$

— The normalized stopping time:

$$\gamma_N = (\tau_N - m)^+/N.$$

We also consider the following value:

$$M_N(m) = \min\{l \geq 0 : \sum_{n=m+l}^{\infty} \mathbf{P}_\infty\{\tau_N = n\} \leq \alpha_N\}.$$

Define the following functions:

— Conditional densities $p_{\infty,n}(\mathbf{z}_n|\mathbf{z}_1^{n-1})$, $n \geq 2$, and $q_{m,n}(\frac{n-m}{N}, \mathbf{z}_n|\mathbf{z}_1^{n-1})$, $n \geq m$;

— Conditional log-likelihood function:

$$\varphi(t, \mathbf{z}_n, \mathbf{z}_1^n) \stackrel{\text{def}}{=} \ln \frac{q_{m,n}(t, \mathbf{z}_n|\mathbf{z}_1^{n-1})}{p_{\infty,n}(\mathbf{z}_n|\mathbf{z}_1^{n-1})}, \ n \geq m.$$

— The Kullback-Leibler information measure.

$$J(n,t) = \mathbf{E}_m \varphi(t, \mathbf{z}_n, \mathbf{z}_1^n) = \int f_{m,n}(t,x) d\mu^{kn}(x) \int \varphi(t,u,x) q_n(t,u|x) d\mu^k(u), \ n \geq$$

Theorem 9.3.1. *Suppose the function $J(n,t)$ is Riemann integrable on $[0,\infty)$ for any $n \geq m$. Then for an arbitrary method of sequential change-point detection dependent on a "large parameter" N:*

$$\mathbf{E}_m \int_0^{\gamma_N} J(n,t)dt \geq \frac{|\ln(\alpha_N M_N(m))|}{N} + O(\frac{1}{N}), \qquad \text{as } N \to \infty. \tag{9.45}$$

The proof of Theorem 9.3.1 is given in Chapter 11 for a more general situation of sequential detection of a gradual change-point.

Now, we formulate the result of this theorem for a sequence of independent scalar r.v.'s.

Suppose $X = (x_1, x_2, \dots)$ is a sequence of scalar independent random variables, such that before the change-point m, their densities (w.r.t. measure μ) are the same, and after the moment m, their densities change with time according to the previous considerations. Namely, in this case, we have

$$\begin{aligned}
&p_{\infty,n}(x_n|x_1^{n-1}) = p_{\infty,n}(x_n) \overset{\text{def}}{=} p_\infty(x),\ n < m,\\
&q_{m,n}(\tfrac{n-m}{N}, x_n|x_1^{n-1}) = q_{m,n}(\tfrac{n-m}{N}, x_n) \overset{\text{def}}{=} q_{m,n}(t,x),\ n \geq m,\\
&\varphi(t, x_n, x_1^n) = \varphi(t, x_n) = \ln \frac{q(\frac{n-m}{N}, x_n)}{p(x_n)} \overset{\text{def}}{=} \varphi(t,x),\\
&J(n,t) = J(t) = \textstyle\int \varphi(t,x) q(t,x) d\mu(x).
\end{aligned}$$

Corollary 9.3.1. *Suppose the function $J(t)$ is Riemann integrable on $[0,\infty)$. Then, for an arbitrary method of sequential change-point detection dependent on a "large parameter" N:*

$$\mathbf{E}_m \int_0^{\gamma_N} J(t)dt \geq \frac{|\ln(\alpha_N M_N(m))|}{N} + O(\frac{1}{N}), \qquad \text{as } N \to \infty. \tag{9.46}$$

Let us discuss why the theorem and the corollary are relevant for the proof of the asymptotic optimality of an arbitrary change-point detection method. Remark that the left-hand side of the a priori inequalities presented in these theorems depends on the normalized delay time in change-point detection (the main reason for the use of this value), while the right-hand side depends on the probability of the false decision. In the sequel, we prove that the left-hand and the right-hand sides tend to certain limits as $N \to \infty$ for all well-known methods of sequential change-point detection. So, we can compare these limit values for a concrete method. If they coincide for a certain choice of the method's parameters, then we say that this method is *asymptotically optimal.*

9.3.2 Rao-Cramer Inequality for Sequential Change-Point Detection

Consider the sequence $\{x_n\}$ of independent random variables with the density function $f_0(\cdot)$ (with respect to a certain σ-finite measure μ) before the change-point $m \geq 1$, and the density function $f(\theta, \cdot)$ after the change-point. Here,
$\theta \in \Theta$ is a certain parameter. Later, the symbols $\mathbf{P}_{\theta m}(\mathbf{E}_{\theta m})$ denote the probability (mathematical expectation) corresponding to the sequence $\{x_n\}$ with the change-point m.

In this and the following subsection, we consider the class of methods $\mathcal{T}_N$ for which the normalized delay time in change-point detection converges almost surely to some constant as $N \to \infty$.

Let τ_N be the moment of the first decision about the presence of a change-point by method A. Here, we consider the rate of convergence of the normalized delay time $\gamma_N = (\tau_N - m)^+/N$ to its limit as $N \to \infty$.

Theorem 9.3.2. *Let the following conditions hold:*

i) θ *belongs to a certain domain* Θ *in the finite-dimensional space.*

ii) The density $f(\theta, z)$ *is differentiable with respect to* θ *for any* z*, the derivative* $f_\theta'(\theta, z)$ *is continuous with respect to both arguments, and there exists Fisher information*

$$\mathbf{J}(\theta) \triangleq \int \left(\frac{\partial}{\partial\theta}\ln f(\theta,x)\right)^2 f(\theta,x)\mu(dx) = \int \frac{\left(f_\theta'(\theta,x)\right)^2}{f(\theta,x)}\mu(dx),$$

with $\sup_{\theta\in\Theta}\mathbf{J}(\theta) < \infty$.

iii) The sequence ρ_N *converges for any* m $\mathbf{P}_{\theta m}$*-a.s. to the deterministic limit* $\gamma(\theta)$*, and the function* $\gamma(\theta)$ *is differentiable;*

iv) For any $m \geq 1$*, there exists the uniform limit with respect to* $\theta \in \Theta$

$$\lim_{N\to\infty}\lim_{u\to\infty} N^{-1}\sum_{k=m+1}^{u}\frac{d}{d\theta}(k-m)\mathbf{P}_{\theta m}\{\tau_N^* = k\} = \gamma'(\theta),$$

where $\tau_N^* = m + (\tau_N - m)^+ \wedge [\gamma(\theta)N]$.

Then the following inequality holds true (the Rao-Cramer inequality):

$$\liminf_{N\to\infty}\mathbf{E}_{\theta m}\left(\sqrt{N}(\gamma_N - \gamma(\theta))\right)^2 \geq \frac{\left(\gamma'(\theta)\right)^2}{\mathbf{J}(\theta)\gamma(\theta)}. \tag{9.47}$$

Proof. By $S(x_i, \theta)$, we denote the random variables $\frac{\partial}{\partial\theta}\ln f_\theta(x_i), i = 1, 2, \ldots$ The continuous differentiability of the density yields $\mathbf{E}_{\theta m}S(x_i,\theta) = 0$.

Since $S(x_i, \theta)$ are independent and $\mathbf{E}_{\theta m}\tau_N^* < \infty$, Wald's identity yields for every $\theta \in \Theta$

$$\mathbf{E}_{\theta m}\sum_{i=m+1}^{\tau_N^*} S(x_i,\theta) = 0$$

and, therefore,

$$\frac{d}{d\theta}\mathbf{E}_{\theta m}\gamma(\theta)\sum_{i=m+1}^{\tau_N^*} S(x_i,\theta) = 0. \tag{9.48}$$

Since $(\tau_N^* - m)/N \overset{\mathbf{P}_{\theta m}}{\longrightarrow} \gamma(\theta)$, the variables $S(x_i\theta)$ are independent and have the finite dispersion according to *ii)*, we can use **Corollary 1.5.1**, which yields for any fixed m and any $\theta \in \Theta$:

$$\frac{1}{\sqrt{N}}\sum_{i=m+1}^{\tau_N^*} S(x_i,\theta) \overset{d}{\longrightarrow} \sqrt{\mathbf{J}(\theta)}W(1). \tag{9.49}$$

Further, for any measurable set $B \in \mathbb{R}^k$,

$$\frac{d}{d\theta}\int_B \prod_{i=1}^{k} f(\theta,z_i)\prod_{i=1}^{k} d\mu(z_i) = \int_B \sum_{i=1}^{k}\frac{\partial}{\partial\theta}f(\theta,z_i)\prod_{i=1}^{k} d\mu(z_i).$$

Therefore,

$$N^{-1}\sum_{k=m+1}^{\infty}(k-m)\frac{d}{d\theta}\mathbf{P}_{\theta m}\{\tau_N^*=k\}=\mathbf{E}_{\theta m}\left((\tau_N^*-m)^+/N)\sum_{i=m+1}^{\tau_N^*}S(x_i,\theta)\right). \tag{9.50}$$

From (9.49) and (9.50), taking into account *iv)*, we write

$$\frac{d}{d\theta}\gamma(\theta)=\frac{d}{d\theta}\lim_{N\to\infty}N^{-1}\sum_{k=m+1}^{\infty}(k-m)\mathbf{P}_{\theta m}\{\tau_N^*=k\}=$$
$$=\mathbf{E}\left(((\tau_N^*-m)^+/N-\gamma(\theta))\sum_{i=m+1}^{\tau_N^*}S(x_i,\theta)\right). \tag{9.51}$$

In virtue of the Schwarz inequality,

$$\mathbf{E}_{\theta m}\left((\tau_N^*-m)^+/N-\gamma(\theta))\sum_{i=m+1}^{\tau_N^*}S(x_i,\theta)\right)\leq$$

$$\sqrt{\mathbf{E}_{\theta m}\left(\sqrt{N}((\tau_N^*-m)^+/N-\gamma(\theta))\right)^2}\sqrt{\mathbf{E}_{\theta m}\left(\frac{1}{\sqrt{N}}\sum_{i=m+1}^{\tau_N^*}S(x_i,\theta)\right)^2}.$$

Therefore, we obtain,

$$\mathbf{E}_{\theta m}\left(\sqrt{N}((\tau_N^*-m)^+/N-\gamma(\theta))\right)^2\geq\frac{\left(\gamma'(\theta)\right)^2}{\mathbf{J}(\theta)\gamma(\theta)}.$$

The required result follows from here immediately. ■

Remark 9.3.1. The most difficult to check is condition *iv)* of the theorem. However, this condition holds if the rate of convergence of the normalized delay time to its limit is exponential, i.e.,

$$\mathbf{P}_{\theta m}\{\tau_N^*=k\}\leq L\exp\left(-(\lambda(\theta)|k-\gamma(\theta)N|)\right),$$

where $\lambda(\theta)$ is a smooth function. But this order of convergence was established in **Subsections 9.2.1–9.2.5** for all sequential change-point detection methods.

Consider our basis Model (9.23) with independent observations $\{\xi(n)\}$. From the proof of the theorem, it is clear that the random variables$\{\xi(n)\}$ should not necessarily be identically distributed. In fact, let $\varphi_{\theta i}(\cdot)$ be the density function of the random variable $\xi(i)$ (with respect to a certain σ-finite measure $\mu(\cdot)$). Put

$$\sigma_i(m,\theta)=\int\left(\frac{(d/d\theta)\varphi_{\theta i}(x-h)}{\varphi_{\theta i}(x)}\right)^2\varphi_{\theta m}(x)d\mu(x)$$

and suppose that there exists the limit

v)

$$\lim_{n\to\infty} \frac{1}{n} \sum_{i=m+1}^{n} \sigma_i(m,\theta) = \mathbf{J}^*(\theta, m).$$

Then, Inequality (9.47) still holds if the value $\mathbf{J}(\theta)$ is changed for $\mathbf{J}^*(\theta, m)$ (it can be established with the help of the generalization of Wald's identity as in 9.3.1).

However, we can make further generalizations. The limit $\gamma(\theta)$ exists (for Model (9.23), and all methods considered above) not only for independent observations $\{\xi(n)\}$, but also for dependent sequences satisfying $(\mathbf{FT})_\mathbf{s}$ conditions. The value of this limit is not connected with characteristics of dependence but only with parameters of the method and the size of the "jump" h. Hence, we obtain the following analog of the Rao-Cramer inequality for Model (9.23) and all methods considered in **Subsections 9.2.1–9.2.5**, if conditions *i)–iii)*, *v)*, and $(\mathbf{FT})_\mathbf{s}$ are satisfied:

$$\liminf_{N\to\infty} \mathbf{E}_{\theta m} \left(\sqrt{N}(\gamma_N - \gamma(\theta))\right)^2 \geq \frac{(\gamma\prime(\theta))^2}{\mathbf{J}^*(\theta, m)\gamma(\theta)} \tag{9.52}$$

9.4 Asymptotic Comparative Analysis of Sequential Change-Point Detection Methods

In the previous subsection, I established *a priori* informational inequalities for sequential change-point problems. These inequalities provide *a priori* informational estimates for the limit value of the normalized delay time in change-point detection and for the rate of convergence of the normalized delay time to its limit.

This approach to the comparative analysis of sequential methods differs essentially from the classic methodology based upon the nonlinear renewal theory and Wald's sequential procedures. The principle of this approach is as follows: parameters of different sequential methods are chosen in such a way that the average time between "false alarms" is asymptotically the same for different compared methods of detection. After that, for a given change $h = h^*$ in the mean value of an observed sequence (corresponding to *a priori* known distribution functions before and after the change-point), it is found which method gives the minimal value of the average delay time in change-point detection. The main drawback of this approach is that the size of a change h^* is almost always *a priori* unknown. Therefore, such comparison of characteristics in a unique point h^* can lead to false conclusions about comparative advantages of different sequential procedures. I show later that the typical situation is as follows: for different values of h^* (the "size" of a change in the mean value), the comparative advantage is gained by different sequential methods of detection.

In this connection, it is useful to introduce the following definitions.

Definition 9.4.1. A nonparametric method of sequential detection is called the *asymptotically* τ *– optimal* for the concrete (but *a priori* unknown) value of a change h, if any of Inequalities (9.45) turns into strict identity for this method.

Definition 9.4.2. The nonparametric method of sequential detection is called the *asymptotically* σ *– optimal* for the concrete (but *a priori* unknown) value of a change h, if Inequality (9.47) turns into strict identity for this method.

An "ideal" method of change-point detection would furnish both identities in (9.45) and (9.47) for any value of a change h. However, this situation is impossible. Therefore, our methodology of the comparative analysis is based upon determination of those diapasons of h, where one of the sequential change-point detection methods gains a comparative advantage.

For the purpose of comparative analysis, it is necessary to find the values $\delta^*(\cdot)$ (or $\delta(\cdot)$) for the concrete methods. For the exponential smoothing method and the family of the "moving sample" methods, we have found these values in **Section 9.2**. For the sake of convenience, we write them again here:

$$\delta^*_{exp} = \frac{C^2_{exp}}{\sigma^2(1-e^{-2})}$$

$$\delta^*_{bd} = \frac{C^2_{bd}\alpha(1-\alpha)}{2\sigma^2}$$

$$\delta^*_{gm} = \frac{C^2_{gm}}{2\sigma^2},$$

where C is the threshold of detection, σ is the parameter from conditions $(\mathbf{SC})_\mathbf{s}$, α is the parameter of the minimax **BD** method.

Below, we find the value of δ for **CUSUM** and **GRSh** methods.

9.4.1 Analysis of the "False Alarm" Probability for CUSUM and GRSh Methods

Let $\{x(n)\}$ be a random sequence defined by (9.23) with $a < 0, h + a > 0$ (recall that the methods **CUSUM** and **GRSh** were considered under these assumptions). Suppose that the conditions $(\mathbf{SC})_\mathbf{s}$ are fulfilled for Model (9.23).

For $0 \le t \le H$, where $H > 0$ is the constant from the Cramer condition, define the function $\varkappa(t)$ as follows:

$$\varkappa(t) = \ln \sup_n \mathbf{E}_\infty \exp\left(tx(n)\right),$$

and consider the equation

$$\varkappa(t) = 0.$$

Since $\mathbf{E}_\infty \exp(tx(n)) = 1 + t\mathbf{E}_\infty x(n) + \frac{t^2}{2}\mathbf{E}_\infty x^2(n) + \ldots$ and $\mathbf{E}_\infty x(n) \equiv \equiv a < 0$, $\sup_n \mathbf{E}_\infty x^2(n) < \infty$, then for small enough $t > 0$ $\varkappa(t) < 0$, and beginning from a certain $t^* > 0$: $\varkappa(t) > 0$.

Theorem 9.4.1. *Suppose the function $\varkappa(t)$ is continuous and has only two zeros: 0 and $t^* > 0$. Let the conditions $(\mathbf{SC})_\mathbf{s}$ be fulfilled. Then, for the method* ***GRSh****, the following relationship holds:*

$$\lim_{z \to \infty} (\ln z)^{-1} |\ln \mathbf{P}_\infty\{\sum_{k=1}^{\infty} \exp(S_k(x)) > z\}| = t^*. \tag{9.53}$$

Now, we formulate the corresponding result about the limit value of the normalized "false alarm" probability for the **CUSUM** method.

Theorem 9.4.2. *Suppose all conditions of* **Theorem 9.4.1** *hold. Then*

$$\delta_{cusum} = \delta_{grsh} = t^*.$$

Now, we give some examples illustrating computation of the limit value of the normalized "false alarm" probability for the **CUSUM** and **GRSh** methods.

Suppose the d.f. of observations $x(n), n = 1, 2 \ldots$ is defined as follows:

$$f(x(n)) = f_0(x(n))\,\mathbb{I}(n < m) + f_\theta(x(n))\,\mathbb{I}(n \geq m), \ n = 1, 2, \ldots \tag{9.54}$$

where m is the change-point and $f_\theta(x)$ belongs to the exponential family of distributions

$$f_\theta(x) = f_0(x) \exp(\theta x - \phi(\theta)), \quad \phi(0) = 0, \quad \theta \in (-\infty, +\infty). \tag{9.55}$$

We assume that the function $\phi(\theta)$ is convex, has a unique nonzero root θ_1, and $\phi(\theta) < \infty$ for any θ. Suppose also that the random sequence $x(n)$ fulfills the ψ-mixing condition.

In virtue of these assumptions, the $(\mathbf{SC})_\mathbf{s}$ conditions are fulfilled, and, therefore, we can use the result of **Theorem 9.4.1**. Then, we obtain

$$a = \mathbf{E}_\infty x(n) = \phi'(0) < 0,$$
$$h - |a| = \mathbf{E}_m x(n) = \phi'(\theta) > 0 \quad \text{for } \theta > \theta_0, \quad \phi'(\theta_0) = 0,$$
$$\delta_{cusum} = \delta_{grsh} = \theta_1.$$

In particular, for the sequence of independent Gaussian random variables with the dispersion σ^2 (and a change in the mathematical expectation), $\delta_{cusum} = \delta_{grsh} = 2|a|/\sigma^2$.

9.4.2 Asymptotic Optimality of Sequential Change-Point Detection Methods

In this section, the asymptotic optimality of the change-point detection methods will be investigated. The analysis of the asymptotic optimality is based upon comparison of the limit characteristics of the normalized delay time in detection and the rate of convergence of the normalized delay time with the *a priori* informational boundaries established in **Theorems 9.3.1 and 9.3.2** (Inequalities (9.45) and (9.47)).

Everywhere in this section, we suppose that an observed random sequence $\{x(n)\}$ has the one-dimensional density satisfying (9.54) and (9.55), and the ψ-mixing condition is fulfilled. Therefore, the conditions $(\mathbf{SC})_\mathbf{s}$ are satisfied.

Besides, we suppose that conditions $(\mathbf{FT})_\mathbf{s}$ are fulfilled (more precisely, we add to $(\mathbf{SC})_\mathbf{s}$ items *i) and iii)* from the condition $(\mathbf{FT})_\mathbf{s}$).

These assumptions allow us to use all results from the previous sections for the analysis of the asymptotic optimality of sequential change-point detection methods.

Consider the functions

$$\Delta(h) \triangleq \gamma_\cdot(h) - \mathbf{I}_{0,1}^{-1}\delta_\cdot(h),$$
$$\Delta^*(h) \triangleq \gamma_\cdot(h) - \mathbf{I}_{0,1}^{-1}\delta_\cdot^*(h).$$

CUSUM and GRSh methods

In the previous, section it was established that $\delta_{cusum} = \delta_{grsh} = \theta_1, \gamma_{cusum} = \gamma_{grsh} = (h - |a|)^{-1} = \left(\phi'(\theta)\right)^{-1}$ for $\theta > \theta_0$, and, as it is easy to check, the Kullback information is equal to $\mathbf{I}_{0,1} \triangleq \mathbf{I}(\theta) = \theta\phi'(\theta) - \phi(\theta) > 0$.

For the Gaussian sequence of independent observations, Model (9.23) takes the form

$$x(n) = a + \xi(n) + h\mathbb{I}(n \geq m),$$

where $\xi(n) \sim \mathcal{N}(0, \sigma^2)$, $a < 0$, $h - |a| > 0$.

In this case, from Theorem 9.3.1, we obtain

$$\frac{1}{h - |a|} \geq \frac{4|a|}{h^2}$$

and, therefore,

$$\Delta(h) = \frac{1}{h - |a|} - \frac{4|a|}{h^2} = \frac{(h - 2|a|)^2}{h^2(h - |a|)}.$$

The point $h = 2|a|$ with $\Delta(h) = 0$ corresponds to the classic Page's test:

$$a = -(\mu_1 - \mu_0)^2/2\sigma^2, \quad h = (\mu_1 - \mu_0)^2/\sigma^2,$$

where μ_0 and μ_1 are the mean values of the diagnostic sequence formed from an initial sample with the use of the likelihood ratio before and after the change-point correspondingly (see the beginning of **Section 9.2**).

In **Theorems 9.2.2 and 9.2.4** it was proved that for any $m \geq 1$

$$\lim_{N\to\infty} \mathbf{E}_m \left(\sqrt{N}(\rho_N - \frac{1}{h-|a|})\right)^2 \leq \frac{\sigma^2}{(h-|a|)^3}. \quad (9.56)$$

On the other hand, from Inequality (9.47), for observations with the density function belonging to the exponential family (9.54)–(9.55) with $\mathbf{E}_\theta x = \phi'(\theta)$, $\mathbf{D}_\theta x = \sigma^2 = \phi''(\theta)$, i.e., for $\gamma(\theta) = 1/\phi'(\theta)$, $\gamma'(\theta) = -\phi''(\theta)/(\phi'(\theta))^2$, $\mathbf{J}(\theta) = \phi''(\theta)$, it follows that for any $m \geq 1$,

$$\liminf_{N\to\infty} \mathbf{E}_{\theta m} \left(\sqrt{N}(\rho_N - \frac{1}{\phi'(\theta)})\right)^2 \geq \frac{\phi''(\theta)}{(\phi'(\theta))^3}. \quad (9.57)$$

From (9.56) and (9.57), we can see that that the equality in the Rao-Cramer estimate is attained for such a sequence. Thus, the **CUSUM** and **GRSh** methods are asymptotically σ-optimal for a wide class of distributions satisfying Models (9.54)–(9.55) of the exponential family.

Exponential smoothing method

From the results obtained above for the exponential smoothing method, it follows that

$$\delta^*(\cdot)_{exp} = C_{exp}^2/[\sigma^2(1-e^{-2})],\ \gamma(h)_{exp} = -\ln(1 - \frac{C_{exp}}{h}).$$

As before, for the case of the Gaussian sequence of observations, $\mathbf{I}_{0,1}(h) = h^2/2\sigma^2$, and for the function $\Delta(h)$, we obtain

$$\Delta(h) = -\ln(1 - \frac{C_{exp}}{h}) - \frac{2C_{exp}^2}{h^2(1-e^{-2})}.$$

Thus, the method of exponential smoothing is not asymptotically τ-optimal.

Now, we analyze the asymptotic σ-optimality for the exponential smoothing method with the following value of the rate of convergence of the normalized delay time:

$$\liminf_{N\to\infty} \mathbf{E}_m[\sqrt{N}(\rho_N - \gamma_{exp})]^2 = \frac{\sigma^2}{2}[1-(1-C_{exp}h^{-1})^2](h-C_{exp})^{-2}.$$

The lower boundary of the Rao-Cramer inequality for the exponential smoothing method in the Gaussian case is equal to $\sigma^2(C_{exp}/h)^2(h-C_{exp})^{-2}\,|\ln(1 - C_{exp}/h)|^{-1}$. This lower boundary is nowhere attainable for the exponential smoothing method.

"Moving sample" methods

BD method

For the **BD** method, it follows from the functional limit theorem (see **Chapter 2**) that for any $n > N$ the process $\sqrt{N}\, Y_N\,([Nt], n)$ weakly converges in $D(\alpha, 1-\alpha)$ as $N \to \infty$ to the process $\sigma[t(1-t)]^{-1}\, W^0\,(t)$, where $W^0\,(t)$ is the standard Brownian bridge. Therefore, $\beta_N \sim g_N$ as $N \to \infty$, where

$$g_N = \mathbf{P}\{\max_{\alpha \le t \le 1-\alpha} \frac{W^0\,(t)}{t(1-t)} \ge \frac{C_{bd}}{\sigma}\sqrt{N}\}.$$

It follows from the results of **Chapter 2** that

$$\overline{\lim_{N\to\infty}}\ \frac{1}{N}\,|\ln\, g_N| = \frac{1}{2}\,(C_{bd}/\sigma)^2\,\alpha\,(1-\alpha).$$

Therefore, for the **BD** method,

$$\Delta\,(h) = \alpha\,\frac{C_{bd}}{|h|} - \frac{C_{bd}^2}{h^2}\,\alpha\,(1-\alpha).$$

We conclude from here that the **BD** method is not asymptotically τ-optimal. Let us now analyze τ- and σ-optimality of the "moving sample" methods in the general situation.

General method

Note that the value $\int_0^1 g^2(t)dt$ is the scale parameter for the general "moving sample" method. Therefore, without loss of generality, we further assume $\int_0^1 g^2(t)dt = 1$. Then, from the results obtained above for the method, it follows that the limit value of the normalized "false alarm" probability is equal to $\delta^*_{gm} = C^2_{gm}/2\sigma^2$, and the limit value of the normalized delay time in detection γ_{gm} is the root of the equation $\int_0^{\gamma_{gm}} g(t)dt = C_{gm}/|h|$.

For the case of the Gaussian stationary d.f. of observations, the main *a priori* inequality (9.45) takes the form: $\gamma_{gm} \ge C^2_{gm}/h^2$. Since $C_{gm}/h = N^{-1}\sum_{k=0}^{[\gamma_{gm}N]} g(k/N) + O(N^{-1})$ as $N \to \infty$, then the last inequality can be written as follows:

$$\gamma_{gm} \ge N^{-2}\left(\sum_{k=0}^{[\gamma_{gm}N]} g(k/N)\right)^2. \tag{9.58}$$

Taking into account that $\int_0^1 g^2(t)dt = 1$, we denote

$$u_i = \frac{1}{\sqrt{N}} g(i/N),\ \sum_{i=1}^{N} u_i^2 = 1.$$

Then, Inequality (9.58) can be written in the following equivalent form:

$$\sqrt{N\gamma_{gm}} - \sum_{i=1}^{[N\gamma_{gm}]} u_i \ge 0. \tag{9.59}$$

This inequality can be used for estimation of the quality of the general "moving sample" method with the fixed set of the coefficients $N^{-1}g(k/N)$. Since the value γ_{gm} depends on h (without loss of generality $0 \le \gamma_{gm} \le 1$), and the size of a change h is *a priori* unknown, the problem of the optimal choice of the weight function $g(t)$ can be formulated as the minimax one: It is necessary to minimize the maximal (with respect to γ_{gm}) value of the left side of the last inequality on the set of the vectors $u = (u_1, \dots, u_N)$, such that $\sum_{i=1}^{N} u_i^2 = 1$.

Denote $n = [N\gamma_{gm}]$. Then, the problem takes the form:

$$\max_{1 \le n \le N} (\sqrt{n} - \sum_{i=1}^{n} u_i) \to \min_{u: \sum_{i=1}^{N} u_i^2 = 1} . \tag{9.60}$$

Theorem 9.4.3. *The optimal coefficients of the "moving sample" method with respect to the minimax criterion are as follows:*

$$g(0) = \cdots = g((k^* - 1)/N) = \sqrt{N}(\sqrt{k^*} - \sqrt{k^* - 1}),$$
$$g(k^*/N) = \sqrt{N}(\sqrt{k^* + 1} - \sqrt{k^*}), \dots, g((N-1)/N) = \sqrt{N}(\sqrt{N} - \sqrt{N-1}), \tag{9.61}$$

where $k^* = [N/e^3]$.

The detailed proof of this theorem is given in Brodsky and Darkhovsky (2000).

Now, consider the question about the asymptotic σ-optimality for the general "moving sample" method.

From **Theorem 9.2.9**, we have the following estimate for the rate of convergence of the normalized delay time

$$\liminf_{N \to \infty} \mathbf{E} \left(\sqrt{N}(\rho_N - \gamma_{gm}(h)) \right)^2 = \frac{\sigma^2}{h^2 g^2 (\gamma_{gm}(h))}.$$

On the other hand, the low boundary from the Rao-Cramer inequality for the general "moving sample" method is equal to $\sigma^2 \left(\gamma_{gm}'(h) \right)^2 / \gamma_{gm}(h)$, where $\gamma_{gm}(h)$ is the root of the equation (suppose that $h > 0$)

$$\int_0^{\gamma_{gm}(h)} g(t) dt = C_{gm}/h. \tag{9.62}$$

Let us show that for any $h > C_{gm} > 0$,

$$\frac{1}{h^2 g^2 (\gamma_{gm}(h))} \ge \frac{\left(\gamma_{gm}'(h) \right)^2}{\gamma_{gm}(h)}. \tag{9.63}$$

Indeed, after differentiating (9.62) with respect to h, we obtain

$g(\gamma_{gm}(h)) = -\left(C_{gm}/h^2\gamma_{gm}^{'}(h)\right)$. Therefore, Inequality (9.63) takes the form (we need to prove it)

$$\frac{h^2}{C_{gm}^2} \geq \frac{1}{\gamma_{gm}(h)}.$$

This relationship holds true in virtue of Schwarz inequality

$$(C_{gm}/h)^2 = \left(\int_0^{\gamma_{gm}(h)} g(t)dt\right)^2 \leq \gamma_{gm}(h) \int_0^1 g^2(t)dt = \gamma_{gm}(h).$$

Thus, the general "moving sample" method is not asymptotically σ-optimal.

9.5 Simulations

In this section, I consider Monte-Carlo tests of the proposed methods. To begin, we ask which issues are new and actual in these methods. The answer is clear: We do not know the future and the parameter h of a possible change, in particular. Therefore, it is reasonable to consider different values of h and characteristics of these methods for different h. Another issue: Comparative analysis of these methods was done under very restrictive conditions on d.f.'s of observations. Here, I can consider dependent and correlated observations with different parameters. The proposed methods are nonparametric versions of well-known tests in sequential change-point detection for univariate models. Therefore, I can consider non-Gaussian d.f.'s of observations.

CUSUM method

The classic CUSUM method is formulated as follows:

$$y_n = (y_{n-1} + \frac{f_1(x_n)}{f_0(x_n)})^+, \qquad y_0 \equiv 0, \quad n = 2, \dots,$$

where the d.f. of observations x_n changes abruptly from $f_0(x_n)$ to $f_1(x_n)$ at the point $n = m$.

The nonparametric version of this test is as follows:

$$x_n = \begin{cases} (x_{n-1} + a + u_n)^+, & x_0 = 0,\ n = 2, \dots, m \\ (x_{n-1} + a + h + u_n)^+, & n > m. \end{cases}$$

Here, $Eu_n \equiv 0$, $h \neq 0$.

This test is designed for detection of possible changes-in-mean of obser- vaions. The parameter h denotes this unknown change-in mean that arrives after the point m.

In tests, we first took $u_n \sim \mathcal{N}(0, 1)$, $a = -1/2$ and the threshold of detection that guarantees the average time between false alarms $T \sim 10000$:

TABLE 9.1
CUSUM Test: Independent Gaussian Observations

h	0.2	0.5	0.8	1.0	1.2	1.5	2.0
w_2	0.040	0	0	0	0	0	0
dn	2537	352.9	13.92	8.64	6.42	4.55	3.21
dn/cm	634	88.2	3.48	2.16	1.60	1.14	0.80

TABLE 9.2
CUSUM Test: Dependent Gaussian Observations

h	0.2	0.5	0.8	1.0	1.2	1.5	2.0
w_2	1.0	1.0	1.0	0.688	0.291	0	0
dn	-	-	-	4757.7	3892.5	1513.4	190.2
dn/cm	-	-	-	432.5	348.4	137.7	17.3

$cm = 4$. The parameter h then varied in the interval $(0, 2]$. In tests, the following values were computed: $w2$ the estimate of the 2nd-type error, dn – the average delay time in change-point detection (in 1000 replications of the test), dn/cm – the normalized delay time in detection. The obtained results are given in Table 9.1.

Then, the following noise sequences were considered:

$$u_n = \rho u_{N-1} + \sigma\sqrt{1-\rho^2}v_n,$$

where $v_n \sim \mathcal{N}(0, 1)$.

Here, the value $\rho = 0.7$ of the correlation of observations in the sequence u_n was taken. Then the threshold $cm = 11.0$ to guarantee $T = 10000$ was assumed. The following results were obtained.

In the third series of tests, non-Gaussian d.f.'s of observations were considered. Here, we give results of tests with uniformely distributed r.v.'s v_n: $\mathcal{U}[0, 1]$, $\sigma = 1, \rho = 0, a = - - 0.85$. In order to guarantee $T = 10000$, the threshold of detection $C = 2.5$ was taken. The obtained results are given in Table 9.3.

TABLE 9.3
CUSUM Test: Independent, Uniformely Distributed Observations

h	0.2	0.5	0.8	1.0	1.2	1.5	2.0
w_2	0.95	0	0	0	0	0	0
dn	4874	16.6	6.11	4.43	3.48	2.71	2.01
dn/cm	1848.6	6.64	2.44	1.77	1.39	1.08	0.80

TABLE 9.4
GRSh Test: Independent Gaussian Observations

h	0.2	0.5	0.8	1.0	1.2	1.5	2.0
w_2	0	0	0	0	0	0	0
dn	1060	25.7	10.4	7.02	5.42	4.09	2.75
dn/cm	265	6.4	2.35	1.63	1.26	0.95	0.69

TABLE 9.5
GRSh Test: Dependent Gaussian Observations

h	0.2	0.5	0.8	1.0	1.2	1.5	2.0
w_2	0.038	0	0	0	0	0	0
dn	2755	133.09	36.83	23.39	17.27	12.3	8.3
dn/cm	234.0	11.09	3.06	1.94	1.44	1.02	0.69

GRSh method

In the second series of tests, the nonparametric version of the GRSh test was taken. First, we considered independent Gaussian observations $u_n \sim \mathcal{N}(0,1)$, $a = -1/2$, and the threshold of detection that guaranteers the average time between false alarms $T \sim 10000$: $cm = \exp(4.0)$. The parameter h then varied in the interval $(0, 2]$. In tests, the following values were computed: $w2$ the estimate of the 2nd-type error; dn, the average delay time in change-point detection (in 1000 replications of the test); dn/cm, the normalized delay time in detection. The obtained results are given in Table 9.4.

Then, the following noise sequences were considered:

$$u_n = \rho u_{N-1} + \sigma\sqrt{1-\rho^2}v_n,$$

where $v_n \sim \mathcal{N}(0,1)$.

Here, the value $\rho = 0.7$ of the correlation of observations in the sequence u_n was taken. Then, the threshold $C = \exp(11.0)$ to guarantee $T = 10000$ was assumed. The results are reported in Table 9.5.

In the third series of tests, uniformely distributed r.v.'s v_n: $\mathcal{U}[0,1]$, $\sigma = 1, \rho = 0, a = -0.85$ were taken. In order to guarantee $T = 10000$, the threshold of detection $C = \exp(2.5)$ was taken. The obtained results are given in Table 9.6.

From the obtained results, we see that the nonparametric GRSh test has distinct advantages over the nonparametric CUSUM method: The avarage delay time in change-point detection for this test is smaller than for the CUSUM test; the characteristic of w_2 is better for the GRSh test. However, theoretical results witness about the asymptotical equivalence of these tests.

MS ("Moving sample") test

In subsequent tests, the nonparametric MS method was taken. First, we

TABLE 9.6
GRSh Test: Independent, Uniformely Distributed Observaions

h	0.2	0.5	0.8	1.0	1.2	1.5	2.0
w_2	0	0	0	0	0	0	0
dn	36.88	9.83	5.82	4.34	3.51	2.98	2.58
dn/cm	15.55	3.93	2.32	1.73	1.40	0.95	1.03

TABLE 9.7
MS Test: 95 and 99 Percent Quantiles of Statistic (2.6)

N	100	200	300	450	500
$\alpha = 0.95$	0.22	0.15	0.11	0.095	0.089
$\alpha = 0.99$	0.24	0.17	0.15	0.11	0.93

computed the threshold of detection. For this purpose, homogenous samples with different sample sizes were taken. In Table 9.7, 95 and 99 percent quantiles of the moving sample statistic are given.

Then, the value of 95 percent quantile was taken as the threshold of detection in tests with nonhomogenous samples. In tests, we computed the following values: w_2 — the frequence of type-2 error, dn — the average delay time in change-point detection (here, $m = 1, 1000$ independent trials of each test); and dn/N = the normalized delay time in change-point detection. The obtained results are given in Table 9.8.

Exponential smoothing method

In experiments, the nonparametric exponential smoothing method was taken:

$$y(n) = (1 - \nu) * y(n - 1) + \nu * u(n),$$

where $u(n) \sim \mathcal{N}(0, 1)$.

First, we computed the threshold of detection C. For this purpose, experiments with homogenous samples were done. In these experiments, we determined the value of the parameter $\nu = \nu(N) = 1/N$, where N is the large

TABLE 9.8
MS Test: Independent Gaussian Observations

N	100	200	300	400	500
C	0.22	0.15	0.11	0.10	0.09
w_2	0.07	0.05	0.03	0.02	0.01
dn	30.21	34.15	35.05	39.5	45.70
dn/N	0.300	0.170	0.116	0.118	0.114

TABLE 9.9
Exponential Smoothing Test: Independent Gaussian Observations

h	0.1	0.2	0.5	0.7	1.0
w_2	0.07	0	0	0	0
dn	756	356.9	129.4	90.3	62.9
dn/N	0.756	0.357	0.129	0.090	0.062

The case $\nu = 0.001$

TABLE 9.10
Exponential Smoothing Test: Independent Gaussian Observations

h	0.1	0.2	0.5	0.7	1.0
w_2	0.8	0.45	0	0	0
dn	697.3	687.1	112.0	43.7	18.3
dn/N	69.7	68.7	11.2	4.37	1.83

The case $\nu = 0.1$

parameter of the method, and computes 95 and 99 percent quantiles of the decision statistic $|y(n)|$. The value of the 95 percent quantile was assumed to be the threshold of detection in experiments with nonhomogenous samples. In these tests, the value of the parameter h was varied in the interval $[0.1; 1.0]$, and for each value of h, the frequency of type-2 error w_2 and the average delay time in change-point detection (dn for the case $m = 1$) were computed. The obtained results are given in Tables 9.9 and 9.10.

First, we took $\nu = 0.001$ and computed $C = 0.06$.

Second, we took $\nu = 0.1$ and computed $C = 0.90$.

From the obtained results, we see that CUSUM and GRSh tests have clear advantages over KS and the exponential smoothing tests. However, more accurate conclusions require additional experiments with these tests.

9.6 Conclusion

The main idea of this chapter is the unified approach to the analysis of parametric and nonparametric sequential change-point detection methods, both for independent and dependent samples of observations. This approach is based upon the methodology of the "large parameter," which enables us to apply asymptotic techniques to retrospective problems similar to those used in **Part I**. For this purpose, I introduce the definitions of the "normalized" delay time in detection and the "false alarm" probability. In terms of these definitions, I analyze characteristics of well-known tests of sequential change-point detection in univariate models: CUSUM and GRSh methods, the method of exponential smoothing, and the "moving widow" tests. Then, on the basis of the a priori inequalities in sequential change-point detection: the a priori inequality for the normalized delay time in detection, the Rao-Cramer type inequality, and the informational inequality for the error of detection — I study the asymptotic optimality properties of sequential methods (so-called τ- and σ-optimality, the definitions of which are given in the text of this chapter). A short review of the most influential publications in the field of sequential change-point detection is given. At the end, I present reluts of a small simulation study of the proposed tests.

10

Sequential Change-Point Detection in Multivariate Models

10.1 Introduction

The problem of sequential detection of change-points in stochastic multivariate systems on the basis of sequential observations has many applications, including detection of changes in parameters of regression equations, testing adequacy of econometric models, fault detection, and isolation in stochastic dynamical systems. There is a extensive statistical and econometric literature dealing with methods of solving these problems.

Page (1954) considered the cumulative sums (CUSUM) test for detection of possible changes in the distribution function (d.f.) of a sequence of independent observations. Girshick and Rubin (1952) proposed a quasi-Bayesian test for solving the same problem.

In 1959, Kolmogorov and Shiryaev proposed the formal statement of the problem of "quickiest detection of spontaneous effects," which was later called the "disorder problem" (see Kolmogorov, Prokhorov, and Shiryaev (1988)). In 1959–1695, Shiryaev found the optimal solution of this problem for the situation of full a priori information on the distribution function of observations and a change-point (Shiryaev (1963)).

In the situation when there is no a priori information on a change-point, Lorden (1971), Pollack (1985), and Moustakides (1986) proved that CUSUM and Shiryaev tests are asymptotically optimal in the problem of sequential detection of an abrupt change in the one-dimensional d.f. of independent observations.

Willsky (1976), and Willsky and Jones (1976) pioneered research into sequential detection of abrupt changes in stochastic dynamical systems. Stochastic "noises" in such systems were assumed to be Gaussian, and change-points were interpreted as instants of spontaneously emerging additive terms in equations of considered systems. For detection of these structural changes, the innovation process of the Kalman filter was used. Different methods generalizing these ideas for sequential change-point detection in stochastic dynamical systems were proposed by Basseville and Benveniste (1983), Basseville and Nikiforov (1993), Nikiforov (1995), and Bansal and Papantoni-Kazakos (1983).

In works of Lai (1995, 1998), the problem of sequential change-point detection in dynamical systems was generalized to the non-i.i.d. case. Lai considers the window-limited generalized likelihood ratio (GLR) schemes and proves their asymptotical optimality in different problems of sequential change-point detection in dynamical systems.

The multidecision change-point detection problem was first considered by Nikiforov (1995) in frames of the detection/isolation research program in stochastic control and surveillance, and then generalized by Lai (2000), and Tartakovsky et al. (2003).

The situation of composite hypotheses both before and after the change-point is considered in modern papers by Mei (2006), and Brodsky and Darkhovsky (2005, 2008). Mei (2006) considers a mixture of distributions before the change-point with known probabilistic mechanism. In papers by Brodsky and Darkhovsky (2005, 2008), parameters of d.f.'s before and after the change-point are assumed to be unknown but nonrandom.

In spite of extensive research into the problem of sequential change-point detection in stochastic dynamical systems, several open problems still exist, and, in particular, the problem of a priori information on observations.

Between the poles of the *full knowledge* (both the probabilistic mechanism of data generation and the specification of a system are known) and the *full ignorance* (neither the probabilistic mechanism of data generation nor the specification of a system is known), there exists the most practically relevant field of *semiparametric* model description in which we know the specification of a stochastic system but the d.f. of observations is unknown to us. Some important examples include the following:

1) The multiple regression models and the systems of simultaneous equations in econometrics. As usual, we know the specification of a model (e.g., the linear regression or the autoregression model) but the d.f.'s of "noise" sequences are unknown to us. The problem consists in sequential detection of structural changes in these models. These structural changes include both abrupt changes in coefficients of equations and new terms in their specification (e.g., new additive factors in econometric equations).

2) Conventional input-output dynamical systems in engineering and control science (the "transfer function" of this system is known, but the d.f. of the "noise" sequence is unknown). The problem consists in sequential detection of spontaneous changes in the transfer function.

Chu, Stinchcombe, and White (1996) considered the problem of monitoring structural changes in coefficients of a linear regression. They used the fluctuation test for sequential detection and diagnosis of abrupt changes in coefficients. These results were followed by Leisch, Hornik, and Kuan (2000). Sequential tests based upon sums of regression residuals were used by Horvath, Huskova, Kokoszka, and Steinebach (2004) for monitoring structural changes. Dynamic econometric models with structural changes were considered by Zeileis, Leisch, Kleiber, and Hornik (2005). The common feature of the works is a regression-based approach to monitirung of structural changes:

It is usually assumed that a "historical period" in model dynamics exists, i.e., an initial time interval when there are no structural changes in coefficients. Using this "historical period", we construct initial regression estimates of coefficients, as well as regression residuals, the estimate of dispersion, etc. Then, on the basis of these estimates, we "monitor" deviations of the current regression model from an initial model, whether via current and initial regression coefficients (fluctuation tests) or via CUSUMs of regression residuals (CUSUM tests based on residuals).

The common drawback of these works is as follows: The quality of proposed tests is analyzed only from the perspective of their limit distributions as the sample volume tends to infinity. Properties of these tests for finite sample volumes are studied only empirically. Moreover, there is no research on optimality and asymptotic optimality of these methods.

In this chapter, a new method for sequential detection of change-points in linear models is proposed. The main performance characteristics of this method are analyzed theoretically for finite sample volumes. Comparison with other well-known methods for sequential detection of structural changes in linear models is carried out via Monte-Carlo tests. Practical applications for the analysis of stability of the German quarterly model of demand for money (1961–1995) and the Russian monthly model of inflation (1994-2005) are considered.

10.2 Model and Assumptions

Model

Consider the following basic specification of the multivariate system with structural changes:

$$Y(n) = \Pi X(n) + \nu_n, \quad n = 1, 2, \dots, \tag{10.1}$$

where $Y(n) = (y_{1n}, \dots, y_{Mn})'$ is the vector of endogenous variables; $X(n) = (x_{1n}, \dots, x_{Kn})'$ is the vector of predetermined variables; and $\nu_n = (\nu_{1n}, \dots, \nu_{Mn})'$ is the vector of errors. $'$ is the transposition symbol.

Recall that the class of predetermined variables $(X(n))$ includes all lagged endogenous variables $(Y(n-1), Y(n-2), \dots)$, as well as all exogenous variables (predictors) for this system.

The $M \times K$ matrix Π changes abruptly at some unknown change-point m, i.e.,

$$\Pi = \Pi(n) = \mathbf{a}I(n \leq m) + \mathbf{b}I(n > m), \quad n = N, N+1, \dots, \tag{10.2}$$

where $\|\mathbf{a} - \mathbf{b}\| > 0$.

Model (10.2) generalizes many widely used regression models, i.e.:
— static and dynamic regression models with multiple predictors
— ARMA models for time series
— systems of simultaneous regression equations in econometrics
— stochastic dynamical systems with fully observed state variables in control theory.

Assumptions

Now, let us formulate assumptions about the random noise process ν_n and predictors $X(n)$ defined on the probability space $(\Omega, \mathfrak{F}, \mathbf{P})$. We suppose that the uniform Cramer's and ψ-mixing conditions are satisfies for these processes.

Let us formulate assumptions about predictors $X(n)$ and noises ν_n. Suppose that predictors $X(n)$ and noises ν_n are continuously distributed and strictly stationary, and the following conditions are satisfied:

1) The vector $X(n) = (x_{1n}, \dots, x_{Kn})'$ is $\mathcal{F}_{n-1}$ measurable.

2) There exists a continuous matrix function $V(t)$, $t \in [0, 1]$, such that for any $0 \le t_1 \le t_2 \le 1$

$$D_N(t_1, t_2) = \frac{1}{N} \sum_{j=[t_1 N]}^{[t_2 N]} X(j)X^{'}(j) \to \int_{t_1}^{t_2} V(t)dt, \quad P - \text{a.s. as } N \to \infty,$$

where $\int_{t_1}^{t_2} V(t)dt$ is the positive definite matrix.

3) The random vector sequence $\{(X(n), \nu_n)\}$ satisfies ψ-mixing and the unified Cramer condition.

4) $\{\nu_n\}$ is a martingale-difference sequence w.r.t. the flow $\{\mathcal{F}_n\}$.

Let us consider some examples.

Example 1.

Consider a multifactor regression model with independent and stationary stochastic predictors and independent Gaussian errors:

$$y(n) = c_0 + c_1 x_1(n) + \cdots + c_k x_k(n) + \nu_n, \quad n = 1, \dots .$$

Suppose that each predictor $x_i(n)$ satisfies the uniform Cramer condition. Then, Conditions 1)–4) are satisfied for this model.

Example 2.

Consider a dynamical autoregressive model AR(1):

$$y_n = \rho y_{n-1} + \nu_n,$$

where $|\rho| < 1$ and ν is a "white noise" random sequence satisfying the Cramer condition.

From basic facts about mixing properties of Markov processes (see, e.g., Bradley (2005)), we know that ψ-mixing condition is satisfied for this model under certain unrestrictive assumptions (e.g., $\psi(k) < 1$ for some $k > 0$). Again, Assumptions 1)–4) are satisfied for this model.

Remark that ψ-mixing condition is imposed here in order to obtain the exponential rate of convergence to zero for type-1 and type-2 error probabilities (see Theorems 10.1 and 10.2 below). Another alternative was to assume α-mixing property, which is always satisfied for aperiodic and irreducible countable-state Markov chains (see Bradley (2005)). Then, we can obtain the hyperbolic rate of convergence to zero for type-1 and type-2 error probabilities.

In general, these assumptions are satisfied in most practical problems of regression analysis, and, in particular, for multifactor regression models with stationary predictors and independent "noises" without "heavy tails," as well as for systems of simultaneous econometric equations with usual conditions.

10.3 Method of Detection

The idea of the proposed method is based upon the "moving window" statistic for sequential detection of a change-point. Suppose the size of this window is defined by a certain large parameter N. For any $n = N, N+1, \ldots$ consider N last vectors of observations $Y(i), X(i), i = n - N + 1, \ldots, n$.

The method of detection is constructed as follows. First, consider the $K \times K$ matrices

$$\mathcal{T}^n(1,l) = \sum_{i=1}^{l} X(i+n-N)X^{'}(i+n-N), \quad l = 1, \ldots, N. \qquad (10.3)$$

Second, consider the $K \times M$ matrices

$$z^n(1,l) = \sum_{i=1}^{l} X(i+n-N)Y^{'}(i+n-N), \quad l = 1, \ldots, N, \qquad (10.4)$$

and, third, consider the decision statistic

$$Y_N^n(l) = \frac{1}{N}(z^n(1,l) - \mathcal{T}^n(1,l)(\mathcal{T}^n(1,N))^{-1} z^n(1,N)), \qquad (10.5)$$

where $l = 1, \ldots, N$, $Y_N^n(N) = 0$ and by definition, $Y_N^n(0) = 0$.

Let us explain why these quantities are relevant for sequential detection of the change-point m. Remark that Statistic (10.5) is the multivariate version of its earlier scalar prototype

$$Z_N^n(l) = \frac{1}{N}(\sum_{i=1}^{l} y_{i+n-N} - \frac{l}{N}\sum_{i=1}^{N} y_{i+n-N}),$$

proposed in Brodsky, and Darkhovsky (2000) for sequential detection of a possible change in the mathematical expectation of a random sequence of scalar observations $y_1, y_2, \ldots$:

$$Ey_i = \begin{cases} 0, & i \leq m \\ h \neq 0, & i > m, \end{cases}$$

where $m > N$ is an unknown change-point. In fact, here, $X(i) = 1$ for any $i \geq 1$, and, therefore, $T^n(1,l) = l$, $l = 1, \ldots, N$. Hence, we obtain statistic (10.5) for this particular case.

For detection of an unknown change-point m in the mathematical expectation of observations, the following stopping time was used (see Brodsky, and Darkhovsky (2000)):

$$v_n = \inf\{n : \max_{[\beta N] \leq l \leq N} |Z_N^n(l)| > C\},$$

where $0 < \beta < 1/2$ and $C > 0$ is the decision bound.

We readily conclude that the mathematical expectation of this statistic is zero for all $l = 1, \ldots, N$ and any $n < m$ and $n > m + N$. For $m \leq n \leq m + N$, the mathematical expectation of the absolute value of the maximum of this statistic over $l = 1, \ldots, N$ is greater than C if $(m - n + N)(n - m)|h| > CN^2$.

So, the limit value γ of the normalized delay time in change-point detection $\gamma_N = v_N/N$ satisfies the following relationship: $\gamma(1-\gamma) = C/|h|$, i.e., γ is the minimal root of the equation $x(1-x) = C/|h|$.

Below, we prove that the multivariate variant of this statistic defined in (10.3.3) behaves like its scalar counterpart, i.e., it is relevant for sequential detection of m.

Remark that the inverse matrix $(\mathcal{T}^n(1,N))^{-1}$ exists if and only if the vectors $x_1^{(n)} = (x_{1n}, \ldots, x_{1(n+1-N)})', \ldots, x_K^{(n)} = (x_{Kn}, \ldots, x_{K(n+1-N)})'$ are linearly independent for any $n \geq N$. For any fixed n, the event of their linear dependence is of probability zero. Therefore, these vectors are a.s. linearly independent and the inverse matrix $(\mathcal{T}^n(1,N))^{-1}$ exists almost surely. This fact is quite sufficient for subsequent theorems in which type-1 and Type-2 error probabilities for the proposed method are considered.

Fix the number $0 < \beta < 1/2$. For detection of the change-point $m > N$, we define the stopping time

$$\tau_N = \inf\{n : \max_{[\beta N] \leq l \leq N} \|Y_N^n(l)\| > C\}, \tag{10.6}$$

where C is a certain decision threshold, $\|A\|$ is the Euclidean norm of the matrix A.

Let us compare this decision rule with other well-known tests proposed for detection of structural changes in regression models. There are two types of tests proposed for detection of structural changes in regression models. The first type consists of *fluctuation tests* based on retrospective estimates of regression coefficients (see Chu et al. (1996)). The second type of tests consists of *residual-based* methods. Here, we can mention tests based on "historical" residuals (Ploberger, and Kramer (1992)), and tests based on "recursive" residuals (Horvath et al. (2004)). All these tests are based on moving windows and variance-covariance matrices of predictors and responses. However, the test proposed in this paper does not use regression estimates, and, therefore is expected to be more robust to possible deviations and inaccuracies in specification of regression models. In comparison with other methods proposed for detection of structural changes in regression models, our test can be easily generalized to different multivariate models (including systems of simultaneous equations).

Another important issue is the volume of a priori information on observations needed for the method proposed in this paper and for other (above considered) tests. All these tests can be called nonparametric or semiparametric, because they do not use a priori information on distributions of noises and predictors. However, the test proposed in this paper does not explicitly use regression parameters and, therefore, can be considered as "less parametric" than other tests.

10.4 Main Results

In the sequel, we denote by $P_0(E_0)$ the measure (mathematical expectation) corresponding to the observed sequence without change-points and by $P_m(E_m)$ — to the sequence with the change-point m. H_0 denotes the hypothesis of statistical homogeneity of observations (no structural changes); H_1 — the hypothesis about the presence of a change-point in the sample.

In this chapter, the following performance characteristics of the proposed method will be used:

1) Probability of type-1 error ("false decision"):

$$\alpha_N = \sup_n P_0\{\max_{[\beta N]\leq l\leq N} \|Y_N^n(l)\| > C\}. \tag{10.7}$$

This characteristic is closely connected with ARL (average run length) to false decision and FAR (false alarm rate) for the case of independent normal observations. However, in the considered situation of dependent multivariate observations with unknown d.f., we can readily construct the exponential upper estimate of the probability of type-1 error (Theorem 1), but any similar result for the traditional ARL and FAR indicators is highly problematic. This is the main motivation for the introduction of a new performance measure α_N.

2) Probability of type-2 error ("missed goal"):

$$\delta_N = P_m\{\max_{m\leq n\leq m+N}\ \max_{[\beta N]\leq l\leq N} \|Y_N^n(l)\| \leq C\}.$$

This characteristic describes the situation when the decision statistic does not exceed the boundary C for a sample with a change-point, i.e., for $m \le n \le m+N$.

3) The normalized delay time in change-point detection:

$$\gamma_N = (\tau_N - m)^+/N, \tag{10.8}$$

where $a^+ = \max(0, a)$.

In the following theorem, the asymptotical behavior of the probability of type-1 error is studied.

Theorem 10.4.1. *Suppose the above assumptions 1), 3), 4) are satisfied. Then, for any $C > 0$ the following exponential upper estimate for the "false alarm" probability holds:*

$$\alpha_N \le \phi_0(C_1) \begin{cases} \exp(-\frac{TNC_1\beta}{4\phi_0(C_1)}), & C_1 > hT \\ \exp(-\frac{NC_1^2\beta}{4h\phi_0(C_1)}), & C_1 \le hT, \end{cases} \tag{10.9}$$

where the constants h, T, and $\phi_0(C_1) \ge 1$ are taken from Cramer's and ψ-mixing condition, respectively, $C_1 = C/(1+\sqrt{K})$.

In the following theorem, we study type-2 error δ_N and the normalized delay time γ_N in sequential change-point detection.

Consider the $K \times K$ matrix

$$A(t) = \int_0^t V(\tau)d\tau, \quad 0 \le t \le 1.$$

Define $I = A(1)$. For any $0 < t \le 1$, the matrix $A(t)$ is positive definite.

For any $0 \le \theta \le 1$, consider the function

$$g(\theta) = \|A(\theta)(E - I^{-1}A(\theta))(\mathbf{a} - \mathbf{b})^{'}\|,$$

where E is the unit matrix $K \times K$.

Evidently, $g(0) = g(1) = 0$. Consider the point $\tilde{\theta}$ of the global maximum of $g(\theta)$ on the segment $[0,1]$ — the root of the equation $E = I^{-1}A(\theta) + A(\theta)I^{-1}$, i.e., $A(\theta) = I/2$. In virtue of the above assumptions, the root of this equation exists and is unique. The function $g(\theta)$ is continuously differentiable by $\theta \in (0,1)$.

Choose the decision threshold $0 < C < g(\tilde{\theta})$. The following theorem holds.

Theorem 10.4.2. *Suppose the above conditions 1)-4) are satisfied and rank$(D) = M$, where $D = (E - I^{-1}A(\theta))(\mathbf{a} - \mathbf{b})^{'}$. Assume also that the sequence $D_N(0,1)$ from condition 2) is uniformely bounded for any $\omega \in \Omega$.*

Denote $d = (g(\tilde{\theta}) - C)/(1 + \sqrt{K})$. *Then, the following exponential upper estimate holds for type-2 error:*

$$\delta_N \leq L_1 \exp(-L_2 d N), \tag{10.10}$$

where constants $L_1 > 0$, *and* $L_2 > 0$ *do not depend on* N.

The relative delay time γ_N *tends almost surely to a deterministic limit as* $N \to \infty$:

$$\gamma_N = \frac{(\tau_N - m)^+}{N} \to \gamma^* \quad P_m - a.s.\ as\ N \to \infty, \tag{10.11}$$

where γ^* *is the minimal root of the equation* $g(t) = C$, $0 < \gamma^* < 1$.

Moreover, for any finite N *and* $0 < \epsilon < 1$, *the following exponential inequality holds* ($v = \epsilon/(1 + \sqrt{K})$):

$$P_m\{|\gamma_N - \gamma^*| > \epsilon\} \leq \mathcal{L}_1 \exp(-\mathcal{L}_2 v N), \tag{10.12}$$

where constants $\mathcal{L}_1 > 0$, *and* $\mathcal{L}_2 > 0$ *do not depend on* N.

The proofs of Theorems 10.4.1 and 10.4.2 are methodologically close to the proof of Theorem 5.4.1 andm therefore, are omitted here. We can see that both 1st- and 2nd-type error probabilities for the proposed method converge to zero exponentially as $N \to \infty$. The normalized (by N) delay time in change-point detection converges to a certain deterministic limit with an exponential rate as $N \to \infty$. In general, such properties characterize an asymptotically optimal method of detection (see, e.g., Brodsky, and Darkhovsky (2000)). Our goal, however, is not to establish the strict asymptotical optimality of the proposed method (and this is surely not the case due to the multivariate character of the problem considered here) but primarily to compare performance characteristics of this method with corresponding measures for other widely used tests, i.e., the *fluctuation test* based on regression estimates (see Chu at al. (1996)), and *CUSUM test* based on regression residuals (see Ploberger, and Kramer (1992)).

10.5 Simulations

In this section, I present results of a simulation study of the proposed method in comparison with other well-known tests for detecting structural changes in model coefficients, i.e.,

– Fluctuation test (Chu et al. (1996));

– CUSUM test based on "historical" OLS residuals (Ploberger, and Kramer (1992)).

– CUSUM test based on recursive residuals (Horvath et al. (2004)).

Below, I brief, describe these methods using original notation. The standard linear regression model used by the authors of these tests is as follows:

$$y_i = x_i^{'}\beta_i + u_i,$$

where at time i, y_i is the observation of the dependent variable, $x_i = (1, x_{i2}, \ldots, x_{ik})^{'}$ is a $k \times 1$ vector of regressors, with the first component usually equal to unity, and β_i is a $k \times 1$ vector of regression coefficients.

The data $i = 1$ to $i = n$ is referred to as the history period, where the regression coefficients are assumed to be constant, and we want to monitor new data from time $n+1$ onwards to test whether any structural change occurs in the monitoring period.

Denote by $\hat{\beta}_i$ the OLS estimate of regression coefficients based on observations from 1 through i. The OLS historical residuals are denoted as $\hat{\epsilon}_i = y_i - x_i^{'}\hat{\beta}_n$ and the recursive residuals as $\tilde{\epsilon}_i = y_i - x_i^{'}\hat{\beta}_{i-1}$. $\hat{\sigma}^2 = \dfrac{1}{n-k}\sum_{i=1}^{n} \hat{u}_i^2$ is the least-square estimate of the common variance of noises.

There are two types of tests proposed for detection of structural changes in regression models. The first type consists of *fluctuation tests* based on retrospective estimates of regression coefficients (RE):

$$Y_n(t) = \frac{i}{\hat{\sigma}\sqrt{n}} Q_{(n)}^{1/2}(\hat{\beta}_i - \hat{\beta}_n),$$

where $1 < t < T$, $i = [k + t(n-k)]$, $Q_{(n)} = X_{(n)}^{'}X_{(n)}/n$. The hypothesis of structural change is accepted if

$$\{\exists t : 1 < t < T, \max_{1 \le j \le k} |Y_n^{(j)}(t)| > b_1(t)\},$$

where the decision boundary is $b_1(t) = [t(t-1)(\lambda^2 + \log(t/(t-1)))]^{1/2}$ and λ determines the significance level of this test.

Another variant of fluctuation tests is based on the moving estimates (ME) of regression coefficients:

$$Z_n(t|h) = \frac{[nh]}{\hat{\sigma}\sqrt{n}} Q_{(n)}^{1/2}(\hat{\beta}_{[nt]-[nh],[nh]} - \hat{\beta}_n),$$

where $\hat{\beta}_{[nt]-[nh],[nh]}$ is the estimate of regression coefficients constructed by observations from the time interval $[nt]-[nh]+1, \ldots, [nt]$, $t \ge h$.

The corresponding decision rule of this test is as follows:

$$\{\exists t : h \le t < T, \max_{1 \le j \le k} |Z_n^{(j)}(t|h)(t)| > c(t)\},$$

where $c(t) = \lambda\sqrt{\log_+ t}$,

The second type of tests consists of *residual-based* methods. The test based on "historical" regression residuals $\hat{\epsilon}_i$, $i = n+1, \dots, n+k$ has the following form:

$$\hat{Q}(n,k) = \sum_{n<j\leq n+k} \hat{\epsilon}_i,$$
$$D_\gamma(n,k) = [n\hat{\sigma}^2]^{-1/2}(\frac{n}{n+k})(\frac{n+k}{k})^\gamma |\hat{Q}(n,k)|.$$

We reject the null hypothesis of no structural change at first time k, such that $D_\gamma(n,k) > c_\alpha(\gamma)$, where α is the error level $(0.01, 0.025, 0.05, 0.10)$ and $0 \leq \gamma < 0.50$. In particular, for $\alpha = 0.05$, $\gamma = 0$, we obtain (Horvath et al. (2004)) $c_\alpha(\gamma) = 2.2365$.

Another variant of residual-based tests is the following test using recursive residuals:

$$\tilde{Q}(n,k) = \sum_{n<i\leq n+k} \tilde{\epsilon}_i,$$
$$\tilde{D}_\alpha(n,k) = [n\hat{\sigma}^2]^{-1/2} |\tilde{Q}(n,k)|.$$

We reject the null hypothesis of no structural change at first time k, such that $\tilde{D}_\alpha(n,k) > [(t+1)(a^2 + \log(1+t))]^{1/2}$, where $\exp(-\frac{a^2}{2}) = \alpha$.

1) Regression models

For this class of Monte-Carlo tests, comparison of the proposed method with other methods of monitoring change-points in linear models was performed. For these purpose, programs written in Matlab were used.

Model

The following regression model was considered:

$$y_i = c_0 + c_1 x_i + \epsilon_i, \quad i = 1, 2, \dots,$$

where $x_i = 2 + \xi_i$ and ϵ_i, $\xi_i \sim \mathcal{N}(0,1)$ are independent Gaussian random sequences.

In order to estimate the FAR, the regression model without structural changes was considered with $c_0 = 0$, and $c_1 = 1$. Then, models with a change-point in the coefficient c_1 were considered.

Method

a) CUSUM test based on "historical" OLS residuals

Parameter $c_\alpha(\gamma) = 2.2365$ of this test was chosen to ensure the FAR $pr = 0.05$. Remark that the empirical false alarm rate of this test is much lower: $pr = 0.02$. It means that the theoretical formula for calculation of the FAR of this test is rather imprecise. The same is true for other well-known tests proposed for monitoring structural changes: Theoretical formulas for determination of decision bounds are obtained from the limiting distributions of these tests. However, the speed of convergence of decision statistics to these limiting distributions is unknown and, therefore, most of these "imposing" formulas are practically useless.

In the following table, I demonstrate results of the Monte-Carlo study of this test in the context of sequential detection of change-points in the

TABLE 10.1
Performance Characteristics of CUSUM Test Based on 'Historical' Residuals

n		25	50	100	200
pr		0.02	0.02	0.015	0.02
$c_1 = 1.5$	w_2	0.004	0	0	0
	$E\tau$	23.9	25.3	29.9	38.4
$c_1 = 1.3$	w_2	0.32	0.04	0.002	0
	$E\tau$	59.0	71.1	65.4	74.3
$c_1 = 1.2$	w_2	0.65	0.36	0.07	0.0
	$E\tau$	68.6	131.4	150.9	159.9

Note: (5000 replications, pr — empirical false alarm rate, w_2 — type-2 error, $E\tau$ — average delay time)

TABLE 10.2
Performance Characteristics of CUSUM Test Based on Recursive Residuals

n		25	50	100	200
pr		0.02	0.02	0.02	0.02
$c_1 = 1.5$	w_2	0.02	0	0	0
	$E\tau$	14.48	15.2	19.34	25.17
$c_1 = 1.3$	w_2	0.40	0.08	0.002	0
	$E\tau$	26.55	37.91	40.87	46.75
$c_1 = 1.2$	w_2	0.71	0.42	0.13	0.0
	$E\tau$	32.4	61.3	83.67	85.08

Note: (5000 replications, pr — empirical false alarm rate, w_2 — type-2 error, $E\tau$ — average delay time)

coefficient c_1. Remark that in the original "monitoring" context, we are not so concerned with the characteristic of the average delay time of change-point detection (we need simply to detect it "sooner or later" within the monitoring period). However, results in Table 10.1 demonstrate that the characteristics w_2 (type-2 error) and $E\tau$ (average delay time) of this test are often inappropriate.

b) CUSUM test based on recursive residuals

In Table 10.2, I demonstrate the corresponding results for the CUSUM test based on recursive residuals. The parameter $a = 1.5$ of this test was chosen in order to ensure the empirial FAR $pr = 0.02$. We see that the characteristic $E\tau$ is significantly better for this test that for the CUSUM test based on "historical" residuals.

c) Fluctuation test

Table 10.3 below contains the corresponding results of Monte-Carlo tests for the fluctuation test based on "historical" regression estimates. The parameter $\lambda = 7.0$ of this test was chosen to ensure the empirical FAR $pr = 0.02$. We

TABLE 10.3
Performance Characteristics of the Fluctuation Test

n		25	50	100	200
pr		0.02	0.02	0.02	0.02
$c_1 = 1.5$	w_2	0.32	0.25	0.004	0
	$E\tau$	21.5	28.4	29.5	31.5
$c_1 = 1.3$	w_2	0.47	0.43	0.40	0.04
	$E\tau$	157.3	182.7	201.26	207.71
$c_1 = 1.2$	w_2	0.93	0.89	0.80	0.55
	$E\tau$	202.2	278.7	345.6	389.7

Note: (5000 Replications, pr — empirical false alarm rate, w_2 — type-2 error, $E\tau$ — average delay time)

see that the characteristics w_2, and $E\tau$ of this test are rather bad, especially for small changes in coefficients.

d) Nonparametric test

Here, I place the corresponding results for the nonparametric test proposed in this paper. First, let us discuss the choice of the decision threshold for this test. For this purpose, the upper estimate of the 1st-type error probability from Theorem 10.4.1 can be used. For independent case, we put $\phi_0(\cdot) = 1$; remark that $h = \sigma^2 \max_{1 \le i \le k} Ex_i^2$ and obtain

$$C = \frac{\sigma(\max_i Ex_i^2)^{1/2}}{\sqrt{N}} \lambda,$$

where σ^2 is the dispersion of ϵ_i and $\lambda > 0$ is the calibration parameter.

This formula includes all factors influencing the choice of the decision threshold. Remark that the calibration parameter λ can be obtained from the "historical" sample without change-points only for one value of N. To check this hypothesis, the following tests were performed. Maximums of decision statistic were computed in k=5000 trials of each experiment, with samples without change-points for different values of the sample volume N. Then, the variation series of these maxima was constructed and the 95 percent and 99 percent quantiles computed. The values of 99 percent quantiles for each value of N were assumed to be the decision thresholds th. In our example $\sigma = 1$, $Ex_i^2 = 5$. Therefore, we obtain the following formula for computation of $\lambda = th\sqrt{N}/2.2361$. The obtained results are reported in Table 10.4.

From these results, I conclude that the calibration parameter λ can be chosen in the diapason $2.0 - 2.1$. For different values of λ from this diapason, the 1st-type error rate slightly varies around $pr = 0.02$.

From these results I see that except for the case $N = 20$, all computed values of λ are in the interval $[2.0; 2.1]$. Remark, however, that precise knowledge of λ is not necessary for practical applications of this test. We can readily

TABLE 10.4
Decision Bounds for the Nonparametric Test

N	20	50	100	200	300	400	500
$p = 0.95$	0.65	0.51	0.32	0.24	0.18	0.16	0.14
$p = 0.99$	0.85	0.65	0.45	0.33	0.27	0.23	0.20
λ	1.7	2.05	2.01	2.08	2.09	2.05	2.00

TABLE 10.5
Performance Characteristics of the Nonparametric Test

N		100	200	300	400
th		0.45	0.33	0.25	0.21
pr		0.021	0.025	0.015	0.025
$c_1 = 1.5$	w_2	0.05	0	0	0
	$E\tau$	18.04	28.4	32.3	35.5
$c_1 = 1.3$	w_2	0.13	0.05	0	0
	$E\tau$	29.0	50.1	53.3	62.1
$c_1 = 1.2$	w_2	0.43	0.36	0.06	0.01
	$E\tau$	44.4	65.6	85.9	90.5

Note: (5000 replications, pr — empirical false alarm rate, w_2 — type-2 error, $E\tau$ — average delay time)

choose $\lambda = 2.0$ and then do minor adjustments of this calibration parameter using "historical" samples.

The essential fact is stability of the calibration parameter λ over different values of N. So, from experiments with "historical" samples, we can estimate th and then compute λ for the volume N^* of these "historical" samples. Then, we can use this λ for all values of N.

In the following series of experiments, regression models with changes in the coefficient c_1 were considered. For each sample volume N and chosen values of the decision threshold th, the estimates of the 1st ("false alarm") and the 2nd-type error probabilities were computed, as well as the average delay time in change-point detection in $k = 5000$ independent trials. The results are reported in Table 10.5.

From these results, I conclude that the proposed test performs quite well even for small changes in coefficients. It is quite competitive with the CUSUM test based on recursive residuals, and is evidently more effective than the fluctuation test and the CUSUM test based on "historical" residuals. In the sequel, I demonstrate that the proposed test is effective also for sequential detection of changes in dynamic regression models and in systems of simultaneous linear regression equations. For these types of regression models, theoretical formulas for decision bounds are imprecise and difficult to obtain. However, we

TABLE 10.6
Decision Bounds of the Nonparametric Test: The Case of Dynamical Regression

N	20	50	100	200	300	400	500
$p = 0.95$	0.73	0.52	0.38	0.28	0.24	0.20	0.18
$p = 0.99$	1.30	0.90	0.63	0.41	0.38	0.32	0.25
λ	2.48	2.71	2.68	2.47	2.80	2.73	2.67

can easily construct empirically robust formulas of the decision threshold for the proposed test and calibrate it using a "historical" sample without change-points.

Now, let us consider the following dynamic regression:

$$y_i = 2 + \rho y_{i-1} + u_i, \quad y_0 = 0, \; i = 1, 2, \dots$$

where $u_i \sim \mathcal{N}(0, 1)$.

Here, I consider the problem of sequential detection of unknown changes in the coefficient ρ.

In the first series of tests, the model without structural changes and the coefficient $\rho = 0.3$ was considered. Again, we begin from the choice of the decision threshold. Using the above formula for it, we determine the calibration constant λ from the following tests. The maximums of decision statistic were computed in k=5000 trials of each experiment for different values of the sample volume N. Then, the variation series of these maxima was constructed, and, the 95 percent and 99 percent quantiles computed. The values of 99 percent quantiles for each value of N were assumed to be the decision thresholds th. In this case, $\sigma = 1$ and $Ey_i^2 = 5.494$. So, we can compute $\lambda = th\sqrt{N}/2.344$. The obtained results are reported in Table 10.6.

From these results, I conclude that the calibration coefficient λ slightly varies in the diapason [2.4; 2.8] depending on the sample volume N. Not surprisingly, for dependent observations y_i, the calibration coefficient is greater than λ from the previous test.

Again, for "historical" samples of the volume N^*, I can estimate th and then compute λ. This λ can be used for any value of N.

Then I consider the regression model with changes in the coefficient ρ. For, each sample volume N and the chosen values of the decision threshold th, the estimates of the 1st-type ("false alarm") and the 2nd-type error probabilities were computed (pr, w_2), as well as the average delay time in change-point detection ($E\tau$) in $k = 5000$ independent trials. The results are reported in Table 10.7.

From these results, I conclude that the proposed test is effective for detection of changes in coefficients of dynamic regression models.

2) System of simultaneous equations (SSE)

TABLE 10.7
Performance Characteristics of the Nonparametric Test

N		20	50	100	200	300
th		1.30	0.90	0.63	0.41	0.38
pr		0.03	0.02	0.03	0.03	0.02
$\rho = 0.7$	w_2	0	0	0	0	0
	$E\tau$	3.39	3.06	3.76	3.97	4.12
$\rho = 0.5$	w_2	0.35	0.18	0.04	0	0
	$E\tau$	9.6	10.5	14.2	18.3	21.3
$\rho = 0.4$	w_2	0.95	0.83	0.74	0.20	0.07
	$E\tau$	24.4	32.7	39.3	50.5	60.5

Note: (Dynamical regression, 5000 replications, pr — empirical false alarm rate, w_2 — type-2 error, $E\tau$ — average delay time)

The following system of simultaneous econometric equations was considered:

$$\begin{aligned} y_i &= c_0 + c_1 y_{i-1} + c_2 z_{i-1} + c_3 x_i + \epsilon_i \\ z_i &= d_0 + d_1 y_i + d_2 x_i + \xi_i \\ x_i &= 0.5 x_{i-1} + \nu_i \\ \epsilon_i &= 0.3 \epsilon_{i-1} + \eta_i, \end{aligned}$$

where $\xi_i, \nu_i, \eta_i,\ i = 1, 2, \ldots$ are independent $\mathcal{N}(0, 1)$ r.v.'s.

So, $(y_i, z_i)'$ is the vector of endogenous variables, x_i is the exogenous variable, and $(1, y_{i-1}, z_{i-1}, x_i)'$ is the vector of predetermined variables of this system.

The dynamics of this system are characterized by the following vector of coefficients: $\mathbf{u} = [c_0\ c_1\ c_2\ c_3\ d_0\ d_1\ d_2]$. The initial stationary dynamics are characterized by the coefficients [0.1 0.5 0.3 0.7 0.2 0.4 0.6].

In the first series of tests of the decision threshold was estimated. For this purpose, the model with the initial set of coefficients $\mathbf{u}$ and without structural changes was used. The empirically robust formula for determination of the decision threshold th has the following form:

$$th = C_0 \sqrt{\frac{N_0}{N}},$$

where N_0, C_0 is the sample volume and the decision bound for the "historical" sample without change-points, respectively.

In order to check this formula, the following Monte-Carlo tests were performed. In 5000 independent trials, the maximums of the decision statistic were computed and the variation series of these numbers constructed. The 95 and 99 percent quantiles are reported in the following table. The 99 percent quantiles were assumed to be the decision thresholds for the corresponding sample volumes. The obtained results are presented in Table 10.8.

TABLE 10.8
Decision Bounds of the Nonparametric Test (SSE Model)

N	20	50	100	200	300	400
$p = 0.95$	0.99	0.67	0.49	0.39	0.30	0.25
$p = 0.99$	1.50	0.85	0.65	0.47	0.38	0.32
th	1.45	0.91	0.65	0.46	0.37	0.32

TABLE 10.9
Performance Characteristics of the Nonparametric Test

N		20	50	100	200
th		1.50	0.85	0.65	0.47
pr		0.02	0.03	0.02	0.03
$d_2 = 0.95$	w_2	0.09	0	0	0
	$E\tau$	3.80	1.71	1.21	1.01
$d_2 = 0.9$	w_2	0.19	0.02	0	0
	$E\tau$	4.83	2.46	1.04	1.10
$d_2 = 0.8$	w_2	0.45	0.15	0.04	0
	$E\tau$	6.52	9.20	13.2	11.2

Note: (SSE model, 5000 replications, pr — empirical false alarm rate, w_2 — type-2 error, $E\tau$ — average delay time)

Suppose the "historical" sample volume was $N_0 = 100$. So, the obtained decision bound is $C_0 = 0.65$. Then, the decision threshold for an arbitrary sample volume can be computed as $th = 6.5/\sqrt{N}$. From Table 10.6, we conclude that empirical values for the decision threshold (for $p = 0.99$) are very close to this formula as N varies from 20 to 400.

In the following series of experiments, the models with changes in the coefficient d_2 were considered. For each sample volume N and the chosen values of the decision threshold th, the estimates of the 1st-type ("false alarm") and the 2nd-type error probabilities were computed, as well as the average delay time in change-point detection in $k = 5000$ independent trials. The results are reported in Table 10.9.

From these results, I again conclude that the smaller the structural change in the coefficients of the considered SSE model, the larger the sample volume for effective detection of this structural change must be.

10.6 Applications

Here, I consider two practical applications of the proposed method: sequential detection of structural changes in the quarterly model of demand for money in

Germany (1961–1990) and in the monthly model of inflation in Russia (1994–2008). For solving these problems, software wtitten in Matlab was used.

10.5.1. Demand for Money in Germany

Lütkepol, Terasvirta, and Wolters (1999) analyzed stability of demand for money in Germany from 1961–1995. The quarterly data on money aggregate M1, index of implicit price deflator of gross national product, real GDP, long-run interest rate of 1960(1)–1995(4) were used. The cointegration and ECM (error correction model) relationships for the demand for money function in the period 1961(1)–1990(2) were constructed, which include the following variables:

$m = \log(M1/PN)$ — the logarithm of real M1 per capita;
$p = \log(P)$ — the logarithm of implicit price deflator;
$y = \log(Y/PN)$ — the logarithm of the real GDP per capita;
R — the nominal interest rate;
N — the size of the population;
$Q1, Q2, Q3$ — quarterly seasonal dummy variables.

The ECM model constructed by the authors has the following form:

$$\begin{aligned}\Delta m_t = \quad & -0.30\Delta y_{t-2} - 0.67\Delta R_t - 1.00\Delta R_{t-1} - 0.53\Delta p_t \\ & -0.12m_{t-1} + 0.13y_{t-1} - 0.62R_{t-1} \\ & -0.05 - 0.13Q1 - 0.016Q2 - 0.11Q3 + \hat{u}_t.\end{aligned}$$

All regression coefficients in this relationship except the intercept are statistically relevant at the error level 1 percent and the determination coefficient is $R^2 = 0.943$.

This model includes the series of residuals of the long-term cointegration relationship

$$e_{t-1} = -0.12m_{t-1} + 0.13y_{t-1} - 0.62R_{t-1},$$

which is stationary by MacKinnon-Davidson criterion.

This model was used by Lütkepol et al. (1999), and Zeileis et al. (2005) for testing structural changes in the full sample of 1960(1)–1995(4).The OLE-based CUSUM test was used and a structural change detected at the point 1990(3).

In our tests, the volume of the moving window $N = 70$ was chosen and $\hat{\sigma} = 0.05$ estimated by the sample 1961(1)–1978(2) using the above specification of the ECM model. So the, estimated decision threshold is $\hat{C} = \hat{\sigma}\lambda(\max_i Ex_i^2)^{1/2}/\sqrt{N} = 0.05\lambda/\sqrt{N}$. Hence, we obtain two decision thresholds: $C_1 = 0.012$ ($\lambda = 2.0$ or 0.05 error level) and $C_2 = 0.016$ ($\lambda = 2.5$ or 0.01 error). Two points of structural changes, $n1 = 52$, and $n2 = 61$, corresponding to the periods 1990(2) and 1992(3) were detected in the whole sample using our method.

Comparing these results with Zeileis et al. (2005), I conclude that the method proposed in this paper is quite well-adapted to sequential detection of structural changes in the real econometric data.

10.5.2. Russian Inflation in 1994–2005

The regression model for the rate of CPI inflation (pi=CPI/100-1) was computed for the period 1994(1)–2008(8) (monthly data, sample volume nn=176) with the following set of predictors:

— inflation expectations (pi(-1));

— the rate of money growth: mu=M2/M2(-1)-1, where M2 is the monetary aggregate M2;

— the rate of growth of the nominal exchange rate of dollar: eps=E/E(-1)-1;

— the rate of growth of electric energy tariffs for population: piel;

— seasonal dummy: Seas.

The regression model obtained with this set of predictors has the following form:

$$\begin{array}{lllll} pi = & 0.0022+ & 0.2734pi(-1)+ & 0.2105piel+ & 0.3547eps+ \\ & (0.214) & (7.781) & (5.353) & (24.852) \\ & 0.1639mu(-6)+ & 0.012Seas- & 0.017Seas(-7) & \\ & (4.877) & (2.312) & (-3.515) & \end{array}$$

The main quality characteristics of this model are as follows: R2 = 0.887; approximation error σ = 0.015; Breusch-Godfrey statistic for higher order residuals autocorrelation AR 1-7 F(7,162)=2.981. All these characteristics are quite good.

The proposed method of monitoring structural changes is based upon the chosen set of predictors for this model (i.e., specification of the model), but the concrete values of regression coefficients are not essential. The choice of the decision bound C (threshold) is very important. For this purpose, the quasistationary subsample 1995(7)–1998(1) of observations was used. The decision bound C computed by this subsample equals $C = 0.0026$. The volume of the "moving window" is $N = 30$. Remark that the estimated value of the decision threshold $\hat{C} = \sigma(\max_i Ex_i^2)^{1/2}/\sqrt{N}) * \lambda$ in this case equals 0.002λ and is very close to the value $C = 0.0026$ computed by the "historical" quasistationary subsample 1995(7)–1998(1) for $\lambda = 1$ corresponding to 10 percent error level.

Two structural changes at the instants $n1 = 1$ and $n2 = 40$ were detected in the whole sample. These changes correspond to two important events in the Russian macroeconomic policy of 1990–2000s: introducing the "currency corridor" in June 1995 and the financial crisis of September 1998. So, the proposed method enables us to detect substantial structural changes in the real econometric data.

10.7 Conclusion

In this chapter, a new method for monitoring structural changes in multivariate stochastic systems is proposed, which enables us to effectively detect changes in parameters of linear regression models by sequential observations. The main performance characteristics of this method are analyzed theoretically for finite sample volumes. In particular, I prove Theorem 10.4.1 about exponential convergence to zero of the 1st-type error as the sample size tends to infinity. In Theorem, 10.4.2 the analogous result is proved for the probability of the 2nd-type error. Comparison with other well-known methods for sequential detection of structural changes in linear models is carried out via Monte-Carlo tests. Practical applications for the analysis of stability of the German quarterly model of demand for money (1961–1995) and the Russian monthly model of inflation (1994–2005) are considered.

11

Early Change-Point Detection

11.1 Introduction

The problem of *early detection* of changes in statistical characteristics of random processes often arises in different practical situations of data processing. The essence of this problem can be presented as follows. Suppose it is a priori known that a "change-point", i.e., an instant of an abrupt change in statistical characteristics of observations is preceded by some period of a slow and gradual trend (accumulation of "faults" and errors of an observed system, a weak drift of its parameters). It is required to detect statistically significant changes not *post factum*, i.e., after an abrupt change of statistical characteristics of data, but *ante factum*, i.e., in a certain instant of a slow trend. In other words, the alarm signal should be raised at some instant of the "transition period" but the number of "false alarms" must be rather small.

Let us give some examples. In ecological monitoring of air and water pollution, gradual negative changes of the quality of water and air caused by slow accumulation of destructive (technological or natural) factors are quite common. Usually, these negative changes are followed by sudden abrupt changes in qualitative characteristics of water or air (e.g., a dam suddenly broken). Detection of the instant of an abrupt change M is useless in this case because the ecological disaster has already happened. So, we need to detect an initial gradual deterioration of qualitative characteristics.

In technological systems, gradual negative changes of regime characteristics are also well known. These negative changes often end by sudden abrupt violations of the system performance (explosion, destruction, etc.). In this case, it is also too late to raise an alarm after the instant M. We need to detect gradual changes that precede abrupt and catastrophic events.

The main goal of this paper is to give the theoretical analysis of such problems. They are different from traditional sequential change-point detection problems. Therefore, we call them the *early change-point detection problems* taking into account their main performance criteria:

— to detect a change-point m as soon as possible; and

— the false alarm rate (FAR) must be rather small, i.e., *false alarms* (erroneous decisions about the presence of a change) must be rather seldom.

It should be noted that the problem of *early detection* was not considered in the statistical literature of 1950–1990. The majority of researchers paid

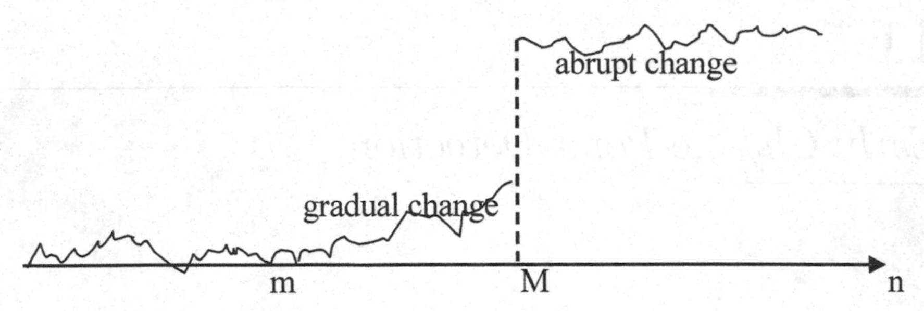

FIGURE 11.1
Abrupt and gradual change-points.

attention only to the "quickest detection" problems, in which the case of an abrupt change of statistical characteristics of data was considered. Optimality and asymptotic optimality of the proposed methods — the cumulative sums (CUSUM), the quasi-Bayesian (Roberts-Shiryaev or GRSh (Girshick-Rubin-Shiryaev)), etc. — was established only for the quickest detection problems, i.e., in the case of abrupt changes of statistical characteristics. However, it is not evident that optimality and asymptotic optimality of these methods will be observed in more general situations of changes in statistical characteristics of data. We demonstrate below that the CUSUM and GRSh methods **are not asymptotically optimal** for any models of changes except abrupt ones (i.e., the classic "change-point" model). In some cases of gradual changes in statistical characteristics of data, the window-limited methods of detection gain comparative advantage.

11.2 Asymptotic Optimality

11.2.1 Univariate Models; Independent Observations

Suppose we observe the following system of independent r.v.'s on the probability space $(\Omega, \mathcal{F}, P)$:

$$X = \{x(1), x(2), \dots\}, \qquad m - \text{ a change-point.}$$

The density function of observation changes at some unknown instant m according to the following pattern:

$$\frac{d}{dz} P\{x(n) \leq z\} = \begin{cases} f(z, 0), & n \leq m \quad \text{or } m = \infty \\ f(z, n-m), & n > m. \end{cases}$$

The problem is to detect the change-point m as soon as possible on condition that "false alarms" (i.e., erroneous decisions about the presence of a change) occur rather seldom.

In this section, we consider the a priori lower bound for a certain performance measure connected with the delay time in change-point detection. For this purpose, consider the following decision rule $d_C(n)$ depending on a large parameter C:

$$d_C(n) = \begin{cases} 1, \text{stop at time } n \text{ and accept } H_1, \\ 0, \text{continue under } H_0. \end{cases}$$

Let $P_\infty(\cdot)$ (E_∞) be the measure (mathematical expectation) corresponding to the sequence without change-points and $P_m(\cdot)$ (E_m) – the measure (expectation) corresponding to the sequence with the change-point m.

Define the following performance characteristics of a change-point detection method:

— the supremal probability of the false decision:

$$\alpha_C = \sup_n P_\infty\{d_C(n) = 1\}.$$

— the stopping time

$$\tau_C = \min\{n : \ d_C(n) = 1\}.$$

— the normalized stopping time:

$$\gamma_C = (\tau_C - m)^+/C.$$

We also consider the following value:

$$M_C = \min\{l : \sum_{n=m+l}^{\infty} P_\infty\{\tau_C = n\} \le \alpha_C\}$$

and the Kullback information measure:

$$J(n) = E_m(\ln \frac{f(z, n-m)}{f(z,0)}), \qquad j(t) = J(m + [tC]), \quad t \ge 0.$$

The following theorem holds.

Theorem 11.2.1. *For an arbitrary method of sequential change-point detection dependent on a large parameter C:*

$$E_m \int_0^{\gamma_C} j(t)dt \ge \frac{|\ln(\alpha_C M_C)|}{C} + O(\frac{1}{C}) \quad \textit{as } C \to \infty.$$

The proof of Theorem 11.2.1 follows from the proof of Theorem 11.2.2, which is given below.

11.2.2 Multivariate Models; Dependent Observations

Suppose $Z = (z_1, z_2, \dots)$ is a sequence of dependent vector-valued observations $z_n = (z_n^1, \dots, z_n^k)$ defined on the probability space $(\Omega, \mathcal{F}, P)$.

By analogy with the previous case of a univariate sequence of independent r.v.'s, define the following performance characteristics of a change-point detection method:

— the supremal probability of the false decision:

$$\alpha_C = \sup_n P_\infty\{d_C(n) = 1\}.$$

— the stopping time;

$$\tau_C = \min\{n : \ d_C(n) = 1\}$$

— the normalized stopping time:

$$\gamma_C = (\tau_C - m)^+/C.$$

We also consider the following value:

$$M_C = \min\{l : \ \sum_{n=m+l}^{\infty} P_\infty\{\tau_C = n\} \leq \alpha_C\}.$$

The following theorem holds.

Theorem 11.2.2. *For an arbitrary method of sequential change-point detection dependent on a large parameter C:*

$$E_m \int_0^{\gamma_C} j(t)dt \geq \frac{|\ln(\alpha_C M_C)|}{C} + O(\frac{1}{C}), \qquad \text{as } C \to \infty,$$

where

$$J(n) = E_m(\ln \frac{f_n(z_n|z_1 \dots z_{n-1})}{f_0(z_n|z_1 \dots z_{n-1})}),$$
$$j(t) = J(m + [tC]), \quad t \geq 0.$$

Proof.

Denote $\tau_N^* = m + (\tau_N - m)^+ \wedge M_N(m)$. Then, taking into account the relationship $\tau_N = \tau_N^*$ on the set $(m \leq \tau_N \leq M_N(m) + m)$ and using Yensen inequality, we obtain

$$\begin{aligned} M_N(m)\alpha_N &\geq \sum_{k=m}^{M_N(m)+m} \mathbf{P}_\infty(\tau_N^* = k) \\ &= \sum_{k=m}^{M_N(m)+m} \mathbf{E}_m\{(\mathbb{I}(\tau_N^* = k)\exp(-\sum_{i=m}^{k} \varphi\left((i-m)/N, \mathbf{z}_i, \mathbf{z}_1^{i-1}\right) \\ &= \mathbf{E}_m \exp\left(-\sum_{i=m}^{\tau_N^*} \varphi((i-m)/N, \mathbf{z}_i, \mathbf{z}_1^{i-1})\right) \\ &\geq \exp\left(-\mathbf{E}_m \sum_{i=m}^{\tau_N^*} \varphi((i-m)/N, \mathbf{z}_i, \mathbf{z}_1^{i-1})\right). \end{aligned} \tag{11.1}$$

Since τ_N^* is the Wald r.v. (i.e., the event $(\tau_N^* \leq n)$ does not depend on the σ-algebra $\sigma\{\mathbf{z}_n, \dots\}$), we can use the following theorem about the sum of a random number of r.v.'s (see Kruglov, and Korolev (1990), p. 98).

$$\mathbf{E}_m \sum_{i=m}^{\tau_N^*} \varphi((i-m)/N, \mathbf{z}_i, \mathbf{z}_1^{i-1}) = \mathbf{E}_m \sum_{i=m}^{\tau_N^*} J\,(i, (i-m)/N)\,. \tag{11.2}$$

From (11.1) and (11.2), we have

$$N^{-1}|\ln(M_N(m)\alpha_N)| \leq N^{-1}\mathbf{E}_m \sum_{i=m}^{\tau_N^*} J\,(i, (i-m)/N)\,. \tag{11.3}$$

Since $\tau_N \geq \tau_N^*$, we obtain for $n \geq m$ from (11.3) and integrability assumption about function $J(\cdot, t)$

$$\mathbf{E}_m \int_0^{\gamma_N} J(n,t)dt \geq \frac{|\ln(\alpha_N M_N(m)|}{N} + O(\frac{1}{N}).$$

Let us discuss why these theorems are relevant for the proof of the asymptotic optimality of an arbitrary change-point detection method. Remark that the left-hand side of the a priori inequalities presented in these theorems depends on the normalized delay time in change-point detection (the main reason for the use of this value), while the right-hand side depends on the probability of the false decision. In the sequel, we prove that the left-hand and the right-hand sides tend to certain limits as $C \to \infty$ for all well known methods of sequential change-point detection. So, we can compare these limit values for a concrete method. If they coincide for a certain choice of method's parameters, then we say that this method is *asymptotically optimal* in a chosen situation.

Now, we study the quality characteristics of well known methods of sequential change-point detection.

11.3 Methods

11.3.1 Univariate Models

In this section, I consider asymptotically optimal methods of early change-point detection for univariate models.

I begin with the simplest model of a gradual change in mean of a univariate time series.

On the probability space $(\Omega, \mathfrak{F}, P)$, consider a univariate sequence of observations $X = \{x_1, x_2, \dots\}$, where

$$x_n = a\,I(1 \leq n \leq m) + h(n-m)\,I(n > m) + \xi(n), \qquad n = 1, 2, \dots$$

and $E\xi(n) = 0$.

For concrete methods, I assume different relationships between a and $h(\cdot)$ (see below).

Sometimes, the hypothesis of a change in mean of observations is no more valid. Often, we can reduce the problem to the case of a change in mean by the corresponding transformation of data. For example, if the volatility of data changes, then we first center our data and after that consider squares of the centered observations. For this transformed sequence of observations, we then apply one of the above considered methods.

However, in many problems, the hypothesis of continuous distribution functions of observations is not correct. For example, if the number of possible states of an observed process is finite and the probabilities of hitting the concrete states are unknown but changing from a certain time, then we cannot directly apply the above methods and must think about modification of them.

Let us consider the following problem. Suppose observations $X = (x_1, x_2, \dots)$ take their values in the finite discrete set of states $S = \{a_1, \dots, a_k\}$ with the following probabilities: for any $j = 1, \dots, k$

$$P\{x_i = a_j\} = \begin{cases} p_j, & \text{for } i \leq m \\ q_j(i) & \text{for } i > m. \end{cases}$$

Probabilities $p_j, q_j(i), j = 1, \dots, k$ are unknown, but the function

$$\delta(n) = \sum_{j=1}^{k} |p_j - q_j(n)| > 0, \qquad n > m$$

does not decrease with respect to (w.r.t.) n.

Both for continuous and discrete distributions of r.v.'s, the problem is to detect the change-point m as soon as possible on condition that "false alarms" are few.

Assumptions

Our assumptions in this section are the uniform Cramer's and ψ-mixing conditions formulated in Chapter 1 (see 1.1).

Continuous distributions

Now, I consider the concrete methods of sequential change-point detection. Our goal here is to study the main performance characteristics of these methods in the situation of early change-point detection, i.e., for gradual changes in statistical parameters of observations.

CUSUM method

This method is based on the following statistic:

$$y_n = (y_{n-1} + x_n - |a|)^+, \qquad y_0 = 0, \qquad d_N(n) = I(y_n > N).$$

Our assumptions about a and $h(\cdot)$ for this method are as follows:

1) *The case of a positive trend function* $h(\cdot)$: $a < 0$ and for any threshold $C > 0$, there exists the root $\gamma > 0$ of the following equation:

$$\int_0^{\gamma} (h(\tau) - |a|)\, d\tau = C.$$

2) *The case of a negative trend function* $h(\cdot)$: $a > 0$ and for any threshold $C < 0$, there exists the root $\gamma > 0$ of the following equation:

$$\int_0^{\gamma} (a + h(\tau))\, d\tau = C.$$

GRSh method

$$R_n = (1 + R_{n-1})e^{x_n - |a|}, \qquad R_0 = 0, \qquad d_N(n) = I(R_n > e^N).$$

The main assumptions about a and $h(\cdot)$ for this method are the same as for CUSUM.

"Window-limited" method

This method is based upon the idea of the "moving window" of observations, i.e., we consider for any $n = N, N+1, \ldots$ the subsample of the last N observations: $X^N(n) = \{x_{n-N+1}, \ldots, x_n\}$. Then, for the same subsample, we define the scale function $g(k/N)$, $k = 1, \ldots, N$, and consider the following statistic:

$$Y_N(n) = N^{-1} \sum_{k=1}^{N} g(\frac{k}{N}) x_{n-k+1}, \qquad n = N, N+1, \ldots.$$

The decision rule for this method is defined as follows:

$$d_N(n) = I(|Y_N(n)| > C), \qquad \text{where} \int_0^1 g^2(t)dt = 1,$$

and $C > 0$ is a certain threshold.

For this method, $a = 0$ (initial observations must be centered) and *in the case of a positive trend function* $h(\cdot)$:
for any $H > 0$ there exists $\gamma > 0$, such that

$$\int_0^{\gamma} h(t)g(t)dt = H;$$

in the case of a negative trend function $h(\cdot)$:

for any $H < 0$, there exists $\gamma > 0$, such that

$$\int_0^\gamma h(t)g(t)dt = H.$$

Discrete distributions; unknown probabilities of states

In order to detect the change-point m, we again use the idea of the "moving window" of the last N observations $X^N(n) = (x_{n-N+1}, \dots, x_n)$, $n = N, N+1, \dots$, where N is a certain large parameter, and consider the following statistic:

$$Y_N^n(l) = \sum_{j=1}^k |n \sum_{i=n-N+1}^n I(x_i = a_j) - N \sum_{i=n-N+1}^{n-N+l} I(x_i = a_j)|/N^2, \quad l = 1, \dots, N,$$

and the decision rule

$$d_N(n) = I\{\max_{1 \le l \le N} Y_N^n(l) > C\},$$

where $C > 0$ is the decision threshold. If $d_N(n) = 1$, then we stop and raise the alarm about a change; if $d_N(n) = 0$, then we continue to collect observations.

For this method, we suppose that for any $0 < C < \frac{k}{4}$, there exists a minimal root $0 < \gamma < 1$ of the following equation:

$$(1-\gamma) \sum_{j=1}^k |\int_0^\gamma (q_j(t) - p_j)dt| = C,$$

where by definition the continuous function $q_j(t)$, $0 < t \le 1$ is such that $q_j(k/N) = q_j(k)$, $k = 1, \dots, N$.

Univariate models; main results

In the following theorem, we study performance characteristics of the CUSUM and GRSh methods. Without loss of generality, let us consider the case of a positive trend function: $a < 0, h(t) > 0$. Recall that for CUSUM, the large parameter is equal to the decision threshold N. For te GRSh method, the large parameter is equal to $\ln N$, where again N is the decision threshold.

Theorem 11.3.1. *(CUSUM, GRSh).*

Suppose Cramer's and ψ-mixing conditions are satisfied for the noise sequence $\xi(n)$. Then,

1) The normalized (by N) probability of the error decision tends to a certain positive limit as $N \to \infty$:

$\delta_N = \frac{|\ln \alpha_N|}{N} \to \delta^*$ *as $N \to \infty$, where δ^* is the minimal nonzero root of the equation*

$$\kappa(t) = \ln \sup_n E \exp(t\xi(n)) = 0.$$

In particular, for the sequence of independent Gaussian random variables with the dispersion σ^2 (and a change in the mathematical expectation), $\delta_{cusum} = \delta_{grsh} = 2|a|/\sigma^2$.

2) The normalized delay time in change-point detection tends almost surely to a certain deterministic limit, i.e., $\gamma_N \to \gamma^*$ P_m-a.s. as $N \to \infty$, where

$$\int_0^{\gamma^*} (h(t) - |a|)dt = 1.$$

The proof of Theorem 11.3.1 is given in the appendix to this chapter. The analogous result for the "window-limited" method is formulated in the following theorem.

Theorem 11.3.2. *("window-limited")*

Suppose the ψ-mixing and Cramer's conditions are satisfied for the noise sequence $\xi(n)$. Then,

i)

$$\lim_{N\to\infty} N^{-1} |\ln \max_{1\le n\le N} \mathbf{P}_\infty (d_N(n) = 1)| = \frac{C^2}{2\sigma^2},$$

where $\sigma^2 = \lim_{N\to\infty} N^{-1} E(\sum_{k=1}^{N} \xi(k))^2$.

ii)

$$\gamma_N \overset{\mathbf{P}_m \ a.s.}{\longrightarrow} \gamma^e_{gm} \quad \textit{as } N \to \infty,$$

where γ^e_{gm} is the minimal root of the equation $\int_0^{\gamma^e_{gm}} h(t)g(t)dt = C$.

The proof of Theorem 11.3.2 repeats the ideas of Theorem 11.3.1 and, therefore, is omitted here.

For the method of change-point detection in the sequence of discrete valued observations, we obtain the following result:

Theorem 11.3.3. *Suppose a sequence of discrete valued r.v.'s is observed that satisfies the ψ-mixing condition. Then,*

i)

$$\lim_{N\to\infty} N^{-1} |\ln \max_{1\le n\le N} \mathbf{P}_\infty (d_N(n) = 1)| = \frac{C^2}{\sigma^2},$$

where $\sigma^2 = \sum_{j=1}^{k} a_j^2 p_j - (\sum_{j=1}^{k} a_j p_j)^2$.

ii)

$$\gamma_N \overset{\mathbf{P}_m \ a.s.}{\longrightarrow} \gamma \quad \textit{as } N \to \infty,$$

where γ is the minimal root of the equation

$$(1-\gamma) \sum_{j=1}^{k} |\int_0^{\gamma} (q_j(t) - p_j)dt| = C.$$

The sketch of the proof of Theorem 11.3.3 is given in the appendix to this chapter.

Univariate models; asymptotic optimality

In this section, the asymptotic optimality of "early detection" methods is studied for the case of the Gaussian sequence $x_1, x_2, \dots,$ with the dispersion $\sigma^2 = 1$ and the trend $h(t)$, $t \geq 0$ in the mathematical expectation of observations.

For the CUSUM and GRSh methods, the *a priori* inequality for a Gaussian sequence has the following form:

$$\int_0^{\gamma_e} \frac{h^2(t)}{2}\, dt \geq \delta^* = 2|a|.$$

On the other hand, the asymptotic limit of the relative delay time in detection for the CUSUM and GRSh methods γ_c satisfies the equation

$$\int_0^{\gamma_c} (h(t) - |a|)\, dt = 1.$$

Hence, in virtue of the Cauchi–Schwartz inequality, we obtain

$$(1 + \gamma_c|a|)^2 \leq \gamma_c \int_0^{\gamma_c} h^2(t)\, dt. \tag{11.4}$$

The property of the asymptotic optimality means

$$\gamma_c = \gamma_e = \gamma \quad \text{for} \quad \int_0^{\gamma} h^2(t) dt = 4|a|.$$

Let us check for which trend function $h(t) \geq 0$ it is possible to satisfy these conditions. Inequality (11.4) takes the form: $(1 - \gamma|a|)^2 \leq 0$, when $\gamma = |a|^{-1}$.

Therefore,

$$\int_0^{|a|^{-1}} h(t)dt = 2, \qquad \int_0^{|a|^{-1}} h^2(t)\, dt = 4|a|,$$

which yields $h(t) \equiv const = 2|a|$.

This case corresponds to the abrupt change in mean from the level $-|a|$ to the level $|a|$. We readily acknowledge here the classic sequential change-point detection problem for the Gaussian sequence: from $\mathcal{N}(0, \sigma^2)$ to $\mathcal{N}(w, \sigma^2)$ with $|a| = |w|/2\sigma^2$.

Let us emphasize this conclusion: *CUSUM and GRSh methods are asymptotically optimal only in the case of abrupt changes* in the mathematical expectation of observations, i.e., in the classic change-point problem. For other trend functions $h(t) \geq 0$, the property of the asymptotic optimality of CUSUM and GRSh methods is violated.

However, the family of "window limited" methods contains the asymptotically optimal procedure for any monotonous trend function $h(t) \geq 0$. In the Gaussian case for "window limited" methods, we can write: $h(1) > C_w$, $\delta^*(\cdot) = C_w^2/2$, $\sigma^2 = 1$, and the a priori inequality takes the following form:

$$\int_0^{\gamma^e} \frac{h^2(t)}{2} \geq \frac{C_w^2}{2}. \tag{11.5}$$

For "window limited" methods:

$$\int_0^{\gamma_m} h(t)g(t)dt = C_w, \tag{11.6}$$

where $\int_0^1 g^2(t)dt = 1$, $0 < \gamma_m \leq 1$.

In virtue of the Cauchi-Schwartz inequality from (11.6), we obtain:

$$\frac{C_w^2}{2} \leq \int_0^{\gamma_m} \frac{h^2(t)}{2}dt \cdot \int_0^{\gamma_m} g^2(t)dt. \tag{11.7}$$

Suppose the equality in (11.5) is attained for a "window limited" method, i.e. for $\gamma_m = \gamma^e = \gamma$. Then, from (11.7), we conclude that $\int_0^{\gamma} g^2(t)dt \geq 1$, and therefore, $\gamma = 1$. So, (11.7) turns into the strict equality that is possible if and only if $g(t) = h(t)/C_w$. From the normalizing condition $\int_0^1 g^2(t)dt = 1$, we obtain $C_w = (\int_0^1 h^2(t)dt)^{1/2}$.

The above considerations allow us to make the following conclusion: The a priori inequality is in the general sharp for any trend function, i.e. the equality in it is attained for a "window limited" method with an appropriately chosen set of parameters. This conclusion leads to the following strategy of *adaptive* change-point detection: Simultaneously with change-point monitoring, we must estimate the trend function $h(t)$ and then adjust the scaling function $g(t)$ for the "window-limited" method.

However, in practice (see Table 11.1 below), CUSUM and GRSh methods gain the comparative advantage over window-limited procedures in most situations of early change-point detection.

11.3.2 Multivariate Models

The above ideas can be generalized to the case of a multivariate model with gradually changing parameters. Below, I consider the most important case: multivariate regression models with gradual changes of coefficients.

Multivariate regression

Let us consider the following model of a multivariate regression with changing parameters:

$$Y(n) = \Pi(n)X(n) + \nu_n, \quad n = 1, 2, \dots .$$

The $M \times K$ matrix Π begins to change at some unknown change-point m, i.e.,

$$\Pi(n) = \mathbf{a}I(n \le m) + \mathbf{b(n)}I(n > m), \quad n = N, N+1, \dots ,$$

where $\|\mathbf{a} - \mathbf{b(n)}\| > 0$.

Assumptions

Suppose that predictors $X(n)$ and noises ν_n are continuously distributed and strictly stationary, and the following conditions are satisfied:

1) The vector $X(n) = (x_{1n}, \dots, x_{Kn})'$ is $\mathcal{F}_{n-1}$ measurable.

2) There exists a continuous matrix function $V(t)$, $t \in [0,1]$, such that for any $0 \le t_1 \le t_2 \le 1$

$$\frac{1}{N} \sum_{j=[t_1 N]}^{[t_2 N]} X(j)X'(j) \to \int_{t_1}^{t_2} V(t)dt, \quad P - \text{a.s. as } N \to \infty,$$

where $\int_{t_1}^{t_2} V(t)dt$ is the positive definite matrix.

3) The random vector sequence $\{(X(n), \nu_n)\}$ satisfies ψ-mixing and the uniform Cramer condition.

4) $\{\nu_n\}$ is a martingale-difference sequence w.r.t. the flow $\{\mathcal{F}_n\}$.

Method

For any $n = N, N+1, \dots$, consider N last vectors of observations $Y(i), X(i), i = n - N + 1, \dots, n$.

First, consider the $K \times K$ matrices:

$$\mathcal{T}^n(1,l) = \sum_{i=1}^{l} X(i+n-N)X'(i+n-N), \quad l = 1, \dots, N.$$

Second, consider the $K \times M$ matrices:

$$z^n(1,l) = \sum_{i=1}^{l} X(i+n-N)Y'(i+n-N), \quad l = 1, \dots, N.$$

Third, consider the decision statistic

$$Y_N^n(l) = \frac{1}{N}(z^n(1,l) - \mathcal{T}^n(1,l)(\mathcal{T}^n(1,N))^{-1} z^n(1,N)),$$

where $l = 1, \dots, N$, $Y_N^n(N) = 0$ and, by definition, $Y_N^n(0) = 0$.

Fix the number $0 < \beta < 1/2$. For detection of the change-point $m > N$, we define the stopping time

$$\tau_N = \inf\{n : \max_{[\beta N] \le l \le N} \|Y_N^n(l)\| > C\},$$

where C is a certain decision threshold, and $\|A\|$ is the Gilbert norm of the matrix A.

The following theorem holds.

Theorem 11.3.4. *Suppose assumptions 1)–2) are satisfied. Then,*

1) for the 1st-type error:

$$P_0\{\max_n \|Z_N(n)\| > C\} \le m_0(C_1) \begin{cases} \exp(-\dfrac{TN\beta C_1}{4m_0(C_1)}), & C_1 > hT \\ \exp(-\dfrac{N\beta C_1^2}{4hm_0(C_1)}), & C_1 \le hT, \end{cases}$$

where $C_1 = C/(1+\sqrt{K})$.

2) for the 2nd-type error, define:

$$S(\theta) = \int_0^{1-\theta} V(\tau)d\tau \cdot I^{-1} \cdot \int_{1-\theta}^{1} V(\tau)(\mathbf{b}(\tau) - \mathbf{a})d\tau$$
$$g(\tilde{\theta}) = \max_\theta g(\theta), \quad g(\theta) = \|S(\theta)\|^2,$$
$$d = (g(\tilde{\theta}) - C)/(1+\sqrt{K}).$$

Then,

$$\delta_N \le m_0(d) \begin{cases} \exp(-\dfrac{TN\beta d}{4m_0(d)}), & d > hT \\ \exp(-\dfrac{N\beta d^2}{4hm_0(d)}), & d \le hT. \end{cases}$$

3) normalized delay time:

$$\gamma_N = \frac{(\tau_N - m)^+}{N} \to \gamma^*, \quad P_m - \ a.s. \ as \ N \to \infty,$$

where γ^* *is the minimal root of the equation* $g(t) = C$.

The proposed method is asymptotically optimal by the order of the performance measures (w.r.t. $N \to \infty$*).*

The proof of this theorem is given in the appendix.

11.4 Simulations

Univariate models

In the first series of tests univariate time series with unknown change-points were studied. I considered both continuous and discrete distribution

TABLE 11.1
Performance Characteristics of CUSUM, GRSh, and WL Tests: Univariate Gaussian Model

k		0.001	0.005	0.01	0.05	0.1	0.5	1.0
CUSUM	$E\tau$	506.36	143.22	85.34	28.84	18.81	7.44	5.09
GRSH	$E\tau$	522.98	146.72	87.74	29.74	19.66	7.92	5.48
WL	$E\tau$	495.8	281.9	224.4	131.38	104.3	61.55	48.8

Note: d.f.'s of observations (5000 replications, $E\tau$ — average delay time)

functions of observations. For the continuous case, the following model of data was studied:

Model

The sequence of Gaussian observations with the d.f. $\mathcal{N}(0,1)$ before the change-point and the linear trend in mean kt, $k > 0$ after the change-point, i.e.,

$$f(x_i) = \begin{cases} \mathcal{N}(0,1), & i \leq m \\ \mathcal{N}(k(i-m),1), & i > m, \end{cases}$$

was considered, where m is an unknown change-point. The problem was to detect m as soon as possible on condition that false alarms occur rather seldomly.

Methods

For solving this problem, the following nonparametric sequential early change-point detection methods were used:

1) **cumulative sums, CUSUM**: $y_n = (y_{n-1} + x_n - |a|)^+$; $a = -0.5$;

2) **Girshick-Rubin-Shiryaev, GRSh**: $y_n = (1+y_{n-1})\exp(x_n - |a|)$; $a = -0.5$;

3) **window-limited, WL**: $y_n = N^{-1} \sum\limits_{i=0}^{N-1} g(i)\, x(n-i)$, $g(i) = h\,i$, $h > 0$.

Parameters of these methods were chosen in order to equalize them by the false alarm characteristics: the frequency of a false decision pr_0 (or the average time between false alarms $E_0\tau$). In experiments, we chose $pr_0 = 0.001$. The following parameters were used:

CUSUM: the threshold of detection: $th = 11.0$;

GRSh: $th = 700000$; and

WL: $N = 400$, $th = 0.048$, $h = 0.001$.

Then, we compared these methods by the characteristic of the average delay time in change-point detection for different values of the parameter $k > 0$ (the slope of the linear trend in the mean values of observations).

The results obtained in 5000 independent trials of each experiment are reported in Table 11.1.

TABLE 11.2
Performance Characteristics of the Proposed Test: Univariate Model

gradual	N	100	200	300	500
th		0.20	0.14	0.11	0.09
pr		0.05	0.03	0.03	0.03
$h = 0.01$	w_2	0.02	0	0	0
	$E\tau$	25.2	26.1	26.4	27.1
$h = 0.005$	w_2	0.42	0.01	0	0
	$E\tau$	84.1	83.6	84.2	86.4
$h = 0.001$	w_2	0.93	0.64	0.02	0
	$E\tau$	234.3	278.9	268.5	279.2

Note: d.f.'s (5000 replications, pr — empirical false alarm rate, w_2 — type-2 error, $E\tau$ — average delay time)

Discrete distributions

In this example of the finite valued alphabet of possible states of an observed random sequence, we suppose that the states are 1, 2, and 3 and the probabilities of these states are as follows: $p_1(n) = 0.2, p_2(n) = 0.3$, and $p_3(n) = 0.5$ before the change-point (i.e. for $n \leq m$) and $q_1(n) = 0.2$, $q_2(n) = 0.3 + h\,(n-m)$, $q_3(n) = 0.5 - h\,(n-m)$, and $n > m$ after the change point m. Here, $h > 0$ is the coefficient of the linear trend in the probabilities of states.

These probabilities were assumed to be unknown for the proposed algorithm. The results obtained are given in Table 11.2.

Multivariate models

Multivariate regression

The following system of simultaneous equations was considered:

$$
\begin{aligned}
y_i &= c_0 + c_1 y_{i-1} + c_2 z_{i-1} + c_3 x_i + \epsilon_i \\
z_i &= d_0 + d_1 y_i + d_2 x_i + \xi_i \\
x_i &= 0.5 x_{i-1} + \nu_i \\
\epsilon_i &= 0.3 \epsilon_{i-1} + \eta_i,
\end{aligned}
$$

where $\xi_i, \nu_i, \eta_i,\ i = 1, 2, \dots$ are independent $\mathcal{N}(0,1)$ r.v.'s.

So, $(y_i, z_i)'$ is the vector of endogenous variables, x_i is the exogenous variable, and $(1, y_{i-1}, z_{i-1}, x_i)'$ is the vector of predetermined variables of this system.

The dynamics of this system are characterized by the following vector of coefficients: $\mathbf{u} = [c_0\ c_1\ c_2\ c_3\ d_0\ d_1\ d_2]$. The initial stationary dynamics are characterized by the coefficients [0.1 0.5 0.3 0.7 0.2 0.4 0.6]. The obtained decision bounds are presented in Table 11.3.

TABLE 11.3
Decision Bounds of the Nonparametric Test (SSE Model)

N	20	50	100	200	300	400
$p = 0.95$	0.99	0.67	0.49	0.39	0.30	0.25
$p = 0.99$	1.50	0.85	0.65	0.47	0.38	0.32
th	1.45	0.91	0.65	0.46	0.37	0.32

TABLE 11.4
Performance Characteristics of the Nonparametric Test (SSE Model)

abrupt	N	20	50	100	200
th		1.50	0.85	0.65	0.47
pr		0.02	0.03	0.02	0.03
$d_2 = 0.95$	w_2	0.09	0	0	0
	$E\tau$	3.80	1.71	1.21	1.01
$d_2 = 0.9$	w_2	0.19	0.02	0	0
	$E\tau$	4.83	2.46	1.04	1.10
$d_2 = 0.8$	w_2	0.45	0.15	0.04	0
	$E\tau$	6.52	9.20	13.2	11.2

gradual ($d_2 = \Delta d$)		$N = 100$
th		0.65
pr		0.02
$\Delta d = 10^{-4}$	w_2	0.23
	$E\tau$	97.4
$\Delta d = 10^{-3}$	w_2	0.02
	$E\tau$	27.8
$\Delta d = 10^{-2}$	w_2	0
	$E\tau$	5.6

Note: 5000 replications, pr — empirical false alarm rate, w_2 — type-2 error, $E\tau$ — average delay time

In the following series of experiments, the models with changes in the coefficient d_2 were considered. For each sample volume N and the chosen values of the decision threshold th, the estimates of the 1st-type ("false alarm") and the 2nd-type error probabilities were computed, as well as the average delay time in change-point detection in $k = 5000$ independent trials. The results are reported in Table 11.4.

11.5 Conclusions

In this chapter, the a priori informational inequalities for the main performance measures in sequential detection of abrupt and gradual changes for univariate and multivariate stochastic models are proved. It is usually assumed that statistical characteristics of observations change instantaneously from one stationary level to another at some unknown points. The optimality and asymptotic optimality of CUSUM, and GRSh and "window-limited" tests were established only under these assumptions. However, in many practically relevant situations of *gradual changes* in statistical characteristics of data, the asymptotic optimality of CUSUM, GRSh, and other well-known tests may be violated. In this chapter, I demonstrate that CUSUM and GRSh tests will be asymptotically optimal in the problem of "early detection" only in the classic situation of an abrupt change from one known density function $f_0(\cdot)$ to another (a priori known) density function $f_1(\cdot)$. The asymptotically optimal methods of early change-point detection in univariate and multivariate stochastic models are proposed. The Monte-Carlo tests are performed for the proposed methods of early change-point detection in univariate and multivariate models.

11.6 Appendix: Proofs of Theorems

Theorem 11.3.1

The proof of i) is given in Chapter 9.

ii)

First, consider the CUSUM method.

Fix $\epsilon > 0$ and consider the event $A = \{\gamma_N > \gamma^e + \epsilon\}$. Denote $\tilde{n} = [N(\gamma^e + \epsilon)]$. Since

$$\begin{aligned} y_{m+1} &= (y_m + x_m)^+ \geq y_m + x_m \geq x_m \quad (y_m \geq 0) \\ y_{m+2} &\geq y_{m+1} + x_{m+1} \geq x_m + x_{m+1} \\ &\vdots \\ y_{m+\tilde{n}} &\geq x_m + x_{m+1} + \cdots + x_{m+\tilde{n}-1}, \end{aligned}$$

we have

$$\begin{aligned} &P_m(A) \leq P_m(y(m+\tilde{n}) < N) \leq P_m\{\sum_{k=m}^{m+\tilde{n}-1} x_k < N\} = \\ &= P_m\{\sum_{k=m}^{m+[\gamma^e N]-1} (h(n) - |a|) + \sum_{k=m+[\gamma^e N]}^{m+\tilde{n}-1} (h(n) - |a|) + \sum_{k=m}^{m+\tilde{n}-1} \xi_k < N\} \\ &\leq P_m\{\sum_{k=m}^{m+\tilde{n}-1} \xi_k < -N \int_{\gamma^e}^{\gamma^e+\epsilon} (h(t) - |a|)\, dt + o(1)\}. \end{aligned}$$

Since $h(t)$ is monotonously increasing, we have $D(\epsilon) = \int\limits_{\gamma^e}^{\gamma^e+\epsilon} (h(t) - |a|)\, dt > 0$.

So, the problem is reduced to the exponential estimate of the probability of the event that the finite sum of dependent random variables exceeds a certain threshold. This exponential estimate follows from results presented in Chapter 1 (see Section 1.4).

$$P_m\{\sum_{k=m}^{m+\tilde{n}-1} \xi_k < -ND(\epsilon) + o(1)\} \leq B_1 \exp(-B_2(\epsilon)N),$$

where the constants B_1, and B_2 do not depend on N.

Now, consider the event

$$B = \{\omega : \gamma_N < \gamma^e - \epsilon\} = \{\omega : m \leq \tau_N \leq m + [N(\gamma^e - \epsilon)]\} \cup \{\omega : \tau_N < m\}.$$

Taking into account that $(\tau_N = k) \subset (d_N(k) = 1)$ and the exponential estimate for the "false alarm" probability, we obtain

$$\begin{aligned} P_m(\tau_N < m) &= P_\infty(\tau_N < m) \leq m \cdot \max_{k<m} P_\infty(d_N(k) = 1) \leq \\ &\leq mC_1 \exp(-C_2 N), \end{aligned}$$

where $C_1, C_2 > 0$.

Denote

$$\begin{aligned} V_m &= y_m, \quad V_{m+1} = y_m + x_{m+1}, \ldots, V_{m+k} = y_m + \sum_{i=1}^{k} x_{m+i}, \\ W_m &= 0, \quad W_{m+1} = \xi_{m+1}, \ldots, W_{m+s} = \sum_{k=m+1}^{m+s} \xi_k, \quad s \geq 1. \end{aligned}$$

Then, for each $s = 0, 1, \ldots$,

$$y_{m+s} = V_{m+s} - \min(0, V_m, V_{m+1}, \ldots, V_{m+s}).$$

Hence, for $s \geq 1$:

$$y_{m+s} \leq y_m + \sum_{i=m}^{m+s-1} (h(i) - |a|) + W_{m+s} - \min(0, W_{m+1}, \ldots, W_{m+s}).$$

Therefore, $(s \geq 1)$:

$$P_m\{y_{m+s} > N\} \leq P_m\{W_{m+s} - \min(0, W_{m+1}, \ldots, W_{m+s}) > \frac{1}{2}(N - \sum_{i=m}^{m+s-1} (h(i) - |a|))\} + P_m\{y_m > \frac{1}{2}(N - \sum_{i=m}^{m+s-1} (h(i) - |a|))\}.$$

Denote $n^* = [N(\gamma^e - \epsilon)]$.

Then,

$$P_m\{\cup_{k=0}^{n^*}(y_{m+k} > N)\} \leq P_m(y_m > N) + \sum_{k=1}^{n^*} P_m(y_{m+k} > N) \leq$$
$$\leq P_m(y_m > N) + \sum_{k=1}^{n^*} P_m\{y_m > \frac{1}{2}(N - \sum_{i=m}^{m+k-1}(h(i) - |a|))\}+$$
$$+ \sum_{k=1}^{n^*} P_m\{\max_{m+1\leq j\leq m+k} W_j > \frac{1}{2}(N - \sum_{i=m}^{m+k-1}(h(i) - |a|))\}.$$

According to the definition of γ^e : $\sum_{i=m}^{m+[\gamma^e N]-1}(h(i) - |a|) = N + o(1)$. Then, $N - \sum_{i=m}^{m+k-1}(h(i) - |a|) > Ng(\epsilon)$ for $k = 1, \ldots, n^*$, $g(\epsilon) > 0$, and, therefore,

$$\begin{aligned} P_m(B) &= P_m(\tau_N < m) + P_m(y_m > N) + n^* P_m\{y_m > \frac{Ng(\epsilon)}{2}\}+ \\ &+ n^* P_m\{\max_{m+1\leq j\leq m+n^*} W_j > \frac{Ng(\epsilon)}{2}\}. \end{aligned}$$

Following considerations used above in the proof of i), we obtain

$$P_m(B) \leq C_1 \exp(-C_2(\epsilon)N),$$

where C_1, and $C_2(\epsilon)$ do not depend on N.

Thus, for every $0 < \epsilon < 1$,

$$\begin{aligned} P_m\{\sup_{k\geq N}|\gamma_k - \gamma^e| > \epsilon\} &\leq \sum_{k=N}^{\infty} P_m\{|\gamma_k - \gamma^e| > \epsilon\} \leq \\ &\leq C_1 \sum_{k=N}^{\infty} \exp(-NC_2(\epsilon)) = \frac{C_1 \exp(-NC_2(\epsilon))}{1 - \exp(-C_2(\epsilon))} \to 0 \\ &\quad as \ N \to \infty, \end{aligned}$$

and, therefore, $\gamma_N \to \gamma^e$ P_m- a.s. as $N \to \infty$.

Now consider the analogous result for the GRSh method applied to dependent observations satisfying Cramer's and ψ-mixing conditions.

We begin with the following representation of the GRSh statistic:

$$R_{m+k} = R_m \exp(S_{m+1}^{m+k}(x) + \sum_{i=1}^{k} \exp(S_{m+i}^{m+k}(x)),$$

where m is the change-point.

Denote $k_0 = 1 + [N\gamma^e]$ and consider the event $\{\tau_N - m > k_0 + [\epsilon N]\}$. We have

$$P_m\{\tau_N - m > k_0 + [\epsilon N]\} \le P_m\{\max_{1\le k\le k_0+[\epsilon N]} R_{m+k} < e^N\} =$$
$$= P_m\{\max_{1\le k\le k_0+[\epsilon N]} (R_m\, e^{S_k(x)} + \sum_{i=1}^{k} e^{S_i(x)}) < e^N\}$$
$$\le P_m\{\max_{1\le k\le k_0+[\epsilon N]} e^{S_k(x)} < e^N\}$$
$$\le P_{\{} \exp(S_{k_0+[\epsilon N]}(\xi)) \exp(\sum_{i=1}^{k_0+[\epsilon N]} (h(i) - |a|)) < e^N\}$$
$$\le P_m\{S_{k_0+[\epsilon N]}(\xi) < -N \int_{\gamma^e}^{\gamma^e+\epsilon} (h(t) - |a|)\, dt + o(1)\}.$$

Since $h(t) - |a|$ is monotonously increasing, we have $D(\epsilon) = \int_{\gamma^e}^{\gamma^e+\epsilon} (h(t) - |a|)dt > 0$. For estimation of the last probability, the exponential estimates for sums of random variables satisfying ψ-mixing conditions can be used (see Section 1.4). Therefore,

$$P_m\{\tau_N - m > k_0 + [\epsilon N]\} \le C_1 \exp(-C_2 D(\epsilon) N),$$

where the constants C_1, and C_2 do not depend on N.

Now, consider the event $\{\tau_N < k_0 - [\epsilon N]\}$ for any $0 < \epsilon < \gamma^e$. Write

$$\begin{aligned} P_m\{\tau_N < k_0 - [\epsilon N]\} &\le P_m\{\max_{1\le k\le k_0-[\epsilon N]} (e^{S_k(x)} R_m + \sum_{i=1}^{k} e^{S_i(x)}) > e^N\} \\ &\le P_m\{\max_{1\le k\le k_0+[\epsilon N]} (R_m\, e^{S_k(x)} > \frac{1}{2} e^N\} \\ &+ P_m\{\max_{1\le k\le k_0-[\epsilon N]} \sum_{i=1}^{k} e^{S_i(x)} > \frac{1}{2} e^N\}. \end{aligned} \tag{11.7}$$

Let us estimate the second term in the right hand of (11.7):

$$P_m\{\max_{1\le k\le k_0-[\epsilon N]} \sum_{i=1}^{k} e^{S_i(x)} > \frac{1}{2} e^N\}$$
$$\le P_m\{\exp(\max_{1\le k\le k_0-[\epsilon N]} S_k(\xi)) \sum_{k=1}^{k_0-[\epsilon N]} e^{\sum_{i=1}^{k}(h(i)-|a|)} > \frac{1}{2} e^N\}$$

Because of the monotonicity of the function $h(i)$:

$$\exp(\sum_{i=1}^{k_0-[\epsilon N]} (h(i)-|a|)) \le \sum_{k=1}^{k_0-[\epsilon N]} e^{\sum_{i=1}^{k}(h(i)-|a|)} \le (k_0-[\epsilon N]) \exp(\sum_{i=1}^{k_0-[\epsilon N]} (h(i)-|a|))$$

and

$$\sum_{i=1}^{k_0-[\epsilon N]} (h(i) - |a|) = N \int_0^{\gamma^e-\epsilon} (h(t) - |a|)dt + o(N).$$

Denote $D(\epsilon) = \int\limits_{\gamma^e - \epsilon}^{\gamma^e} (h(t)) - |a|)dt$. Then,

$$P_m\{\exp(\max_{1\leq k\leq k_0-[\epsilon N]}(S_k(\xi)) \sum_{k=1}^{k_0-[\epsilon N]} e^{\sum_{i=1}^{k}(h(i)-|a|)} > \frac{1}{2}e^N\}$$
$$\leq P_m\{\max_{1\leq k\leq k_0-[\epsilon N]} S_k(\xi) > ND(\epsilon) + o(N)\}.$$

The last probability is estimated as earlier for the CUSUM method. So

$$P_m\{\max_{1\leq k\leq k_0-[\epsilon N]} \sum_{i=1}^{k} e^{S_i(x)} > \frac{1}{2}e^N\} \leq C_1\exp(-C_2(\epsilon)N),$$

where the constants C_1, and C_2 do not depend on N.

Now, let us estimate the first probability in the right hand of (11.6.1). For any $0 < \delta < 1$, write

$$P_m\{\max_{1\leq k\leq k_0-[\epsilon N]} R_m e^{S_k(x)} > \frac{1}{2}e^N\} \leq P_m\{R_m > \frac{1}{2}e^{N\delta}\}$$
$$+ P_m\{\max_{1\leq k\leq k_0-[\epsilon N]} e^{S_k(x)} > e^{N(1-\delta)}\},$$

The second probability in the right hand is estimated as above. The exponential estimate for the first probability is obtained as earlier for the "false alarm" probability.

So, for the GRSh method,

$$P_m\{|\tau_N - k_0| > [\epsilon N]\} \leq C_1\exp(-C_2(\epsilon)N),$$

where the constants C_1, and C_2 do not depend on N.

Proof of Theorem 11.3.3

Recall that Theorem 11.3.3 refers to the case of discrete valued observations $X = (x_1, x_2, \ldots)$, satisfying the ψ-mixing condition. Suppose that the set of all possible values of observations is $S = (a_1, \ldots, a_k)$ and

$$P\{x_i = a_j\} = \begin{cases} p_j, & i \leq m \\ q_j(i), & i > m, \end{cases}$$

where the vectors $p_1, \ldots, p_k$ and $q_1(i), \ldots, q_k(i)$ differ in a certain sense for any $i > m$ (precise formulations are given below).

Then, we have under the measure P_0 (no change-points):

$$I(x_i = a_j) = p_j + \eta_j(i),$$

where $E_0\eta_j(i) = 0$ (the law of large numbers).

Therefore,

$$P_0\{\max_{[\beta N]\leq l\leq N} Y_N^n(l) > C\}$$
$$\leq \sum_{l=[\beta N]}^{N}\sum_{j=1}^{k} P_0\{|l\sum_{i=n-N+1}^{n}\eta_j(i) - N\sum_{i=n-N+1}^{n-N+l}\eta_j(i)|/N^2 > C\}.$$

Since the d.f.'s of the r.v.'s $\eta_j(i)$ do not depend on n, the right-hand side of the last inequality can be estimated from above like in Brodsky, and Darkhovsky (2000), i.e. using the exponential upper estimates for the sums of centered r.v.'s satisfying the Cramer and ψ-mixing conditions. Proceeding this way, we obtain

$$P_0\{\max_{[\beta N]\le l\le N} Y_N^n(l) > C\} \le kN\, m_0(C) \begin{cases} \exp(-\frac{HN\beta C}{4m_0(C)}), & C > gH \\ \exp(-\frac{N\beta C^2}{4gm_0(C)}), & C \le gH, \end{cases}$$

where for any $\epsilon > 0$, the number $m_0(\epsilon) \ge 1$ is defined from the following condition: $\psi(l) \le \epsilon$ for $l \ge m_0(\epsilon)$, and g, H are taken from the uniform Cramer condition. Now, let us consider the 2nd-type error:

$$P_m\{\max_{m\le\le N+m}\ \max_{[\beta N]\le l\le N} Y_N^n(l) \le C\}.$$

Denote

$$\Lambda = \sup_{0<\gamma<1} (1-\gamma)\sum_{j=1}^{k} |\int_0^{\gamma} (q_j(t) - p_j)dt|.$$

We suppose that $C < \Lambda$ and denote $\delta = \Lambda - C$. Then,

$$\begin{aligned} & P_m\{\max_{m\le\le N+m}\ \max_{[\beta N]\le l\le N} Y_N^n(l) \le C\} \\ & \le \sum_{l=[\beta N]}^{N} \sum_{j=1}^{k} P_0\{|l \sum_{i=n-N+1}^{n} \eta_j(i) - N \sum_{i=n-N+1}^{n-N+l} \eta_j(i)|/N^2 > \delta\} \\ & \le kN\, m_0(\delta) \begin{cases} \exp(-\frac{HN\beta\delta}{4m_0(\delta)}), & \delta > gH \\ \exp(-\frac{N\beta\delta^2}{4gm_0(\delta)}), & \delta \le gH. \end{cases} \end{aligned}$$

Now, let us consider the normalized delay time $\gamma_N = (\tau - m)^+/N$. Suppose γ^* is the minimal root of the equation

$$(1-\gamma)\sum_{j=1}^{k} |\int_0^{\gamma} (q_j(t) - p_j)dt| = C.$$

Then, for any $0 < \epsilon < 1$,

$$P_m\{|\gamma_N - \gamma^*| > \epsilon\} \le kN\, m_0(\epsilon) \begin{cases} \exp(-\frac{HN\beta\epsilon}{4m_0(\epsilon)}), & \epsilon > gH \\ \exp(-\frac{N\beta\epsilon^2}{4gm_0(\epsilon)}), & \epsilon \le gH. \end{cases}$$

It follows from here that $\gamma_N \to \gamma^*$ as $N \to \infty$ P_m—almost surely.

Proof of Theorem 11.3.4
The proof of 1) is given in Chapter 10.
2)
In this section, we consider type-2 error

$$\delta_N = P_m\{\max_{m\le n\le m+N}\ \max_{[\beta N]\le l\le N} \|Y_N^n(l)\| \le C\}.$$

This characteristic describes the situation when the decision statistic does not exceed the boundary C for a sample with a change-point, i.e., for $m \le n \le m+N$.

For $m \le n = m+[\theta N] \le m+N$, denote $\zeta_N^\theta(t) = Y_N^{m+[\theta N]}([Nt])$, $0 \le \theta \le 1$. Then, $\zeta_N^\theta(t) = A_1 + A_2$, where

$$A_2 = \frac{1}{N}(\sum_{i=1}^{[Nt]} X(i^*)\nu_{i^*}^{'} - T^n(1,[Nt])(T^n(1,N))^{-1}\sum_{i=1}^{N} X(i^*)\nu_{i^*}^{'}), \quad i^* = i+n-N,$$

but the first term A_1, in this case, $(m \le n \le m+N)$, is not zero, and for $0 \le t \le 1-\theta$ equals

$$A_1 = \frac{1}{N}(\sum_{i=1}^{[Nt]} X(i^*)X^{'}(i^*)\mathbf{a}^{'} - T^n(1,[Nt])(T^n(1,N))^{-1}$$
$$\times(\sum_{i=1}^{N-[\theta N]} X(i^*)X^{'}(i^*)\mathbf{a}^{'} + \sum_{i=N-[\theta N]+1}^{N} X(i^*)X^{'}(i^*)\mathbf{b}^{'}(i^*)).$$

In virtue of assumption 2 of this theorem, A_1 tends a.s. to the following matrix for $0 \le t \le 1-\theta$:

$$A_1 \to \int_0^t V(\tau)d\tau\,(\int_0^1 V(\tau)d\tau)^{-1}\int_{1-\theta}^1 V(\tau)(a-b(\tau))^{'}d\tau.$$

Define

$$S(\theta) = \int_0^{1-\theta} V(\tau)d\tau \cdot I^{-1} \cdot \int_{1-\theta}^1 V(\tau)(\mathbf{b}(\tau)-\mathbf{a})d\tau$$
$$g(\tilde{\theta}) = \max_\theta g(\theta), \quad g(\theta) = \|S(\theta)\|^2,$$
$$d = (g(\tilde{\theta}) - C)/(1+\sqrt{K}).$$

Then, we can prove that

$$\delta_N \le m_0(d)\begin{cases} \exp(-\dfrac{TN\beta d}{4m_0(d)}), & d > hT \\ \exp(-\dfrac{N\beta d^2}{4hm_0(d)}), & d \le hT. \end{cases}$$

If we consider the normalized delay time in change-point detection, then

$$\gamma_N = \frac{(\tau_N - m)^+}{N} \to \gamma^*, \quad P_m - \text{ a.s. as } N \to \infty,$$

where γ^* is the minimal root of the equation $g(t) = C$.

12

Sequential Detection of Switches in Models with Changing Structures

12.1 Introduction

This chapter is aimed at sequential detection of switches in models with changing structures. The problem can be formulated as follows. Suppose we observe an output of an object that can function in two different regimes with some large and small probability, respectively. In the univariate case, the density function of observations $x_1, x_2, \dots$ can be modeled as follows:

$$f(x_i) = (1 - \epsilon) f_0(x_i) + \epsilon f_1(x_i), \tag{12.1}$$

where $\mathbf{E}_0 x_i = 0$, $\epsilon = 0$ in the "normal" regime; $\mathbf{E}_1 x_i \neq 0$, $0 < \epsilon < 1/2$ in the "outliers" regime.

Suppose we receive observations sequentially, and until some unknown instant $m \geq 1$, we have $\epsilon = 0$, i.e., the observed object is in the "normal" regime. However, beginning from the instant m, the regime of this object changes, i.e., $0 < \epsilon < 1/2$ (in other words, certain abnormal observations may appear). Our goal is to detect the change-point m as soon as possible on condition that false alarms are few. Moreover, in the case that the alternative $\epsilon > 0$ is accepted, we need to classify all obtained observations into subsamples of "normal" and "outliers" data.

In other terms early violations of the "normal" state of an observed object can be considered as some unusual observations that appear quite rarely and irregularly. Our goal is to detect such initial outliers and make necessary corrections at the early stage of a possible disorder.

In the multivariate case, we can consider sequential regressions with abnormal observations, the multivariate sequential classification problems, etc.

Methodologically, the considered problem is the sequential variant of traditional classification problems, e.g., the problem of splitting mixtures of probabilistic distributions, the problem of constructing regressions with abnormal observations, etc. A short review of methods proposed for retrospective detection and estimation of switches in models with changing structures was given in Chapter 4.

The main result of this chapter is the method proposed for estimation of m in model of type (12.1) and the method of classification of observations. Moreover, we establish a priori informational inequalities for such problems. These inequalities enable us to make a conclusion about the asymptotic optimality of the proposed method.

The structure of this chapter is as follows. In Section 12.2, I give the problem statement and formulate the method for solving this problem. In Section 12.3, I study the main statistical characteristics of the proposed method: type-1 and type-2 errors, and the delay time in change-point detection. In Section 12.4, I prove the a priori informational inequalities for the problem of sequential change-point detection in the considered statistical context. These inequalities can be used for the study of the asymptotic optimality of the proposed method. In Section 12.5, I give results of the experimental study of the proposed method for different stochastic models with switching regimes and in comparison with cumuletive sums (CUSUM) and Girshick-Rubin-Shyriaev (GRSh) methods. Main proofs are given in the appendix.

12.2 Problem Statement and Detection Method

On the probability space $(\Omega, \mathcal{F}, \mathbf{P})$ a sequence $X = \{x_i\}_{i=1}^{\infty}$ of r.v.'s is observed with the following density function (d.f.) of observations:

$$f(x_i) = \begin{cases} f_0(x_i), & i \leq m \\ f_\epsilon(x_i) \stackrel{\text{def}}{=} (1-\epsilon) f_0(x_i) + \epsilon f_1(x_i), & i > m, \end{cases} \tag{12.2}$$

where $m > 1$ is an unknown change-point; $0 \leq \epsilon < 1/2$, $\int x f_0(x) dx = 0$, $\int x f_1(x) dx = h \neq 0$, $f_0(x_i)$ is symmetric w.r.t. zero.

Below, we denote by $\mathbf{P}_0(\mathbf{E}_0)$, $\mathbf{P}_m(\mathbf{E}_m)$ measure (mathematical expectation) of the sequence X under the condition $\epsilon = 0$ (no outliers, no change-point) and under the condition $\epsilon > 0$, $h \neq 0$, i.e., the change-point at the instant m.

Our main problems are as follows:

1. Detect the change-point m as soon as possible on condition that false alarms are few; and

2. Classify the whole obtained sample of observations into subsamples of ordinary ($f_0(\cdot)$) and abnormal observations.

Let us discuss which methods could be effective for solving these problems. Of course, we can apply the CUSUM or GRSh procedures. However, since the parameter ϵ can be rather small, the average delay time in change-point detection for these methods will be large for most cases.

We use the idea of the moving window of the last N observations. In what follows, we suppose that $m = N + k$, $k > 0$, and for any fixed $k > 0$, consider the family (w.r.t. N) of probability measures $\mathbf{P}_m = \{\mathbf{P}_{N+k}\}$.

For each $n = N, N+1, \dots$, define $X_n^N = \{x_i\}_{i=n-N+1}^n$ and do as follows:
1) Compute the estimate of the mean value

$$\theta_N^n = \frac{1}{N} \sum_{i=n-N+1}^{n} x_i.$$

Evidently,

$$\mathbf{E}\theta_N^n = \begin{cases} 0, & \text{if } n \le m \\ \frac{(n-m)^+}{N}\epsilon h, & \text{if } m < n \le (m+N) \ / \\ \epsilon h, & \text{if } n > m+N. \end{cases}$$

2) Fix the numbers $0 < \kappa < B$ and parameter $b \in \mathbb{B} \stackrel{\text{def}}{=} [\kappa, B]$ and classify observations as follows: If an observation falls into the interval $(\theta_N^n - b, \theta_N^n + b)$, then we place it into the subsample of ordinary observations, otherwise, place it into the subsample of abnormal observations.

Then, for each $b \in \mathbb{B}$, we obtain the following decomposition of the sample X_n^N into two subsamples:

$$\begin{aligned} X_1^n &= \{\tilde{x}_{i_1}^n, \tilde{x}_{i_2}^n, \dots, \tilde{x}_{i_{N_1}}^n\}, \quad |\tilde{x}_{i_k}^n - \theta_N^n| < b,\ k = 1, 2, \dots, N_1 \\ X_2^n &= \{\hat{x}_{j_1}^n, \hat{x}_{j_2}^n, \dots, \hat{x}_{j_{N_2}}^n\}, \quad |\hat{x}_{j_k}^n - \theta_N^n| \ge b,\ k = 1, 2, \dots, N_2, \end{aligned}$$

where $N_1 = N_1(b, n)$, $N_2 = N_2(b, n)$ the sizes of the subsamples X_1^n and X_2^n, respectively, $N = N_1 + N_2$.

The parameter b is chosen so that the sub-samples X_1^n and X_2^n are separated in the best way. For this purpose, consider the following statistic:

$$\Psi_N(b, n) = \frac{1}{N^2} \left(N_2 \sum_{k=1}^{N_1} \tilde{x}_{i_k}^n - N_1 \sum_{k=1}^{N_2} \hat{x}_{i_k}^n \right).$$

3) Define the decision threshold $C > 0$ and the stopping time

$$\tau_N = \inf\{n : \sup_{b \in \mathbb{B}} |\Psi_N(b, n)| > C\}.$$

At each moment $n > N$, we take the solution: if $J^n \stackrel{\text{def}}{=} \sup_{b \in \mathbb{B}} |\Psi_N(b, n)| \le C$, then we accept the hypothesis $\mathbf{H}_0$ (*no change-point*) and continue to collect observations; if, however, $J^n > C$, then the hypothesis $\mathbf{H}_0$ is rejected, we take additional N observations after τ_N, and construct estimates of the parameters ϵ and h.

Remark that our primary goal is to separate ordinary and abnormal observations in the sample. Evidently, classification errors must be small, and, therefore, we have to require some kind of convergence of the estimate $\hat{\epsilon}_N, \hat{h}_N$ to its true value ϵ, h.

The estimates of ϵ and h, if the functional form of d.f.'s $f_0(\cdot)$, $f_1(\cdot)$ is known a priori, are formulated in the proof of Theorem 4.3.2.

We will show that the estimates ϵ_N^* and h_N^* tend almost surely to the true values ϵ and h as $N \to \infty$. The subsample of abnormal observations is $X_2^{\tau_N+N}$.

The quality of the proposed method is characterized by

1) type-1 error: the supremal probability of the false decision under $\mathbf{P}_0$:

$$e_N^{(1)} = \sup_n \mathbf{P}_0\{\sup_{b\in\mathbb{B}} |\Psi_N(b,n)| > C\};$$

2) type-2 error:

$$e_N^{(2)} = \mathbf{P}_m\{\max_{m\le n\le m+N} \sup_{b\in\mathbb{B}} |\Psi_N(b,n)| \le C\}.$$

3) the normalized delay time in change-point detection:

$$\gamma_N = \frac{(\tau_N - m)^+}{N}.$$

12.3 Main Results

In this section, I suppose that the uniform Cramer's and ψ-mixing conditions are satisfied for the observed sequence.

The main results are formulated in the following theorems. Below, we denote by the same symbols C, N_0, L_1, and L_2 probably different constants that do not depend of N.

First, I study the supremal probability of a false decision for the proposed method.

Theorem 12.3.1. *Let $\epsilon = 0$. Suppose the d.f. $f_0(\cdot)$ is symmetric w.r.t. zero and bounded. Then, for any $0 < \kappa < B$, there exist $C > 0, N_0 > 1, L_1 > 0, L_2 > 0$, such that for $N > N_0$, the following estimate holds:*

$$\sup_n \mathbf{P}_0\{\sup_{b\in\mathbb{B}} |\Psi_N(b,n)| > C\} \le L_1 \exp(-L_2 N).$$

The proof of Theorem 12.3.1 repeats the main ideas of the proof of Theorem 4.3.1 (retrospective case) and therefore is omitted here.

Second, I study the probability of type-2 error.

For formulation of the following theorem, consider the function

$$G(b,t) \stackrel{\text{def}}{=} (1-t)c + td - t\epsilon h\left((1-t)\alpha + t\beta\right),$$

where

$$c = \int_{t\epsilon h-b}^{t\epsilon h+b} x f_0(x)dx,\ d = \int_{t\epsilon h-b}^{t\epsilon h+b} x f_\epsilon(x)dx,\ \alpha = \int_{t\epsilon h-b}^{t\epsilon h+b} f_0(x)dx,\ \beta = \int_{t\epsilon h-b}^{t\epsilon h+b} f_\epsilon(x)dx.$$

Evidently, the function $G(b,t)$ is continuous on $U \stackrel{\text{def}}{=} \mathbb{B} \times [0,1]$, and, therefore, its module attains its maximum. Put $Q \stackrel{\text{def}}{=} \max_{(b,t)\in U} |G(b,t)|$.

Theorem 12.3.2. *Suppose $\epsilon > 0$, $h \neq 0$, and the density function f_ϵ is bounded. Then:*

1) for any $0 < \kappa < B$, there exist $C > 0, N_0 > 1, L_1 > 0, L_2 > 0$ such that for any fixed $k > 0$, and $m = N + k$

$$\mathbf{P}_m\{ \max_{m \le n \le m+N} \sup_{b \in \mathbb{B}} |\Psi_N(b,n)| \le C\} \le L_1 \exp(-L_2 N).$$

2) for any sufficiently small $0 < \beth$, any fixed $k > 0$ and $m = N + k$

$$\mathbf{P}_m\{|\gamma_N - \gamma^*| > \beth\} \le L_1 \exp(-L_2 N),$$

where γ^* is the minimal root of the equation $\max_{b \in \mathbb{B}} |G(b,t)| = 1.2C$ and $0 < C < 1/4Q$.

Remark 12. 3. 1. I consider the family (w.r.t. N for the fixed $k > 0$) of probability measures $\mathbf{P}_m = \{\mathbf{P}_{N+k}\}$ (due to the use of a moving window). Consider a sequence of r.v.'s $\{\zeta_N\}$ and the set of elementary events $\mathbb{A} \stackrel{\text{def}}{=} \{\omega :$ $\{\zeta_N\}$ *не сходится* к случайной величине $\zeta\}$. In analogy with the standard definition, we say that the sequence $\{\zeta_N\}$ *converges almost surely w.r.t. the family* $\mathbf{P}_m = \{\mathbf{P}_{N+k}\}$ *for any fixed* $k > 0$, if

$$\lim_{N \to \infty} \mathbf{P}_{N+k}\mathbb{A} = 0.$$

Taking into account this definition and sufficient conditions of a.s. convergence analogous to the standard situation, section 2 of this theorem can be reformulated as follows:

$$\gamma_N \to \gamma^*$$

almost surely w.r.t. the family $\mathbf{P}_m = \{\mathbf{P}_{N+k}\}$

for any fixed $k > 0$.

■

12.4 Asymptotic Optimality

In this section, I formulate and prove a general result concerning a wide enough class of sequential methods in scheme of type (12.2.1) for a sequence of independent random variables. From this result follows the asymptotic optimality of the method proposed in Section 12.2.

So, let us consider a sequence of i.r.v.'s $X = \{x_n\}_{n=1}^{\infty}$ and suppose that the d.f. of these r.v.'s is described by scheme (1), where m is the change-point.

In this section, I denote by $\mathbf{P}_{m,\epsilon}(\mathbf{E}_{m,\epsilon})$ the measure (mathematical expectation) corresponding to the sequence with the change-point m and the parameter $\epsilon > 0$. By $\mathbf{P}_0(\mathbf{E}_0)$, I denote the measure (mathematical expectation) of the sequence without a change-point.

All reasonable methods of change-point detection (and corresponding stopping times) must have a "large parameter" L; if L increases, then the probability of the false decision (i.e., stopping at a moment when there is no change) tends to zero. Typically, this parameter coincides with the threshold (for example, in the case of the CUSUM), but this is not always the case.

Denote by $d(L, n)$ a decision function of a change-point detection method with the large parameter, L (i.e., a function measurable with respect to the natural flow of σ-algebras generated by the observations), such that $d(L, n) = 1$ (resp. $d(L, n) = 0$) corresponds to the decision that n is the change-point (resp. n is not the change point).

In particular, for the CUSUM method, in the case of independent random variables, such that the density function changes from $f_0(\cdot)$ to $f_1(\cdot)$ the decision function is as follows:

$$\mathbb{D}(L, n) = \begin{cases} 1, & \max\limits_{1 \le s \le n} \sum\limits_{u=s}^{n} \ln \dfrac{f_1(x_u)}{f_0(x_u)} > L \\ 0, & \text{otherwise.} \end{cases} \tag{12.3}$$

The stopping time, corresponding to the decision function $d(L, n)$ is denoted by τ_L, i.e., $\tau_L = \inf\{n : d(L, n) = 1\}$.

Now, let us define the *supremal probability of a false decision* (SPFD) $\alpha(\tau_L)$ for the s.t. τ_L, generated by a decision function $d(L, n)$:

$$\alpha(\tau_L) \stackrel{\text{def}}{=} \sup_n \mathbf{P}_0\{d(L, n) = 1\}.$$

Notice that SPFD, generally speaking, is not a false alarm probability.

Let $M = \{\tau_L : \mathbf{P}_0\{d(C, n) = 1\} \le \rho \quad \forall n\}$. It is known (see Brodsky and Darkhovsky, 2000) that

$$\inf_{\tau_L \in M} \mathbf{E}_0 \tau_L \sim (2\rho)^{-1}$$

as $\rho \to 0$.

Consider the following class of detection methods (and the corresponding mathematical expectations):

$$\mathfrak{M} = \{\tau_L\colon \liminf_{L \to \infty} \frac{|\ln \alpha_L|}{L} \ge R > 0,\ \gamma_L \stackrel{\text{def}}{=} \frac{(\tau_L - m)^+}{L} \to \gamma(\epsilon)\, \mathbf{P}_m \text{ a.s. as } L \to \infty\}.$$

We suppose that the function $\gamma(\cdot)$ is differentiable, $\gamma'(\cdot) > 0$, and in some neighborhood $\mathcal{E}$ of the point ϵ there exists the second derivative $\gamma''(\cdot)$.

Let us discuss these assumptions. The assumption about the exponential rate of convergence to zero for error probabilities is quite mild and requires the proper choice of the large parameter L. Remark that for the CUSUM method, the large parameter L is equal to the decision threshold. For other known methods of sequential change-point detection, the choice of a large parameter is also rather simple (see, e.g., Brodsky and Darkhovsky (2000)) and gives the exponentially decreasing α_L.

The assumption $\gamma_L \to \gamma(\epsilon)$ is also quite common. Due to the nonlinear renewal theory (see Woodroofe (1982)), it is usually satisfied for most sequential testing methods. For known sequential change-point detection methods, it is also satisfied (see Brodsky and Darkhovsky (2000)).

The following theorem holds.

Theorem 12.4.1. *Let* $\int \ln\left(\frac{f_{\epsilon'}(x)}{f_\epsilon(x)}\right) f_{\epsilon'}(x)dx < \infty$ *for* $\epsilon' \in \mathcal{E}$. *Then, for any s.t.* $\tau_L \in \mathfrak{M}$ *and any* $0 < \delta < \max\left(1, 2/(1+v)^2\right)$, *the following inequality holds:*

$$\liminf_L \, L^{-1} \ln \mathbf{P}_{m,\epsilon}\{|\gamma_L - \gamma(\epsilon)| > \delta\} \geq -\frac{\gamma(\epsilon)J(\epsilon)}{[\gamma'(\epsilon)]^2}\, \delta^2\, (1+v)^2,$$

where $v = \dfrac{\max_{u\in\mathcal{E}} |\gamma''(u)|}{[\gamma'(\epsilon)]^2}, \quad J(\epsilon) = \displaystyle\int \frac{(f_1(x) - f_0(x))^2}{f_\epsilon(x)}\, dx.$

Proof.

Consider any stopping time $\tau_L \in \mathfrak{M}$ corresponding to the sample $X^L = \{x(1), \ldots, x(\tau_L)\}$ obtained up to the first decision about the presence of a change-point $d_L(\cdot) = 1$. By definition, $\tau_L = m + [\gamma_L L]$.

Let $\lambda_L = \mathbb{I}\{|\gamma_L - \gamma(\epsilon)| > \delta\}$. Denote by $f(X^L, \epsilon)$ the likelihood function for the sample X^L and the parameter ϵ. Then, for any $d > 0$ and $\epsilon' \in \mathcal{E}$, we have:

$$\begin{aligned}
\mathbf{P}_{m,\epsilon}\{|\gamma_L - \gamma(\epsilon)| > \delta\} &= \mathbf{E}_{m,\epsilon}\lambda_L \geq \mathbf{E}_{m,\epsilon}(\lambda_L \mathbb{I}\left(\frac{f(X^L, \epsilon')}{f(X^L, \epsilon)} < e^d\right)) \geq \\
&\geq e^{-d}\mathbf{E}_{m,\epsilon'}\left(\lambda_L \mathbb{I}(\frac{f(X^L, \epsilon')}{f(X^L, \epsilon)} < e^d)\right) \geq \\
&\geq e^{-d}\left(\mathbf{P}_{m,\epsilon'}\{|\gamma_L - \gamma(\epsilon)| > \delta\} - \mathbf{P}_{m,\epsilon'}\left\{\frac{f(X^L, \epsilon')}{f(X^L, \epsilon)} \geq e^d\right\}\right).
\end{aligned} \tag{12.4}$$

Let us choose ϵ' as follows:

$$\epsilon' = \epsilon + \frac{\delta + \eta}{\gamma'(\epsilon)} \stackrel{\text{def}}{=} \epsilon + s, \tag{12.5}$$

where $\eta \equiv v\delta = \dfrac{\delta}{[\gamma'(\epsilon)]^2} \max_{u\in\mathcal{E}} |\gamma''(u)|$.

Then, the Taylor expansion gives

$$|\gamma(\epsilon^{'}) - \gamma(\epsilon)| = \delta + \eta + \frac{\gamma^{''}(u)}{2} \frac{(\delta+\eta)^2}{[\gamma^{'}(\epsilon)]^2} > \delta, \quad u \in \mathcal{E}, \tag{12.6}$$

for $0 < \delta < \max(1, 2/(1+v)^2)$.

For any method from the considered class $\mathfrak{M}$, $\gamma_L \to \gamma(\epsilon')\mathbf{P}_{m,\epsilon'}$-a.s. as $L \to \infty$. Therefore, taking into account our choice of $\epsilon^{'}$ and the inequality $|\gamma(\epsilon^{'}) - \gamma(\epsilon)| > \delta$ (see (12.6)), we have

$$\mathbf{P}_{m,\epsilon'}\{|\gamma_L - \gamma(\epsilon)| > \delta\} \to 1, \quad \text{as} \quad L \to \infty.$$

For independent observations, we can write

$$\mathbf{P}_{m,\epsilon'}\left\{f(X^L,\epsilon^{'})/f(X^L,\epsilon) \geq e^d\right\} = \mathbf{P}_{m,\epsilon'}\left\{\sum_{i=1}^{[\gamma_L L]} \ln \frac{f_{\epsilon'}(x_i)}{f_\epsilon(x_i)} \geq d\right\}. \tag{12.7}$$

If we choose (for any $\kappa > 0$)

$$d = L\left(\kappa + \gamma(\epsilon) \int f_{\epsilon'}(x) \ln \frac{f_{\epsilon'}(x)}{f_\epsilon(x)} dx\right),$$

then the right hand of (12.7) converges to zero as $L \to \infty$ in virtue of the law of large numbers (from our assumptions and the following property $\gamma_L \to \gamma(\epsilon') \quad \mathbf{P}_{m,\epsilon'}$-a.s.).

Then, from (12.4), (12.7), and our choice of $\epsilon^{'}$, we obtain (since $\kappa > 0$ is arbitrary):

$$\liminf_L L^{-1} \ln \mathbf{P}_{m,\epsilon}\{|\gamma_L - \gamma(\epsilon)| > \delta\} \geq -\gamma(\epsilon) \int f_{\epsilon'}(x) \ln \frac{f_{\epsilon'}(x)}{f_\epsilon(x)} dx. \tag{12.8}$$

From (12.5), we obtain

$$f_{\epsilon'}(x) = (1-\epsilon-s)f_0(x) + (\epsilon+s)f_1(x) = f_\epsilon(x) + s(f_1(x) - f_0(x)), \tag{12.9}$$

and so

$$\ln \frac{f_{\epsilon'}(x)}{f_\epsilon(x)} = \ln(1 + s\frac{f_1(x) - f_0(x)}{f_\epsilon(x)}) \leq s\,\frac{f_1(x) - f_0(x)}{f_\epsilon(x)}. \tag{12.10}$$

Then, from (12.9) and (12.10),

$$\int \left(\ln \frac{f_{\epsilon'}(x)}{f_\epsilon(x)}\right) f_{\epsilon'}(x) dx \leq s^2\, J(\epsilon) \tag{12.11}$$

The statement of the theorem now follows from (12.8) and (12.11).

Theorem 12.4.1 is proved. ■

Corollary

From Theorem 12.4.1 and Theorem 12.3.2, it follows that the proposed method is asymptotically optimal by the order of convergence

TABLE 12.1
α-Quantiles of the Proposed Statistic

N	50	100	300	500	800	1000	1200
$\alpha = 0.95$	0.1681	0.1213	0.0710	0.0534	0.044	0.0380	0.037
$\alpha = 0.99$	0.1833	0.1410	0.1105	0.0814	0.0721	0.0615	0.0390

12.5 Simulations

Consider the following model of the d.f. of independent observations:

$$f(x_i) = \begin{cases} f_0(x_i), & i \leq m \\ (1-\epsilon)f_0(x_i) + \epsilon f_0(x_i - h), & i > m, \end{cases} \tag{12.12}$$

where $f_0(\cdot) = \mathcal{N}(0,1), 0 \leq \epsilon < 1/2$.

We tried to study performance characteristics of the proposed method and to compare them with the analogous characteristics of the CUSUM method. Remark that the proposed method makes it possible to estimate the parameter ϵ and to classify the whole obtained sample of observations into subsamples of ordinary and abnormal observations. This property is totally absent for the CUSUM test.

In this experiment, we first studied critical thresholds of the decision statistic for the proposed method computed. For this purpose, for homogenous samples ($\epsilon = 0$), α-quantiles of the decision statistic $\max_{b\in\mathbb{B}} |\Psi_N(b, n_0)|$ were computed for some fixed $n_0 > N$ ($\alpha = 0.95, 0.99$). The results obtained in 5000 independent trials of each experiment are presented in Table 12.1.

The quantile value for $\alpha = 0.99$ was chosen as the critical threshold C in experiments with nonhomogenous samples (for $\epsilon \neq 0$). For different sample sizes in 5000 independent trials of each test, the estimate of type-2 error w_2 (i.e., the frequency of the event $\max_{m \leq n \leq m+N} \max_b |\Psi_N(n, b)| < C$ for $\epsilon > 0$) and the average delay time in change-point detection (dn) were computed. The results are presented in Table 12.2.

CUSUM test

The obtained results were compared with the analogous results for the following CUSUM test:

$$\begin{aligned} y_n &= (y_{n-1} + x_n - a)^+, \qquad y_0 = 0, \\ d_n &= \mathbb{I}(y_n > C), \end{aligned}$$

where $C > 0$ is the decision threshold, $a = -0, 5$, $x_n, n = 1, 2, \ldots$ is an initial sequence of observations with the d.f. (12.12),where $h = 2$, $\epsilon = 0, 1$.

Such results for the CUSUM method may seem strange, but we should not forget that the mathematical expectation of observations after the change-point is 0.2 in this example, and, therefore, (for the best choice of the CUSUM

TABLE 12.2
Statistical Characteristics of the Proposed Method

$\epsilon = 0.1$	h=2.0			
N	300	500	800	1000
C	0.11	0.08	0.07	0.06
pr_1	0.03	0.03	0.03	0.03
w_2	0.03	0.01	0.01	0.01
dn	42.5	65.4	91.4	143.6

TABLE 12.3
Statistical Characteristics of the CUSUM Method

$\epsilon = 0.1$	h=2.0			
C	10.2	10.9	11.2	11.4
pr_1	0.04	0.03	0.03	0.03
w_2	0.03	0.01	0.01	0.01
dn	945.3	1339.5	1425.4	1541.4

test parameters), the average delay time is about 100. However, empirically, outliers arrive quite irregularly, and, therefore, empirical delay times can be even larger. We see that the proposed test substantially exceeds the CUSUM test in the considered case.

12.6 Conclusions

The sense of the problem considered in this chapter can be formulated as follows. Suppose we receive observations sequentially and at some unknown time m statistical properties of data change. Initially, there can appear certain outliers with different statistical characteristics. We must detect the instant m as soon as possible, i.e., to make a decision that statistical characteristics of data changed. In other terms, we should classify outliers differently, i.e., to solve the typical problem of the discriminant analysis in its fully probabilistic statement. However, in existing statements of the problems of discriminant analysis, each new observation is instantly classified to a certain existing class of observations. The notions of "classification error" and "delay time" in classification are totally absent in these statements. In this chapter, I demonstrate that even in the simplest binary, case these notions naturally emerge due to the probabilistic description of the data.

For the nonparametric test proposed in this chapter, the exponential upper estimates for type-1 and type-2 errors can be proved (see Theorems 12.3.1 and 12.3.2), and the limit of the normalized delay time in change detection can be given. I also prove the asymptotic optimality of the proposed test (by the order of convergence to the true values of parameters) and compare it empirically with the CUSUM test constructed for this case. Both theoretical and experimental results witness to the high efficiency of the proposed test.

12.7 Appendix: Proofs of Theorems

Proof of Theorem 12.3.2.

Below, we need two auxiliary statements.

Suppose $X = \{x_n\}$, $Y = \{y_n\}$ are random sequences satisfying the uniform Cramer's condition and the ψ-mixing condition, and $\mathbf{E}x_n = \mathbf{E}y_n \equiv 0$.

Lemma 12.7.1. *Under the above formulated assumptions*

$$i)\mathbf{E}|N^{-1}\sum_{k=1}^{N} x_k| \to 0,$$

$$ii)\mathbf{E}|N^{-1}\sum_{k=1}^{N} x_k|^2 \to 0,$$

$$iii)\mathbf{E}|N^{-2}\sum_{k=1}^{N} x_k \sum_{k=1}^{N} y_k| \to 0$$

as $N \to \infty$.

Proof. Let us prove the first statement

$$\mathbf{E}|N^{-1}\sum_{k=1}^{N} x_k| = \int_0^{\infty} \mathbf{P}\{|N^{-1}\sum_{k=1}^{N} x_k| > x\}dx.$$

Using the exponential upper estimate from Section 1.4, we obtain for an arbitrary $\epsilon > 0$

$$\mathbf{E}|N^{-1}\sum_{k=1}^{N} x_k| = \int_0^{\epsilon} \mathbf{P}\{|N^{-1}\sum_{k=1}^{N} x_k| > x\}dx + \int_{\epsilon}^{\infty} \mathbf{P}\{|N^{-1}\sum_{k=1}^{N} x_k| > x\}dx \leq$$
$$\leq \epsilon + \int_{\epsilon}^{\infty} A(\epsilon)\exp(-B(x)N)dx \leq \epsilon + O(1/N).$$

Since $\epsilon > 0$ is arbitrary, the formulated statement is proved.

In analogy,

$$\mathbf{E}|N^{-1}\sum_{k=1}^{N} x_k|^2 = \int_0^\infty \mathbf{P}\{|N^{-1}\sum_{k=1}^{N} x_k|^2 > x\}dx = \int_0^\infty \mathbf{P}\{|N^{-1}\sum_{k=1}^{N} x_k| > \sqrt{x}\}dx =$$
$$= \int_0^\epsilon \mathbf{P}\{|N^{-1}\sum_{k=1}^{N} x_k| > \sqrt{x}\}dx + \int_\epsilon^\infty \mathbf{P}\{|N^{-1}\sum_{k=1}^{N} x_k| > \sqrt{x}\}dx \le$$
$$\le \epsilon + \int_\epsilon^\infty A(\epsilon)\exp(-B(\sqrt{x})N)dx \le \epsilon + O(1/N).$$

The second statement is proved.
Further,

$$\mathbf{E}|N^{-2}\sum_{k=1}^{N} x_k \sum_{k=1}^{N} y_k| = \int_0^\infty \mathbf{P}\{|N^{-1}\sum_{k=1}^{N} x_k\ N^{-1}\sum_{k=1}^{N} y_k| > x\}dx$$
$$\le \int_0^\infty \mathbf{P}\{|N^{-1}\sum_{k=1}^{N} x_k| > \sqrt{x}\}dx + \int_0^\infty \mathbf{P}\{|N^{-1}\sum_{k=1}^{N} y_k| > \sqrt{x}\}dx$$
$$= \int_0^\epsilon \mathbf{P}\{|N^{-1}\sum_{k=1}^{N} x_k| > \sqrt{x}\}dx$$
$$+ \int_\epsilon^\infty \mathbf{P}\{|N^{-1}\sum_{k=1}^{N} x_k| > \sqrt{x}\}dx + \int_0^\epsilon \mathbf{P}\{|N^{-1}\sum_{k=1}^{N} y_k| > \sqrt{x}\}dx$$
$$+ \int_\epsilon^\infty \mathbf{P}\{|N^{-1}\sum_{k=1}^{N} y_k| > \sqrt{x}\}dx$$
$$\le 2\epsilon + 2\int_\epsilon^\infty A(\epsilon)\exp(-B(\sqrt{x})N)dx \le 2\epsilon + O(1/N).$$

The third statement is proved.

■

Now, let us fix certain $b \in \mathbb{B}$, $t \in [0,1]$. Put $n = m + [tN]$ and denote $t_N = (n-m)^+/N$. Remark that $t_N \to t$ uniformly w.r.t. $t \in [0,1]$ with the rate $O(1/N)$.

Lemma 12.7.2. *For any fixed $b \in \mathbb{B}$, $t \in [0,1]$, $k > 0$, and $m = N + k$, $n = m + [tN]$, the following relationship holds*

$$\lim_{N\to\infty,} \mathbf{E}_m \Psi_N(b,n) = G(b,t)$$

where $G(b,t)$ was defined in Theorem 12.3.2.

Proof. Since $n = m + [tN]$, then $\mathbf{E}_m \theta_N^n = t_N \epsilon h = t\epsilon h + O(1/N)$.
Put

$$\theta_N^n = t\epsilon h + \lambda_n, \tag{12.13}$$

where $\lambda_n = O(1/N) + z_n$, $\mathbf{E}_m z_n = 0$.

Consider the first term in relationship for$\Psi_N(b,n)$. We obtain

$$\mathbf{E}_m\left\{1/N\sum_{k=[tN]-N+m+1}^{m+[tN]} x_k\mathbb{I}(|x_k-\theta_N^n|<b)\right\}=$$
$$=\left\{\frac{N-[Nt]}{N(N-[Nt])}\sum_{k=[tN]-N+m+1}^{m}\mathbf{E}_m(x_k\mathbb{I}(|x_k-\theta_N^n|<b))+\right. \tag{12.14}$$
$$\left.+\frac{[Nt]}{N[Nt]}\sum_{k=m+1}^{m+[tN]}\mathbf{E}_m(x_k\mathbb{I}(|x_k-\theta_N^n|<b))\right\}.$$

For $[tN]-N+m+1\le k\le m$, we have

$$\mathbf{E}_m(x_k\mathbb{I}(|x_k-\theta_N^n|<b))=\mathbf{E}_m\int_{\theta_N^n-b}^{\theta_N^n+b} xf_0(x)dx=\int_{t\epsilon h-b}^{t\epsilon h+b} xf_0(x)dx+$$
$$+\mathbf{E}_m\left\{\int_{t\epsilon h+b}^{t\epsilon h+b+\lambda_n} xf_0(x)dx-\int_{t\epsilon h-b}^{t\epsilon h-b+\lambda_n} xf_0(x)dx\right\}. \tag{12.15}$$

Since the d.f. $f_0(\cdot)$ is bounded by the constant L, then integrals in the right hand of (12.15) are bounded by the value $L\,(t\epsilon h+b+O(1/N)+|z_n|)\,(O(1/N)+|z_n|)$. Therefore, for the proof of convergence of the left hand of (12.14) to the deterministic value, it is sufficient to prove that $\mathbf{E}_m|z_n|$ and $\mathbf{E}|z_n|^2$ tend to zero as $N\to\infty$. It follows, however, from i) and ii) of Lemma 12.7.1.

Analogous considerations are valid for the second term in (12.14). We need only to change $f_0(\cdot)$ for $f_\epsilon(\cdot)$ and use the boundedness of this d.f.

Taking into account all these considerations, we obtain

$$\lim_{N\to\infty}\mathbf{E}_m\left\{1/N\sum_{k=[tN]-N+m+1}^{m+[tN]} x_k\mathbb{I}(|x_k-\theta_N^n|<b)\right\}=$$
$$=(1-t)\int_{t\epsilon h-b}^{t\epsilon h+b} xf_0(x)dx+t\int_{t\epsilon h-b}^{t\epsilon h+b} xf(x)dx. \tag{12.16}$$

Now, let us consider the second additive term in an obtained relationship for the main statistic. This term can be written as follows:

$$\mathbf{E}_m\left\{\left(\frac{N_1(b,n)}{N}\right)\left(\frac{1}{N}\sum_{k=[tN]-N+m+1}^{m+[tN]} x_k\right)\right\}. \tag{12.17}$$

In virtue of Cramer's and ψ-mixing conditions, it follows that the first multiplicative term in (12.17) converges almost surely w.r.t. the family $\mathbf{P}_m=\{\mathbf{P}_{N+k}\}$ with the exponential rate to the value

$$(1-t)\int_{t\epsilon h-b}^{t\epsilon h+b} f_0(x)dx+t\int_{t\epsilon h-b}^{t\epsilon h+b} f_\epsilon(x)dx. \tag{12.18}$$

Since

$$\mathbf{E}_m\left(\frac{N_1(b,n)}{N}\right) = \mathbf{E}_m\left\{1/N \sum_{k=[tN]-N+m+1}^{m+[tN]} \mathbb{I}(|x_k - \theta_N^n| < b)\right\} =$$

$$= \left\{\frac{N-[Nt]}{N(N-[Nt])} \sum_{k=[tN]-N+m+1}^{m} \mathbf{E}_m(\mathbb{I}(|x_k - \theta_N^n| < b)) + \right.$$

$$\left. + \frac{[Nt]}{N[Nt]} \sum_{k=m+1}^{m+[tN]} \mathbf{E}_m(\mathbb{I}(|x_k - \theta_N^n| < b))\right\}.$$

and the corresponding integrals are Lipshitz functions of their upper limits (in virtue of a bounded d.f.), we can make considerations analogous to estimation of (12.15) and then use Lemma 12.7.1. Then, we obtain that the mathematical expectation of the first multiplicative term in (12.17) also converges to (12.18).

The second multiplicative term in (12.17) converges almost surely with the exponential rate to its mathematical expectation $t\epsilon h$ in virtue of the same Cramer's and ψ-mixing conditions. Then, the product in brackets in (12.17) converges almost surely w.r.t. the family $\mathbf{P}_m = \{\mathbf{P}_{N+k}\}$ to the value

$$[(1-t)\alpha + t\beta]\, t\epsilon h. \tag{12.19}$$

Now, we can use iii) of Lemma 12.6.1 in order to obtain

$$\lim_{N\to\infty} \mathbf{E}_m\left\{\left(\frac{N_1(b,n)}{N}\right)\left(\frac{1}{N}\sum_{k=[tN]-N+m+1}^{m+[tN]} x_k)\right)\right\} = [(1-t)\alpha + t\beta]\, t\epsilon h. \tag{12.20}$$

From (12.16) and (12.20), we obtain the result of this lemma.

■

Now, let us consider the proof of the theorem.

Item 1.

Consider the main statistic:

$$\Psi_N(b,n) = \left(N \sum_{i=1}^{N_1(n,b)} \tilde{x}_i^n - N_1(n,b) \sum_{i=1}^{N} x_i\right) / N^2.$$

We need to prove that there exists $C > 0$, such that for sufficiently large N, the probability

$$\mathbf{P}_m\{\max_{m\le n\le m+N} \sup_{b\in\mathbb{B}} |\Psi_N(b,n)| < C\} \stackrel{\text{def}}{=} \mathbf{P}_m(\cdot)$$

tends to zero exponentially.

Since for any $b \in \mathbb{B},\ n \in [m, m+N]$,

$$\mathbf{P}_m(\cdot) \le \mathbf{P}_m\{|\Psi_N(b,n)| < C\}, \tag{12.21}$$

then it suffices to prove that the right hand of (12.21) exponentially converges to zero for certain *fixed* $b \in \mathbb{B}$, $t \in [0,1]$, and $n = m + [tN]$.

From Lemma 12.7.2, it follows that for $n = m + [tN]$,

$$\mathbf{E}_m \Psi_N(b, n) = G(b, t) + O(1). \tag{12.22}$$

Evidently, the function $G(b, y)$ is continuous and is not equal to zero for all (b, t). Therefore, there exists a pair $(b^*, t^*) \in \mathbb{B} \times [0, 1]$, for which the function $|G(b,t)|$ attains its maximum $Q = |G(b^*, t^*)| = \max_{(b,t) \in U} |G(b,t)|$.

Now, let us use one simple inequality. Suppose ξ is random variable $\mathbf{E}\xi = a$. Then,

$$\mathbf{P}\{|\xi| \leq |a|/4\} \leq \mathbf{P}\{|\xi - a| > |a|/2\} \tag{12.23}$$

From Inequality (12.23), it follows that for the proof of item 1 of this theorem, it suffices to demonstrate that the probability

$$\mathbf{P}_m\{|\Psi_N(b^*, n^*) - \mathbf{E}_m \Psi_N(b^*, n^*)| > |\mathbf{E}_m \Psi_N(b^*, n^*)|/2\}$$

exponentially converges to zero as $N \to \infty$.

But from (12.22) it follows that for large enough N,

$$\begin{aligned} &\mathbf{P}_m\{|\Psi_N(b^*, n^*) - \mathbf{E}_m \Psi_N(b^*, n^*)| > |\mathbf{E}_m \Psi_N(b^*, n^*)|/2\} \leq \\ &\leq \mathbf{P}_m\{|\Psi_N(b^*, n^*) - \mathbf{E}_m \Psi_N(b^*, n^*)| > |G(b^*, t^*)|/4\}. \end{aligned} \tag{12.24}$$

But the probability in the right hand of (12.24) exponentially converges to zero, because we proved earlier such convergence of the main statistic to its mathematical expectation. Therefore, if $0 < C \leq |G(b^*, t^*)|/4$, then from (12.24) it, follows that

$$\mathbf{P}_m(\cdot) \to 0$$

with the exponential rate. The theorem is proved.

Item 2.

As to the proof of 2), we first observe that for all $\gimel > 0$:

$$\mathbf{P}_m\{|\gamma_N - \gamma^*| > \gimel\} = \mathbf{P}_m\{\gamma_N > \gamma^* + \gimel\} + \mathbf{P}_m\{\gamma_N < \gamma^* - \gimel\},$$

where γ^* is the minimal root of the equation $\max_{b \in \mathbb{B}} |G(b,t)| = 1.2C$, and $0 < C \leq 1/4 \max_{(b,t) \in U} |G(b,t)| = 1/4Q$.

By definition of the stopping time τ_N and the variable γ_N, we obtain

$$\begin{aligned} &\mathbf{P}_m\{\gamma_N > \gamma^* + \gimel\} \leq \mathbf{P}_m\{\max_{m \leq n \leq m + [(\gamma^* + \gimel)N]} \sup_{b \in \mathbb{B}} |\Psi_N(b, n)| \leq C\} \\ &\leq \mathbf{P}_m\{|\Psi_N(\tilde{b}, m + [(\gamma^* + \gimel)N])| \leq C\}. \end{aligned} \tag{12.25}$$

Here, $\tilde{b} \in \mathbb{B} = \arg\max_{b \in \mathbb{B}} |G(b, \gamma^* + \gimel)|$.

From Lemma 12.7.2, it follows that

$$\lim_{N\to\infty} \mathbf{E}_m \Psi_N(\tilde{b}, m + [(\gamma^* + \gimel)N]) = G(\tilde{b}, \gamma^* + \gimel) \qquad (12.26)$$

From (12.25), we obtain

$$\mathbf{P}_m\{\gamma_N > \gamma^* + \gimel\} \leq \mathbf{P}_m\{|\Psi_N(\tilde{b}, m + [(\gamma^* + \gimel)N]) - G(\tilde{b}, \gamma^* + \gimel)| \\ > |G(\tilde{b}, \gamma^* + \gimel)| - C\}. \qquad (12.27)$$

Let us choose $\gimel > 0$ so small that the following inequality is true (recall that by definition of γ^*, the following relationship holds $\max_{b\in\mathbb{B}} |G(b, \gamma^*)| = 1.2C$):

$$\left|\max_{b\in\mathbb{B}} |G(b, \gamma^*)| - \max_{b\in\mathbb{B}} |G(b, \gamma^* + \gimel)|\right| = \left|\max_{b\in\mathbb{B}} |G(b, \gamma^*)| - |G(\tilde{b}, \gamma^* + \gimel)|\right| \\ \leq 0.1C \qquad (12.28)$$

Then, taking into account the choice of $\tilde{b}$ from (12.25), (12.27), and (12.28) we obtain

$$\mathbf{P}_m\{\gamma_N > \gamma^* + \gimel\} \leq \mathbf{P}_m\{|\Psi_N(\tilde{b}, m + [(\gamma^* + \gimel)N]) - G(\tilde{b}, \gamma^* + \gimel)| \\ > \max_{b\in\mathbb{B}} |G(b, \gamma^*)| - 1.1C\}. \qquad (12.29)$$

Recall that $\max_{b\in\mathbb{B}} |G(b, \gamma^*)| = 1.2C$. Therefore from (12.7.31), we have

$$\mathbf{P}_m\{\gamma_N > \gamma^* + \gimel\} \leq \mathbf{P}_m\{|\Psi_N(\tilde{b}, m + [(\gamma^* + \gimel)N]) - G(\tilde{b}, \gamma^* + \gimel)| > 0.1C\}. \qquad (12.30)$$

But the right hand of inequality (12.29) tends to zero exponentially by N (here, we estimate deviation of the normed sum of r.v.'s satisfying Cramer's and ψ-mixing conditions from its mathematical expectation).

In analogy, we obtain the exponential upper estimate for the probability $\mathbf{P}_m\{\gamma_N < \gamma^* - \gimel\}$ for small enough $\gimel > 0$.

Item 2 of theorem is proved.

Corollary

Suppose $(b^*, t^*) \in \arg \max_{(b,t)\in U} |G(b, t)|$. Define $n^* = \tau_N + [t^* N]$ and consider the following system:

$$N_1(n^*, b^*)/N = (1 - t^*) \int_{t^*\epsilon h - b^*}^{t^*\epsilon h + b^*} f_0(x)dx + t^* \int_{t^*\epsilon h - b^*}^{t^*\epsilon h + b^*} f_\epsilon(x)dx$$
$$\max_{b\in\mathbb{B}} |\Psi_N(b, n^*)| = |G(t^*, b^*)|.$$

In virtue of Theorem 12.3.2, solutions of this system (w.r.t. (ϵ, h)) are consistent estimates of (ϵ, h) (see Borovkov (1998)).

13

Sequential Detection and Estimation of Change-Points

13.1 Introduction

In this chapter, the problem of sequential detection and estimation of change-points is considered. The sense of this problem is as follows: In practice, after raising an alarm signal about a change, it is often required to divide the whole obtained sequential sample into subsamples of observations before and after an unknown change-point. In this chapter, asymptotically optimal methods are proposed for sequential detection and estimation of a change-point in this problem.

Although this problem was already discussed in earlier works by Girshick and Rubin (1952), Duncan (1974), and Robbins and Siegmund (1972), in subsequent research, it was almost completely neglected. The main bulk of results in sequential change-point analysis refers to sequential change-point detection (see, e.g., Shiryaev (1961), Page (1954), Lorden (1971), Pollak (1985), and Moustakides (1986)). Sequential estimation was understood mainly as the problem of estimation of a parameter θ of the d.f. $f(x, \theta)$ of observations after sequential detection. For example, Lorden and Pollak (2005) propose a nonanticipating scheme for estimation of an unknown papameter of the d.f. of observations after a change-point, which develops the initial ideas of Robbins and Siegmund (1972). However, in 1990–2000s, some methods for sequential change-point detection and estimation were proposed (Srivastava and Wu (1999), Ding (2003)). These methods have a substantial drawback: The proposed change-point estimates are biased. In the paper by Srivastava and Wu (1999), a sequential method for detection and estimation of a change-point in the Wiener process was proposed that gives asymptotically biased estimates in most cases. Gombay (2003) considers a general situation of change-point detection and estimation in independent sequences and also proposes a biased estimate of a change-point.

However, I can mention many practical problems where some more accurate estimates of a change-point are needed. For example, in quality control problems after detection of a change-point, it is often required to call back a portion of the defect production, and, for this purpose, we need to estimate an unknown change-point. In control science, it is usually required to readjust a control law and to reidentify a system after detection of a change.

So, the problem is twofold: first, to quickly detect a change-point; and second, to estimate the detected change-point as precisely as possible. In order to achieve these two goals that often contradict each other, we need to develop methodology that combines sequential and retrospective approaches to change-point detection and estimation. In this chapter, I propose such methodology based upon the idea of a "large parameter".

This large parameter must provide an interface between sequential and retrospective stages of the problem. Below, I give a detailed description of this parameter for the proposed method.

The quality of retrospective change-point detection methods is characterized by the probabilities of the 1st-type and the 2nd-type errors, as well as the probability of the error of estimation. Asymptotics of these probabilities are thoroughly investigated in the theory of retrospective change-point detection methods (see, e.g., Brodsky and Darkhovsky, 2000).

However, in sequential detection problems, quite different performance measures are used: the "average run length" (ARL) and the average time between false alarms. In 1950–2000s, an essential methodological gap between retrospective and sequential methods of change-point analysis was formed. But the statistical phenomenon — a change-point, i.e., a change in statistical characteristics of observations — is the same in retrospective and sequential problems that differ only by the method (retrospective or online) of information collecting. This consideration leads to the idea that qualitative characteristics and performance criteria in retrospective and sequential change-point problems must be methodologically close to each other. The goal of this chapter is to propose such comparable performance measures for the concrete problem of change-point detection and estimation.

13.2 Sequential Change-Point Detection

In this chapter, I consider the nonparametric approach to change-point detection and estimation. This means that the d.f. of observations both before and after a change-point are unknown to us. There are two distinctive features of any nonparametric setup: I consider the problem of a change in a certain integral statistical characteristic of observations (e.g., the mean value, dispersion, etc.), or we construct some score functions of observations (e.g., the empirical d.f. with weights). Here, I try the first way and consider the *basic problem* of sequential detection of a change in the mean value of observations. It is possible to argue that many general formulations of sequential detection problems can be reduced to this basic statement. Let us demonstrate it for the classic parametric statement of a sequential change-point detection problem.

Suppose the d.f. of sequential observations x_i, $i = 1, 2, \dots$ changes at some unknown point m from $f_0(x_i)$ to $f_\theta(x_i)$, where the parameter θ of the d.f. is also unknown. CUSUM and GRSh tests are based upon the likelihood ratios $\ln \frac{f_\theta(x_i)}{f_0(x_i)}$. For example, for the CUSUM test, the decision statistic is

$$z_i = (z_{i-1} + \ln \frac{f_\theta(x_i)}{f_0(x_i)})^+,$$

where $z_0 = 0$.

Remark that $E_0 \ln \frac{f_\theta(x_i)}{f_0(x_i)} < 0$ and $E_\theta \ln \frac{f_\theta(x_i)}{f_0(x_i)} > 0$. So, the mean value of the sequence $\ln \frac{f_\theta(x_i)}{f_0(x_i)}$ changes from a negative to a positive value at the change-point m. Since the parameter θ of the d.f. is unknown, the parametric situation of sequential change-point detection is essentially reduced to the following problem.

On the probability space $(\Omega, \mathcal{F}, P)$, let us consider the following model of observations:

$$x(n) = a + hI(n > m) + \xi_n, \tag{13.1}$$

where m is an unknown change-point, and $h \geq \delta > |a|$ is an unknown change in the mathematical expectation of a centered random sequence ξ_n ($E\xi_n = 0$, $E\xi_n^2 = \sigma^2$). Remark that usually $a < 0$ (the mean value of observations before a change-point) and $h > 0$ (the mean value of observations after a change-point). The parameter $\delta > |a|$ (a minimal detected change in mean) plays an important role in the sequel.

Suppose ξ_i satisfies the uniform Cramer and ψ-mixing conditions.

For **sequential detection of the change-point** m, let us consider the following the *nonparametric CUSUM method*:

$$y_n = (y_{n-1} + x(n))^+, \quad y_0 \equiv 0, \tag{13.2}$$

where $b^+ = \max(b, 0)$.

The decision function of the CUSUM method is $d_N(n) = I(y_n > N)$, where N is the threshold of detection.

The choice of this method is motivated by its asymptotical optimality in sequential change-point detection (see Brodsky and Darkhovsky (2008)).

Denote by P_0, P_m measures corresponding to a sequence of observations without a change-point and with the change-point m, respectively. We introduce the following performance measures of sequential change-point detection:

— the supremal probability of a "false decision":

$$\alpha_N = \sup_n P_0\{d_N(n) = 1\}. \tag{13.3}$$

This characteristic is closely connected with ARL to false decision and FAR (false alarm rate) for the case of independent normal observations. However, for dependent observations satisfying Cramer's and ψ-mixing conditions, we can readily estimate α_N, but to obtain any similar estimates for ARL and FAR measures is very difficult. This is the main motivation for introduction of this new performance measure.

— the stopping time

$$\tau_N = \inf\{n : \ d_N(n) = 1\}. \tag{13.4}$$

Remark that the threshold N plays the role of a "large parameter" for the CUSUM method: the probability of a false decision about a change-point (type-1 error) tends to zero as $N \to \infty$. The following theorem holds.

Theorem 13.2.1. *Suppose the sequence of observations satisfies the uniform Cramer and ψ-mixing conditions. Then, for any $a < 0$ and any $m > 0$:*

$$\alpha_N \leq L \max(\exp(-\frac{N|a|}{H}), \exp(-\frac{(N+|a|)T}{4})), \tag{13.5}$$

where $L = (1 - e^{-y})^{-1}$, $y = \min(\frac{a^2}{4H}, |a|T)$.

For any $h > |a|$, $\epsilon > 0$ and $\gamma_N = (\tau_N - m)^+/N$:

$$\lim_{N \to \infty} N^{-1} \ln P_m\{|\gamma_N - \frac{1}{h - |a|}| > \epsilon \ |(\tau_N > m)\} \leq -\frac{\epsilon^2}{2\sigma^2}(h - |a|)^3. \tag{13.6}$$

From this theorem, it follows that the normalized delay time γ_N tends a.s. to $(h - |a|)^{-1}$ as $N \to \infty$ with the rate of convergence estimated in (13.6).

The proof of Theorem 13.2.1 is given in the appendix.

13.3 Retrospective Estimation

For **estimation of a detected change-point**, we use the retrospective sample of the last M observations: $\{x(\tau_N - M + 1), \dots, x(\tau_N)\}$. The volume M of this sample can be obtained from (13.6):

$$M = \frac{N}{\delta - |a|} + \sigma\sqrt{2N}\frac{|\ln \alpha|^{1/2}}{(\delta - |a|)^{3/2}}. \tag{13.7}$$

This choice of M guarantees that the retrospective sample "covers" the true change-point m with the confidence probability $1 - \alpha$ for any $0 < \alpha < 1$. We can choose $\alpha = \exp(-\beta N)$ for $\beta > 0$ and then $M = N(\frac{1}{\delta - |a|} + \frac{\sigma(2\beta)^{1/2}}{(\delta - |a|)^{3/2}})$.

So $P_m(H_\beta) \geq 1 - \exp(-\beta N)$, where H_β is the hypothesis that the retrospective sample contains the true change-point m.

For estimation of the change-point m, the following statistic is used:

$$T_M(n) = \sqrt{\frac{n(M-n)}{M}}\,(\frac{1}{n}\sum_{i=1}^{n} x(i+\tau_N - M) - \frac{1}{M-n}\sum_{i=n+1}^{M} x(i+\tau_N - M)) \tag{13.8}$$

for $n = 1, \ldots, M$.

Then, the estimate of the change-point m can be constructed as follows: $\hat{m} = \hat{n} + \tau_N - M$, where $\hat{n}$ is the minimal point of the set $arg\max_n |T_N(n)|$.

Define the following values: $\hat{\rho}_M = 1 - \frac{(\tau_N - m)^+}{M}$ and $\tilde{\rho}_M = \frac{\hat{n}}{M}$. From Theorem 1 it follows that

$$\hat{\rho}_M = 1 - \frac{(\tau_N - m)^+}{M} \to \rho(h) \equiv 1 - \frac{\delta - |a|}{h - |a|}(1 + \frac{(\delta - |a|)^{1/2}}{\sigma(2\beta)^{1/2}})\ P_m\text{-a.s. as } N \to \infty.$$

Then, the following theorem holds.

Theorem 13.3.1. *Suppose the sequence of observations satisfies the uniform Cramer and ψ-mixing conditions. Then, for any $0 < \epsilon < \rho(h)$:*

$$P_m\{|\tilde{\rho}_M - \rho(h)| > \epsilon | H_\beta \cap \{\tau_N > m\}\} \leq L_1 \exp(-L_2 \epsilon^2 M), \tag{13.9}$$

where the constants $L_1, L_2 > 0$ do not depend on M.

From this theorem, it follows that the proposed estimate $\hat{m}$ will be in the $[\epsilon M]$ neighborhood of the true change-point m with the probability increasing to 1 as $M \to \infty$ for any $0 < \epsilon < 1$.

The proof of Theorem 13.3.1 is given in the appendix.

13.4 Simulations

In this section, results of a simulation study of the proposed methods are presented. The following examples were studied:

1) Change-in-mean: independent Gaussian r.v.'s with an unknown change in the mathematical expectation

2) Change-in-dispersion: independent Gaussian r.v.'s with an unknown change in dispersion

1) Change-in-mean

The sequence of Gaussian observations $X = (x(1), x(2), \ldots)$ with the change-point $m = 1000$ was modeled. Both m and a change in mean h $(N(0,1) \to N(h,1))$ were unknown to the proposed algorithm.

For detection of a change-point, the nonparametric CUSUM test was used:

$$y_n = (y_{n-1} + x(n) + a)^+ > N, \quad y_0 = 0.$$

TABLE 13.1
Characteristics of the Proposed Method: Change in Mean

h		0.55	0.6	0.8	1.0	1.5	2.0	2.5
CUSUM	$E_m\tau - m$	113.1	82.1	36.4	23.3	12.2	8.2	5.9
	$\sigma(\tau - m)$	88.5	61.5	18.4	12.5	3.8	2.9	1.3
K-S	Δ	79.3	45.5	13.9	13.5	4.6	3.9	2.1

The parameters of this test were chosen as follows: $N = 12$, $a = -0.5$. The supremal probability of a false decision for these parameters is $\exp(-2|a|N/\sigma^2) = \exp(-12) = 6.14 * 10^{-6}$ and corresponds to the average time before a "false alarm" $E_0\,\tau = 162750$.

Let $\tau = \inf\{n \geq 1 : d_N = 1\}$ be the instant when the change was detected. Then, $E_m\tau - m$ is the average delay time in change-point detection. Besides the average delay time, we compute the value $\sigma(\tau - m)$ — the square root of the dispersion of the delay time.

At the 2nd stage, we estimate the change-point using the retrospective sample of the last M observations. The choice of M is made by formula (13.7) for $\alpha = 0.05$. Here, $\beta = 0.25$, and for $N = 12, a = -0.5$ and $\delta = 0.55$, we obtain $M = 1000$.

In experiments, the following measure of the estimation error was computed:

$$\Delta = \left(\frac{1}{k-1}\sum_{i=1}^{k}(\hat{m}_k - m)^2\right)^{1/2},$$

where $k = 5000$ is the number of independent trials.

The results obtained are reported in Table 13.1.

2) Change-in-dispersion

In this group of tests, the following model of observations was considered: the sequence of i.r.v.'s $X = (x(1), x(2), ...)$ with the change-point $m = 1000$. The d.f. of observations changes at the instant $m = 1000$: $N(0,1) \to N(0,(1+h)^2)$. Both m and h were unknown to the algorithm.

For detection of the change-point m, the nonparametric CUSUM test was used:

$$y_n = (y_{n-1} + x^2(n) + a)^+ > N, \quad y_0 = 0.$$

The parameters of this test were chosen as follows: $N = 20$, $a = -1.25$. The supremal probability of a false decision for these parameters is $\exp(-10) = 4.54 * 10^{-5}$ and corresponds to the average time before a "false alarm" $E_0\,\tau = 22026$.

The choice of M is made by formula (13.7) for $\alpha = 0.05$. Here, $\beta = 0.25$, and for $N = 20, a = -1.25$, and $\delta = 1.69$, we obtain $M = 150$. Results are presented in Table 13.2.

TABLE 13.2
Characteristics of the Proposed Method: Change in Dispersion

h		0.3	0.4	0.5	0.7	0.9	1.0
CUSUM	$E_m\tau - m$	65.7	37.8	26.7	15.9	10.8	9.4
	$\sigma(\tau - m)$	54.3	28.3	18.9	10.4	7.1	6.1
K-S	Δ	70.9	37.1	24.5	19.3	12.5	11.6

13.5 Asymptotical Optimality

The asymptotical optimality of the CUSUM method for sequential detection of a change-point in the sense of different criteria (see Lorden (1971), Lai (1988), and Pollak (1985)) is a well-known fact. In contrast to these results and criteria of optimality (the *1st order optimality* or τ-optimality; see Chapter 9), in this paper, we are primarily interested in another kind of optimality of the CUSUM method and in another criterion of the asymptotical optimality (the *2nd order optimality* or σ-optimality) of sequential methods. The details of this approach are presented below.

We consider a sequence of independent r.v.'s $X = \{x_1, x_2, \dots\}$ with the d.f.

$$f(x_n) = \begin{cases} f_0(x_n), & n \le m \text{ or } m = \infty \text{ no change point} \\ f_\theta(x_n), & n > m, \end{cases}$$

where $\theta \in \Theta$ is a certain parameter.

At any step of decision making, we test the null hypothesis H_0 of no change in the d.f. of observations against the alternative H_1 of a change occurred in this d.f. Suppose a certain method with the decision function $d_C(\cdot)$ (depending on some "large parameter" C) is used:

$$d_C(\cdot) = \begin{cases} 0, & \text{assume } H_0 \text{ and continue} \\ 1, & \text{assume } H_1 \text{ and stop.} \end{cases}$$

In the sequel, we denote measures and expectations for the sequence with the change-point m and the parameter θ of the d.f. by $P_{\theta m}$ and $E_{\theta m}$, respectively. Measures and expectations referring to sequences without change-points are denoted by P_0 and E_0.

Consider the following characteristics:

1) Supremal probability of a false decision:

$$\alpha_C = \sup_n P_0\{d_C(n) = 1\}, \tag{13.10}$$

2) Stopping time τ_C and the normalized delay time γ_C:

$$\tau_C = \inf\{n : d_C(n) = 1\}, \quad \gamma_C = (\tau_C - m)^+/C. \tag{13.11}$$

As before (see, e.g., Chapter 12), we consider the class of methods with certain upper constraints on the error probabilities. In particular, we consider the class of methods Γ_θ with the exponential rate of convergence to zero of the supremal probability of false decision:

$$\liminf_{C\to\infty} \frac{|\ln \alpha_C|}{C} \geq L > 0 \qquad (\Gamma_\theta - 1).$$

Another feature of the class Γ_θ concerns the normalized delay times in change-point detection γ_C. We consider methods for which this sequence tends in probability $P_{\theta m}$ to some deterministic limit $\gamma(\theta)$, i.e.,

$$\gamma_C \to \gamma(\theta), \qquad \text{in } P_{\theta m} \text{ as } C \to \infty \qquad (\Gamma_\theta - 2),$$

and the finite derivatives $\gamma'(u) \neq 0$, $\gamma''(u)$ in some neighborhood of $u = \theta$ exist.

Besides these properties of methods belonging to the class Γ_θ, we need yet the following technical assumptions:

1) $\frac{\partial^i}{\partial\theta^i} f_\theta(x)$ exists and is finite $P_{\theta m}$-a.s. for any θ and $i = 1, 2$;

2) $\int |\frac{\partial^i}{\partial\theta^i} f_\theta(x)|dx < \infty$ for any θ and $i = 1, 2$;

3) the Fisher information $I(\theta) = \int \frac{(f_\theta'(x))^2}{f_\theta(x)} dx$ exists for any θ.

The following theorem holds.

Theorem 13.5.1. *Suppose assumptions 1)–3) are satisfied. Then, for the methods from the class Γ_θ, the following asymptotical inequality holds true for $\epsilon \to 0$:*

$$\lim_{C\to\infty} C^{-1} \ln P_{\theta m}\{|\gamma_C - \gamma(\theta)| > \epsilon | \tau_C > m\} \geq -\frac{\epsilon^2(1 + o(1))}{2[\gamma'(\theta)]^2} \gamma(\theta) I(\theta). \tag{13.12}$$

The proof of Theorem 13.5.1 repeats the main ideas of the proof of Theorem 12.4.1 and, therefore, is omitted here.

A method of sequential change-point detection is called the *2nd order asymptotically optimal* if the equality sign in (13.12) is attained for this method as $C \to \infty$. The nonparametric CUSUM test is the 2nd order asymptotically optimal.

The 2nd order asymptotical optimality in sequential detection is closely connected with the asymptotical optimality of change-point estimation at the 2nd stage of the proposed two-stage procedure.

Now, let us prove the a priori low inequality for the error of change-point estimation. Suppose a method from the class Γ_θ was used at the 1st stage of change-point detection, which gives the normalized delay time $\gamma_C = (\tau_C - m)^+/C$ and the corresponding change-point parameter $\hat{\rho}_C = 1 - \gamma_C$ at the 2nd stage of the retrospective change-point estimation. The sequence $\hat{\rho}_C$ tends to the limit $\rho(\theta)$ in probability $P_{\theta m}$ as $C \to \infty$. Suppose some method of the retrospective change-point estimation is used at the 2nd stage (below, we specify requirements to this method) that gives the estimate $\tilde{\rho}_C$ of the limiting change-point parameter $\rho(\theta)$. For any $0 < \epsilon < 1$, we need to estimate from below the probability $P_\theta\{|\tilde{\rho}_C - \rho(\theta)| > \epsilon\}$.

The following theorem holds.

Theorem 13.5.2. *For* $0 < \epsilon < 1$ *and any fixed* m,

$$P_\theta\{|\tilde{\rho}_C - \rho(\theta)| > \epsilon | H_\beta \cap (\tau_C > m)\} \geq L_1 \exp(-L_2 \epsilon^2 C), \qquad (13.13)$$

where positive constants L_1 *and* L_2 *do not depend on* C.

The proof of Theorem 13.5.2 is given in the appendix.

13.6 Conclusion

In this chapter, a new two-stage method is proposed for sequential detection and estimation of a change-point. The basis problem statement is purely nonparametric: at the first stage, a nonparametric version of the CUSUM method is used for sequential detection of a change-point, and at the second stage, we use a modified Kolmogorov-Smirnov statistic for estimation of the detected change-point. A distinctive feature of the proposed two-stage method: It provides the estimate of a change-point without the asymptotical bias (for example, earlier methods by Srivastava and Wu (1999) and Gombay (2003) do not achieve this). Moreover, we prove that this method is asymptotically optimal in the sense that the theoretical lower bounds for the errors of change-point detection and estimation are asymptotically attained for this method.

The proposed method is remarkable for two reasons. First, it can serve as a clear illustration and example of the methodology of a *large parameter* that essentially unify the sequential and retrospective approaches to change-point problems. In this chapter, I demonstrated that the volume of the retrospective sample M must be connected with the threshold N in sequential change-point detection. An explicit relationship between the large parameters N and M is obtained from new performance measures in sequential detection and retrospective estimation problems that are defined in this paper.

Second, the proposed method can provide a background for more advanced methods in sequential estimation. In fact, many modern approaches to sequential estimation of the d.f. parameter are based upon the initial idea of Robbins and Siegmund (1972): to construct the estimate $\hat{\theta}_n$ of the d.f. parameter on the basis of the first $n-1$ observations $x_1, \dots, x_{n-1}$. However, when a change-point m occurs, such estimates react too slowly to a sudden change in the d.f. parameter θ. Why not detect a change-point m first and then readjust our sequential estimation procedure beginning from the estimate $\hat{m}$ of a change-point? This path was not yet tried, but it seems to be quite promising in many practical problems. Its clear advantage: We can eliminate the asymptotical bias in $\hat{\theta}_n$ and escape time-consuming correction procedures.

13.7 Appendix: Proofs of Theorems

Proof of Theorem 13.2.1

Denote $S_n(x) = \sum_{i=1}^{n} x(i)$. CUSUM statistics have the following form:

$$y_n = S_n(x) - \min_{0 \le j \le n} S_j(x).$$

Therefore,

$$\begin{aligned} P_0\{d_N(n) = 1\} &= P_0\{S_n(x) - \min_{0 \le j \le n} S_j(x) > N\} \\ &= P_0\{\max_{0 \le j \le n} \sum_{k=j+1}^{n} x(k) > N\}. \end{aligned}$$

After time transformation $k \to (n-k+1)$, $k = 1, \dots, n$ and introduction of new variables $\zeta(k) = \xi_{n-k+1}$, $k = 1, \dots, n$, we obtain

$$\begin{aligned} \alpha_N &= \sup_n P_0\{\max_{1 \le k \le n} (-|a|k + \sum_{i=1}^{k} \zeta(i)) > N\} \\ &\le \sup_n \sum_{k=1}^{n} P_0\{\sum_{i=1}^{k} \zeta(i) > N + k|a|\}. \end{aligned}$$

Remark that the transformed variables $\zeta(k)$ still satisfy Cramer's and ψ-mixing conditions. Since $E\zeta(k) = 0$, Cramer's condition is equivalent to the following: There exists $H > 0$, such that

$$Ee^{t\zeta(i)} \le \exp(t^2 H), \quad \text{for} \quad 0 \le t \le T.$$

Denote $Z_k = \sum_{i=1}^{k} \zeta(i)$. We wish to obtain the exponential upper estimate for $|Z_k|$. Such estimates were obtained first by Petrov (1965) for i.i.d. case and then by Brodsky and Darkhovsky (2000) (for Cramer's and mixing conditions, see Chapter 1). In our case, this estimate has the following form:

$$P_0\{|Z_k| > N + k|a|\} \le \begin{cases} \exp\left(-\dfrac{(N+k|a|)^2}{4kH}\right), & |a| + \dfrac{N}{k} \le HT \\ \exp\left(-\dfrac{(N+k|a|)T}{4}\right), & |a| + \dfrac{N}{k} > HT. \end{cases}$$

Therefore, for α_N, we can write

$$\alpha_N \leq \begin{cases} \exp(-\frac{N|a|}{2H}) \cdot \sup_n(\sum_{k=1}^{n} \exp(-[\frac{N^2}{4kH} + \frac{ka^2}{4H}])), & N \leq k(HT - |a|) \\ \exp(-\frac{NT}{4}) \cdot \sup_n(\sum_{k=1}^{n} \exp(-\frac{k|a|T}{4})), & N > k(HT - |a|). \end{cases}$$

Finally, for α_N, we obtain the following upper estimate:

$$\alpha_N \leq L \max(\exp(-\frac{N|a|}{H}), \exp(-\frac{(N + |a|)T}{4})),$$

where $L = (1 - e^{-y})^{-1}$, $y = \min(\frac{a^2}{4H}, |a|T)$.

Now let us consider the proof of (6). A new content in (6) is the condition $\tau_N > m$. Remark that the unconditional counterpart of (6) was proved in Brodsky and Darkhovsky (2000):

$$\lim_{N \to \infty} N^{-1} \ln P_m\{|\gamma_N - \frac{1}{h - |a|}| > \epsilon\} = -\frac{\epsilon^2}{2\sigma^2}(h - |a|)^3.$$

Denote $A_\epsilon = \{|\gamma_N - \frac{1}{h - |a|}| > \epsilon\}$. Then,

$$P_m\{A_\epsilon | \tau_N > m\} = \frac{P_m\{A_\epsilon \cap (\tau_N > m)\}}{P_m\{\tau_N > m\}} = \frac{P_m\{A_\epsilon \cap (\tau_N > m)\}}{1 - P_m\{\tau_N \leq m\}} \leq \frac{P_m\{A_\epsilon\}}{1 - m\alpha_N}.$$

In virtue of exponential Estimate (13.5) for α_N, for any finite m, we obtain (13.6).

Proof of Theorem 13.3.1

First, let us consider the probability of the event $\{\sup_{M \geq k} |\tilde{\rho}_M - \rho(h)| > \epsilon\}$.

Write

$$\begin{aligned} & P_m\{\sup_{M \geq k} |\tilde{\rho}_M - \rho(h)| > \epsilon\} \\ = \; & P_m\{\sup_{M \geq k} |\tilde{\rho}_M - \rho(h)| > \epsilon, \sup_{M \geq k} |\hat{\rho}_M - \rho(h)| \leq \frac{\epsilon}{2}\} \\ + \; & P_m\{\sup_{M \geq k} |\tilde{\rho}_M - \rho(h)| > \epsilon, \sup_{M \geq k} |\hat{\rho}_M - \rho(h)| > \frac{\epsilon}{2}\} \\ \leq \; & P_m\{\bigcup_{M \geq k} (\{|\tilde{\rho}_M - \hat{\rho}_M| > \frac{\epsilon}{2}\} \cap \{|\hat{\rho}_M - \rho(h)| \leq \frac{\epsilon}{2}\})\} \\ + \; & P_m\{\sup_{M \geq k} |\hat{\rho}_M - \rho(h)| > \frac{\epsilon}{2}\}. \end{aligned}$$

From Theorem 13.2.1, it follows that the last probability in the right hand tends to zero as $k \to \infty$:

$$P_m\{\sup_{M \geq k} |\hat{\rho}_M - \rho(h)| > \frac{\epsilon}{2}\} \leq L_1 \exp(-L_2 \epsilon^2 k),$$

where positive constants L_1 and L_2 do not depend on M.

Consider the first term in the right hand:

$$P_m\{\bigcup_{M\geq k}(\{|\tilde{\rho}_M - \hat{\rho}_M| > \frac{\epsilon}{2}\}, \{|\hat{\rho}_M - \rho(h)| \leq \frac{\epsilon}{2}\})\}$$
$$\leq \sum_{M\geq k} P_h\{|\tilde{\rho}_M - \hat{\rho}_M| > \frac{\epsilon}{2}, |\hat{\rho}_M - \rho(h)| \leq \frac{\epsilon}{2}\}.$$

Write

$$P_m\{|\tilde{\rho}_M-\hat{\rho}_M|>\frac{\epsilon}{2}, |\hat{\rho}_M-\rho(h)| \leq \frac{\epsilon}{2}\} = \sum_{i:|r_i-\rho(h)|\leq \frac{\epsilon}{2}} P_m\{|\tilde{\rho}_M-r_i| > \frac{\epsilon}{2}, \hat{\rho}_M = r_i\}.$$

Consider the event $A_i = \{\hat{\rho}_M = r_i\}$. For $\omega \in A_i$, the following decomposition of the statistic $Y_M(\cdot)$ into the deterministic and stochastic component holds:

$$Y_M(n) = \phi_{r_i}(n) + \eta_{r_i}(n),$$

where the deterministic function $\phi_{r_i} = E_m(Y_M(n)|A_i)$ is equal to

$$\phi_{r_i}(n) = \begin{cases} h \quad r_i\sqrt{\dfrac{M-n}{n}}, & \text{for } [r_iM] < n \leq M-1 \\ h \quad (1-r_i)\sqrt{\dfrac{n}{M-n}}, & \text{for } 1 \leq n \leq [r_iM]. \end{cases}$$

Without loss of generality, suppose $h > 0$. Then, $\phi_{r_i}(n)$ has a unique maximum at the point $[r_iM]$:

$$\max_{1\leq n\leq M-1} \phi_{r_i}(n) = h\sqrt{r_i(1-r_i)}.$$

For the random process $\eta_{r_i}(n)$, we have for any $\delta > 0$:

$$\begin{aligned} & P_m\{\max_{r_i,n} |\eta_{r_i}(n)| > \delta\} \\ \leq \; & P_m\{\max_{1\leq n\leq M-1} \sqrt{\frac{n(M-n)}{M^2}}\, |\frac{1}{n}\sum_{i=1}^{n} \xi_i - \frac{1}{M-n}\sum_{i=n+1}^{M} \xi_i| > \delta\} \\ \leq \; & C_1 \exp(-C_2\delta M), \end{aligned}$$

where the last inequality follows from the exponential estimates for sums of random variables satisfying Cramer's and ψ-mixing conditions (Section 1.4).

Consider the change of time $n \to n/M$ and the linear interpolation of the functions $\phi_{r_i}(\cdot)$ and $\eta_{r_i}(\cdot)$ by discrete points n/M on the interval $t \in [0, 1]$.

The function $\phi_{r_i}(t)$ satisfies the reversed Lipschitz condition at the point r_i:

$$|\phi_{r_i}(r_i) - \phi_{r_i}(t)| \geq B(r_i)|t - r_i|,$$

where $B(r_i) = \frac{1}{2}h\max((\frac{r_i}{1-r_i})^{1/2}, (\frac{1-r_i}{r_i})^{1/2})$.

Hence, for $\rho(h) > \epsilon$:

$$\begin{aligned}
&P_m\{|\tilde{\rho}_M - \hat{\rho}_M| > \frac{\epsilon}{2}, \hat{\rho}_M = r_i\} \\
&\leq P_m\{\phi_{r_i}(r_i) - \phi_{r_i}(\tilde{\rho}_M) > B(r_i)\frac{\epsilon}{2}, \phi_{r_i}(\tilde{\rho}_M) + \eta_{r_i}(\tilde{\rho}_M) > \phi_{r_i}(r_i) + \eta_{r_i}(r_i)\} \\
&\leq P_m\{\eta_{r_i}(\tilde{\rho}_M) - \eta_{r_i}(r_i) > B(r_i)\frac{\epsilon}{2}\} \\
&\leq P_m\{\max_{r_i,n} |\eta_{r_i}(n)| > \min_{r_i} B(r_i)\frac{\epsilon}{4}\} \\
&= P_m\{\max_{r_i,n} |\eta_{r_i}(n)| > \frac{h\epsilon}{8} \max((\frac{\rho(h) - \epsilon/2}{1 - \rho(h) + \epsilon/2})^{1/2}, (\frac{1 - \rho(h) + \epsilon/2}{\rho(h) - \epsilon/2})^{1/2})\} \\
&\leq C_1 \exp(-C_2 \epsilon h M).
\end{aligned}$$

Therefore,

$$P_m\{|\tilde{\rho}_M - \hat{\rho}_M)| > \epsilon, |\rho_M - \rho(h)| < \frac{\epsilon}{2}\} \leq \epsilon M C_1 \exp(-C_2 \epsilon M)$$

and

$$P_m\{\bigcup_{M \geq k} (\{|\tilde{\rho}_M - \hat{\rho}_M| > \frac{\epsilon}{2}\} \cap \{|\hat{\rho}_M - \rho(h)| \leq \frac{\epsilon}{2}\})\} \leq C_1 \exp(-C_2 \epsilon k),$$

where the constants C_1 and C_2 do not depend on M.

Finally, we obtain

$$P_m\{\sup_{M \geq k} |\tilde{\rho}_M - \rho(h)| > \epsilon\} \leq C_1 \exp(-C_2 \epsilon k) + L_1 \exp(-L_2 \epsilon^2 k),$$

and, therefore, $\tilde{\rho}_M \to \rho(h)$ P_m-a.s. as $M \to \infty$.

Now, consider the probability

$$\begin{aligned}
&P_m\{|\tilde{\rho}_M - \rho(h)| > \epsilon | H_\beta \cap (\tau_N > m)\} \\
&\leq \frac{P_m\{|\tilde{\rho}_M - \rho(h)| > \epsilon\}}{P_m\{H_\beta \cap (\tau_N > m)\}} = \frac{P_m\{|\tilde{\rho}_M - \rho(h)| > \epsilon\}}{1 - P_m\{\bar{H}_\beta \cup (\tau_N \leq m)\}} \\
&\leq \frac{P_m\{|\tilde{\rho}_M - \rho(h)| > \epsilon\}}{1 - P_m(\bar{H}_\beta) - P_m(\tau_N \leq m)} \leq \frac{P_m\{|\tilde{\rho}_M - \rho(h)| > \epsilon\}}{1 - e^{-\beta N} - m\alpha_N} \\
&\leq L_1 \exp(-L_2 \epsilon^2 M).
\end{aligned}$$

Theorem 13.3.1 is proved.

Proof of Theorem 13.5.2
Let $0 < \delta < \epsilon$. We have

$$P_\theta\{|\tilde{\rho}_C - \rho(\theta)| > \epsilon\} \geq P_\theta\{|\hat{\rho}_C - \rho(\theta)| > \epsilon + \delta\} - P_\theta\{|\tilde{\rho}_C - \hat{\rho}_C| > \delta\}.$$

The first term in the right hand can be estimated from below as in Theorem 13.3.1:

$$P_\theta\{|\hat{\rho}_C - \rho(\theta)| > \epsilon + \delta\} \geq L_1 \exp(-L_2(\epsilon + \delta)^2 C),$$

where the positive constants do not depend on C.

For the 2nd term, we obtain

$$P_\theta\{|\tilde{\rho}_C - \hat{\rho}_C| > \delta\} \le \sum_r P_\theta\{(|r - \tilde{\rho}_C| > \delta) \cap (\hat{\rho}_C = r)\}$$
$$= \sum_r P_\theta\{|r - \tilde{\rho}_C| > \delta | \hat{\rho}_C = r\} \cdot P_\theta\{\hat{\rho}_C = r\}.$$

The conditional probability $P_\theta\{|r - \tilde{\rho}_C| > \delta | \hat{\rho}_C = r\}$ is exactly the error of the retrospective change-point estimation when the change-point parameter is equal to $0 < r < 1$ and the d.f. of independent observations changes from $f_0(x)$ to $f_\theta(x)$ at $n_0 = [rC]$. From Brodsky and Darkhovsky (2000), we know that the asymptotically optimal methods for this problem have the following order of the change-point estimation:

$$P_\theta\{|r - \tilde{\rho}_C| > \delta | \hat{\rho}_C = r\} \le L_1 \exp(-L_2 \delta C),$$

where the positive constants L_1 and L_2 do not depend on C and r.

For the class R_θ of these methods, we obtain from here:

$$P_\theta\{|\tilde{\rho}_C - \hat{\rho}_C| > \delta\} \le L_1 \exp(-L_2 \delta C).$$

Let $\delta = \epsilon/2$. Then,

$$P_\theta\{|\tilde{\rho}_C - \rho(\theta)| > \epsilon\} \ge L_1 \exp(-L_2 \epsilon^2 C).$$

Now, let us consider the conditional probability

$$P_\theta\{|\tilde{\rho}_C - \rho(\theta)| > \epsilon | H_\beta \cap (\tau_C > m)\} \ge P_\theta\{(|\tilde{\rho}_C - \rho(\theta)| > \epsilon) \cap H_\beta \cap (\tau_C > m)$$
$$\ge P_\theta\{|\tilde{\rho}_C - \rho(\theta)| > \epsilon\} - P_\theta(\bar{H}_\beta) - m\alpha_C.$$

Since $P_\theta(\bar{H}_\theta) \le \exp(-\beta C)$, $\alpha_C \le \exp(-LC)$, for any fixed m, we obtain for small enough ϵ:

$$P_\theta\{|\tilde{\rho}_C - \rho(\theta)| > \epsilon | H_\beta \cap (\tau_C > m)\} \ge L_1 \exp(-L_2 \epsilon^2 C),$$

where the positive constants L_1 and L_2 do not depend on C. From Theorem 13.2.1 and the last inequality, we conclude that the proposed method is asymptotically optimal by the order of the retrospective change-point estimation.

General Conclusions

In the introduction to this book, the following main goals were formulated:

1. To consider retrospective and sequential problems of detection of nonstationarities.

2. To prove optimality and asymptotic optimality of proposed methods.

3. To consider parametric and nonparametric methods of detection of nonstationarities.

4. To consider different types of changes: abrupt changes in statistical characteristics ("change-points" or instants of "structural changes"); gradual changes in statistical characteristics (deterministic and stochastic trends or "unit roots"); purely random and disappearing changes in statistical characteristics ("outliers", switches in coefficients of stochastic models).

5. To consider changes in parameters of univariate and multivariate nonstationary stochastic models.

6. To consider hypothesis testing and change-point detection methods.

In this book, all these goals were achieved.

Different statements of the retrospective change-point problem with the special sections devoted to the asymptotically optimal choice of parameters of decision statistics and a priori lower bounds for performance efficiency in multiple change-points problems are studied in Part I.

Retrospective detection and estimation problems for stochastic trends are considered. A new method for retrospective detection of a stochastic trend is proposed, and the problem of discrimination of a stochastic trend and a structural change hypotheses is considered.

Retrospective detection and estimation of switches in models with changing structures is studied in Part I. The case of outliers in univariate models is analyzed and the asymptotically optimal method for detection of these outliers is proposed. I analyze here methods for splitting mixtures of probabilistic distributions (nonparametric problem statement).

Problems of retrospective detection and estimation of change-points in multivariate models are considered. I propose methods for detection of changes in parameters of multivariate stochastic models (including multifactor regressions, systems of simultaneous equations, etc.) and prove that probabilities of type-1 and type-2 errors for these methods converge to zero as the sample size tends to infinity. Then, the asymptotic lower bounds for the probability of the error of estimation for the case of dependent multivariate observations in situations of one change-point and multiple change-points are proved.

Applied problems of the retrospective change-point detection for different financial models, including GARCH and SV models, as well as Copula models, are considered. I propose methods for the retrospective detection of structural changes and give experimental results demonstrating their efficiency.

In Part II, sequential problems of change-point detection are considered. Sequential hypotheses testing problems are studied in a separate chapter.

Here, we present new prior inequalities in sequential problems of hypotheses testing and asymptotically optimal one-sided and multisided tests. Results of computer simulation of sequential tests are given.

Sequential problems of change-point detection for univariate models are considered. First, I analyze performance characteristics of well-known parametric and nonparametric sequential tests. Prior lower bounds for performance efficiency of sequential tests are proved, including the a priori lower estimate for the delay time, the Rao-Cramer type inequality for sequential change-point detection methods, and the lower bound for the probability of error in the normalized delay time estimation. Then, I give results of the asymptotic comparative analysis and asymptotic optimality of sequential change-point detection methods.

Sequential change-point detection problems in multivariate linear systems are considered. I consider a multivariate model in which coefficients of equations can change unexpectedly. A new method for change-point detection in such a system is proposed: Type-1 and type-2 error probabilities and the normalized delay time in change-point detection are studied. The prior informational lower bound for performance efficiency in change-point detection problems for multivariate systems is proved, and the asymptotic optimality of the proposed method is demonstrated. Results of computer experiments with the proposed method for different types of multivariate systems are given.

Early change-point detection are considered is a separate chapter. Methods of early change-point detection are proposed, and their statistical characteristics, including type-1 and type-2 error probabilities and the normalized delay time in detection, are studied. I prove that the proposed methods are asymptotically optimal, i.e., the lower informational bounds for performance efficiency are attained for them. Results of experimental testing of proposed methods are given.

Sequential detection of switches in models with changing structures is studied in Part-II. These problems include decomposition of mixtures of probabilistic distributions (sequential context), classification of multivariate observations and classifications of observations in regression models with randomly changing coefficients of regression equations. I propose nonparametric methods for detection of switches and analyze their statistical characteristics (type-1 and type-2, error probabilities, the normalized delay time in detection). Then, I prove the a priori informational lower bound for sequential problems of detection of switches and demonstrate that the proposed method is asymptotically optimal by the order of detection.

Problems of sequential detection and estimation of change-points are considered. The sense of these problems is as follows: After sequential detection of a change-point, I need to estimate the true point of a change. This problem is actual, because all sequential methods have a certain delay time in change-point detection. I propose a two-stage method for solving this problem (first stage — sequential detection; second stage — retrospective estimation of this change-point using a moving window of observations) and analyze its statistical characteristics. The asymptotic optimality of this method is proved. Results of computer simulation of this method are given.

So, this book has the following new features:

— Change-points of different types in multivariate models are considered.

— Informational inequalities for the main characteristics of performance efficiency of retrospective and sequential methods are proved.

— New problems of retrospective and sequential detection of random switches in univariate and multivariate nonstationary stochastic models are studied.

I hope that the future development of these features will be useful at the modern stage of research in change-point detection for nonstationary stochastic models.

tistical characteristics. The asymptotic optimality of this method is proved. Results of computer simulations of this method are given.

So, this book has the following new features:

- Change-points of different types in multivariate models are considered.
- Informational inequalities for the main characteristics of performance efficiency of retrospective and sequential methods are proved.
- New problems of retrospective and sequential detection of random switches in univariate and multivariate nonstationary stochastic models are studied.

I hope that the future development of these ideas will be useful in the modern state of research in retrospective detection for nonstationary stochastic models.

Bibliography

[1] Anderson, T., and D. Darling. 1952. Asymptotic theory of certain goodness-of-fit criteria based on stochastic processes. *Annals of Mathematical statistics,* 23:193–212.

[2] Andrews, D.W.K. 1993. Tests for parameter instability and structural change with unknown change point. *Econometrica,* 61:821–856.

[3] Andrews, D.W.K., and W. Ploberger. 1994. Optimal tests when a nuisanse parameter is present only under the alternative. *Econometrica,* 62:1383–1414.

[4] Ang, A., and G. Bekaert. 2002. International asset allocation with regime shifts, *Review of Financial Studies,* 15:1137–1187.

[5] Antoch, J., Huskova M., and Z. Praskova. 1997. Effect of dependence on statistics for determination of change. *J. of Statistical Planning and Inference,* 60:291–310.

[6] Asai, M., M. McAleer, and J. Yu. 2006. Multivariate stochastic volatility. A review. *Econometric Reviews,* 25:145–175.

[7] Aubin, J.-P., and I. Ekeland. 1984. *Applied Nonlinear Analysis.* NY:Springer.

[8] Aue, A., L. Horvath, P. Kokoszka, and J. Steinebach. 2008. Monitoring shifts in mean: Asymptotic normality of stopping times. *Test* 17:513–530.

[9] Bagshaw, M., and R.A. Johnson. 1975. The influence of reference values and estimated variance on the ARL of CUSUM test. *J. Roy. Statist. Soc. B.,* 37:413–420.

[10] Bagshaw, M., and R.A. Johnson. 1977. Sequential procedures for detecting parameter changes in a time series model. *J. Amer. Statist. Assoc.,* 72:593–597.

[11] Bai, J., R. Lumsdaine, and J. Stock. 1998. Testing for and dating common breaks in multivariate time series. *Review of Economic Studies,* 65:395–432.

[12] Bai, J., and P. Perron. 1998. Estimating and testing linear models with multiple structural changes. *Econometrica,* 66:47–78.

[13] Banerjee, A., R. Lumsdaine, and J. Stock. 1992. Recursive ans sequential tests of the unit-root and trend-break hypotheses: Theory and international evidence. *J. of Business and Economic Statistics,* 10:271–287.

[14] Banzal, R.K., and P. Papantoni-Kazakos. 1983. An algorithm for detecting a change in a stochastic process. *IEEE Trans. on Inf. Theory,* 20:709–723.

[15] Bartlett, M.S. 1946. The large-sample theory of sequential tests. *Proc.Camb. Phil. Soc.,* 42:239–244.

[16] Basseville, M., and A. Benveniste. 1983. Design and comparative study of some sequential jump detection algorithms for digital signals. *IEEE Trans. on A.S.S.P.,* ASSP-31: 3.

[17] Basseville, M., and A. Benveniste. 1983. Sequential detection of abrupt changes in spectral characteristics of digital signals. *IEEE Trans. on Inf. Theory,* 20:709–723.

[18] Basseville M., and A. Benveniste. 1986. *Detection of abrupt changes in signals and dynamic systems.* NY:Springer.

[19] Basseville, M. and I. Nikiforov. 1993. *Detection of abrupt changes: Theory and applications.* NY:Prentice-Hall.

[20] Basu, A.P., J.K. Ghosh, and S.N. Joshi. 1986. On estimating change-point in a failure rate. *Statistical Decision Theory and Related Topics,* 2:239–252.

[21] Baum, L., T. Petri, G. Soules, and N. Weiss, 1980. A maximization technique occurring in the statistical analysis of probabilistic functions of markov chains. *Annals of Mathematical Statistics,* 41:164–171.

[22] Berkes, I., L. Horvath, and P. Kokoszka. 2003. GARCH processes: Structure and estimation. *Bernoulli,* 9:201–207.

[23] Billingsley, P. 1969. *Convergence of probability measures.* NY:Wiley.

[24] Bhattacharya, P.K., and F. Frierson Jr. 1981. A nonparametric control chart for detecting small disorders. *Ann. Statist.,* 9:544–554.

[25] Bhattacharya, P.K., and P.J. Brockwell. 1976. The minimum of an additive process with applications to signal estimation and storage theory. *Z. Wahrsch. verw. Gebiete,* 37:51–75.

[26] Bhattacharya, G.K., and B. Johnson. 1968. Nonparametric tests for shifts at an unknown time point. *Ann. Math. Statist.,* 23:183–208.

[27] Bhattacharya, P.K. 1987. Maximum likelihood estimation of a change-point. *J. Multivar. Analysis,* 23:183–208.

[28] Bhattacharya, P.K. 1994. Some aspects of change-point analysis. In *Change-Point Problems, IMS Lecture Notes – Monograph Series,* ed. Carlstein, E., H.G. Muller, and D. Siegmund (eds.), 23:28–56, Hyward, California.

[29] Bhattacharya, P.K., and P.A. Zhao. 1994. A rank cusum procedure for detecting small changes in a symmetric distribution. In *Change-Point Problems, IMS Lecture Notes – Monograph Series,* Carlstein E., H.G. Muller, and D. Siegmund (eds.). 23:28–56, Hyward, California.

[30] Bollerslev, T. 1986. Generalized autoregressive conditional heteroscedasticity. *J. of Econometrics,* 51:307–327.

[31] Bollerslev, T., R.F. Engle, and D.B. Nelson. 1994. ARCH models. In *Handbook of Econometrics, 2,* Z. Grillihes, and M.D. Intrilligator (eds.), 2959-3038. North Holland: Elsevier.

[32] Borovkov, A.A. 1989. *Mathematical statistics.* Moscow: Nauka *(in Russian).*

[33] Borovkov A.A. 1998. Asymptotically optimal decisions in change-point problems. *Probability theory and its applications,* 43:625–654.

[34] Box, G.E.P., and G.C. Tiao. 1965. A change in level of a nonstationary time series. *Biometrika,* 52:181–192.

[35] Bradley, R. 2005. Basic properties of strong mixing conditions: A survey and some open questions. *Probability Surveys,* 2:107–144.

[36] Brant, R. 1990. Comparing classical and resistant outlier rules. *JASA,* 85:1083–1090.

[37] Breymann, W., and A. Dias, P. Embrechts. 2003. Dependence structures for multivariate high-frequency data in finance. *Quantitetive Finance,* 3:1–14.

[38] Brodsky, B., and B. Darkhovsky. 1993. *Non-Parametric Methods in Change-Point Problems,* North-Holland: Kluwer.

[39] Brodsky, B., and B. Darkhovsky. 2000. *Non-Parametric Statistical Diagnosis: Problems and Methods.* North-Holland: Kluwer.

[40] Brodsky, B.E., and B.S. Darkhovsky. 2005. Asymptotically optimal methods of change-point detection for composite hypotheses. *Journal of Statistical Planning and Inference,* 133:123–138.

[41] Brodsky, B.E., and B.S. Darkhovsky. 2008. Sequential change-point detection for mixing random sequences under composite hypotheses. *Statistical Inference for Stochastic Processes,* 11:35–54.

[42] Brodsky, B.E., and B.S. Darkhovsky. 1990. Comparative study of non-parametric sequential change-point detection methods. *Theory of Probability and Its Applications,* 35:655–668.

[43] Brodsky, B.E., and B.S. Darkhovsky. 1990. Asymptotic analysis of some estimates in the aposteriori change-point problem. *Teor. Veroyatn. Primen,* 35:551–557 *(Russian).*

[44] Brodsky, B.E., and B.S. Darkhovsky. 1989. A nonparametric method for detection of switching times for two random sequences. *Avtom. Telemekh.,* 10:66–75 *(Russian).*

[45] Brodsky, B.E. and B.S. Darkhovsky. 1993. A posteriori detection of multiple change-points of a random sequence. *Avtom. Telemekh.,* 10:62–67.

[46] Brodsky, B.E., and B.S. Darkhovsky. 1983. On the quickiest detection of the change-point of a random sequence. *Avtom. Telemekh.,* 10:101–108. *(Russian).*

[47] Brodsky, B.E. 1996. Method of decomposition of mixtures of probabilistic distributions. *Avtom. Telemekh.,* 2:76–88.

[48] Brodsky, B.E. 1995. Asymptotically optimal methods in sequential change-point detection, Part 1, 2. *Avtom. Telemekh.,* 9, 10: 60–72, 50–59.

[49] Brodsky, B.E., and B.S. Darkhovsky. 2008. Minimax methods for multi-hypothesis sequential testing and change-point detection problems. *Sequential Analysis,* 27:141–173.

[50] Brodsky, B.E., and B.S. Darkhovsky. 2008. Minimax sequential tests for multiple composite hypotheses, I. *Probability Theory and Its Applications,* 52:565–579.

[51] Brodsky B.E., B.S. Darkhovsky. Minimax sequential tests for multiple composite hypotheses, II. *Probability Theory and Its Applications,* 53:1–12

[52] Brodsky, B.E. 2009. Sequential change-point detection in linear models. *Sequential Analysis,* 28:1–28.

[53] Brodsky, B.E. 2010. Sequential detection and estimation of change-points. *Sequential Analysis,* 29, 2.

[54] Brodsky, B., and H. Penikas, and I. Safaryan. 2012. Copula structural shift identification. *HSE Preprints: Financial Economics,* 2:1–20.

[55] Brodsky, B.E. 2012. Sequential change-point detection in state-space models. *Sequential Analysis,* 31:145–171.

[56] Brodsky, B.E., and B.S. Darkhovsky. 2012. Structural changes and unit roots in non-stationary time series. *Journal of Statistical Planning and Inference,* 142:327–335.

[57] Brodsky, B.E., and B.S. Darkhovsky. 2013. Asymptotically optimal methods of early change-point detection. *Sequential Analysis,* 32:158–181.

[58] Brodsky, B.E., and B.S. Darkhovsky. 2015. Sequential detection of switches in models with changing structures. *Stochastics, DOI 10.1080 17442508.2015 1089349.*

[59] Brown, R.L., J. Durbin, and J.M. Evans. 1975. Techniques for testing the constancy of regression relationships over time. *J.Royal Statist. Soc. Ser. B,* 37:149–192.

[60] Bulinskii, A.V. 1989. *Limit theorems under weak dependence conditions. Moscow: Nauka (in Russian).*

[61] Campos, J., and N.R. Ericsson, and D.F. Hendry. 1996. Cointegration tests in the presence of structural breaks. *J. of Econometrics,* 70:187–220.

[62] Carey, V., C. Wager, E. Walters, and B. Rosner. 1997. Resistant and test-based outlier rejection: Effects on Gaussian one- and two-sample inference. *Technometrics,* 39:320–330.

[63] Carlstein, E. 1988. Nonparametric change-point estimation. *Ann. Statist.,* 16:188–197.

[64] Carnero, M., D. Pena, and E. Ruiz. 2004. Persistence and kurtosis in GARCH and stochastic volatility models. *J. of Financial Econometrics,* 2:319–342.

[65] Chan, L., and K. Arjun. 2012. Parametric Statistical Change-Point Analysis. Birkhauser: Springer.

[66] Chatterjee, S., and A.S. Hadi. 1988. *Sensitivity Analysis in Linear Regression.* NY:Wiley.

[67] Chernoff, H., and S. Zacks. 1864. Estimating the current mean of a normal distribution which is subject to changes in time. *Ann. Math. Statist.,* 35:999–1028.

[68] Cherubini, U., E. Luciano, and W. Vecciato. 2004. *Copula methods in finance.* NY:Wiley.

[69] Cho, H., and H. Fryzlewitz. 2012. Multiscale and multilevel technique for consistent seqmentation of nonstationary time series. *Statistica Sinica,* 22:207–229.

[70] Christiano, L.J. 1992. Searching for breaks in GNP. *J. of Business and Economic Statistics,* 10:237–250.

[71] Chu, C., M. Stinchcombe, and H. White. 1996. Monitoring structural change. *Econometrica,* 64:1045–1065.

[72] Cerra, V., and S. Saxena. 2005. Did Output Recover from the Asian Crisis? IMF Staff Papers, 52:1–23.

[73] Clark, P.K. 1973. A subordinated stochastic process model with finite variance for speculative prices. *Econometrica,* 41:135–155.

[74] Cosslett, S., and L. Lee. 1985. Serial correlation in discrete variable models. *Journal of Econometrics,* 27:79–97.

[75] Cox, D.R. 1963. Large sample sequential tests for composite hypotheses. *Sankhya,* 25:5–12.

[76] Cox, C.P., and T.D. Roseberry. 1966. A large sample sequential test, using concomitant information, for discrimination between two composite hypotheses. *J. Amer. Statist. Assoc.,* 61:357–367.

[77] Csörgő, M., and L. Horváth. 1987. Nonparametric tests for the change-point problem. *J. of Statist. Plan. Inf.,* 17:1–9.

[78] Csörgő, M., and L. Horváth. 1988. Nonparametric methods for the change-point problems. In *Handbook of Statistics V.7,* Krishnaiah, P.R., and C.R. Rao (eds.), 403-425 Elsevier Science Publ. North–Holland.

[79] Csörgő, M., and L. Horváth. 1987. Detecting change in a random sequence. *J. of Multivar. Analysis,* 23:119–130.

[80] Chu, C., K. Hornik, and C. Kuan. 1985. MOSUM tests for parameter constancy. *Biometrika,* 82:603–617.

[81] Cobb, W.G. 1978. The problem of the Nile: Conditional solution to a change-point problem. *Biometrika,* 65:243–251.

[82] Crowder, W.G. 1987. A simple method for studying run-length distributions of exponentially weighted moving average charts. *Technometrics,* 29:401–407.

[83] Cui, S., and Y. Sun. 2004. Checking for the gamma family distribution under the marginal proportional frailty model. *Statistica Sinica,* 14:249–267.

[84] Dai, Q., K. Singleton, and Y. Wei. 2003. Regims Shifts in a Dynamic Term Structure Model of U.S. Treasury Bonds. Working Paper, Stanford University.

[85] Darkhovsky, B.S. 1976. Nonparametric method for detection of the change-point of a random sequence. *Teor. Veroyatn. Primen.*, 21:180–184.

[86] Darkhovsky, B.S. 1985. Nonparametric method for estimation of homogeneity intervals of a random sequence. *Teor. Veroyatn. Primen.*, 30:796–799.

[87] Darkhovskii, B.S., and B.E. Brodskii. 1979. Identification of the change-time of the random sequence. In *Proceedings of the 5th IFAC-IFORS symposium on identification and system parameter estimation.* Düsseldorf-Darmstadt.

[88] Darkhovsky, B.S., and B.E. Brodsky. 1980. Aposteriori change-point detection of a random sequence. *Teor. Veroyatn. Primen.*, 25:635–639. *(Russian).*

[89] Darkhovsky, B.S., and B.E. Brodsky. 1987. Nonparametric method of the quickiest detection of a change in the mean value of a random sequence. *Teor. Veroyatn. Primen.*, 32:703–711. *(Russian).*

[90] Darkhovsky, B.S., 1984. On two problems of estimation the moments of changes in probabilistic characteristics of a random sequence. *Teor. Veroyatn. Primen.*, 29:464–473. *(Russian).*

[91] Darkhovsky, B.S. 1995. Retrospective change-point detection in some regression models *Teor. Veroyatn. Primen.*, 4:898–903.

[92] Davig, T. 2004. Regime-switching debt and taxation. *Journal of Monetary Economics,* 51:837–859.

[93] Davis M.H.A. 1975. The application of nonlinear filtering to fault detection in linear systems. *IEEE Trans. Autom. Control,* AC-20:257–259.

[94] Davis, R., T. Lee, and G. Rodriguez-Yam. 2006. Structural break estimation for nonstationary time series. *JASA,* 101:223–239.

[95] Deshayes, J., and D. Picard. 1986. Off-line statistical analysis of change-point models using nonparametric and likelihood methods. In *Lect. Notes in Control and Inf. Sciences,* Allgower, F., and M. Morary (eds.), 77:103–168. Springer.

[96] van Dobben de Bruyn, D.S. 1968. *Cumulative sum tests: Theory and practice.* L.: Griffin.

[97] Dickey, D., and W. Fuller. 1979. Distribution of the estimators for autoregressive time series with a unit root. *J. of American Statistical Asiciation,* 74:427–431.

[98] Dickey. D., and W. Fuller. 1981. Likelihood ratio statistics for time series with a unit root. *Econometrica,* 49:1057–1072.

[99] Diebold, F., and A. Inoue. 2001. Long memory and regime switching. *J. of Econometrics,* 105:131–159.

[100] Ding, K. 2003. A lower confidence bound for the change-point after a sequential CUSUM test. *Journal of Statistical Planning and Inference,* 115:311–326.

[101] Dragalin, V. 1988. Asymptotic solutions to the change-point problem with an unknown parameter. In *Stat. Probl. Upr.* 83:47–51. Institute of Mathematics and Cybernetics. Vilnius: Lithuanian Academy of Sciences *(Russian).*

[102] Dragalin, V.P., and A.A. Novikov. 1988. Asymptotic solution of the Kiefer-Weiss problem for processes with independent increments. *Theory of Probab. and Its Appl.,* 32:617–627.

[103] Dragalin, V.P., and A.A. Novikov. 1999. Adaptive sequential tests for composite hypotheses. *Review of Applied and Industrial Mathematics* 6:387–398.

[104] Dragalin, V.P., A.G. Tartakovsky, and V. Veeravalli. 1999. Multihypothesis sequential probability ratio tests, I: Asymptotic optimality. *IEEE Trans. Inform. Theory,* 45:2448–2461.

[105] Dragalin, V.P., A.G. Tartakovsky, and V. Veeravalli. 2000. Multihypothesis sequential probability ratio tests, II: Accurate asymptotic expansions for the expected sample size. *IEEE Trans. Inform. Theory,* 46:1366–1383.

[106] Dobric, J., and F. Schmid. 2005. Testing goodness-of-fit parametric families of copulas: Application to financial data. *Communications in Statistics: Simulation and Computation,* 34:1053–1068.

[107] Dudley, R.M. 1973. Sample functions of the Gaussian process. *Ann. Probab.,* 1:66–103.

[108] Dümbgen L. 1991. The asymptotic behavior of some nonparametric change-point estimators. *Annals of Statistics,* 19:1471–1495.

[109] Duncan A.J. 1974. *Quality control and industrial statistics.* NY:Irwin.

[110] Dvoretzky, A., J. Kiefer, and J. Wolfowitz. 1956. Asymptotic minimax character of the sample distribution function and of the classical multinomial estimator. *Annals of Mathematical Statistics,* 27:642–669.

[111] Elliott, G., T. Rothenberg, and J. Stock. 1996. Efficient tests for an autoregressive unit root. *Econometrica,* 64:813–836.

[112] Engle, R.F. 1982. Autoregressive conditional heteroscedasticity and estimates of the variance of UK inflation. *Econometrica,* 50:987–1008.

[113] Ferger, D. 1995. Nonparametric tests for nonstandard change-point problems. *Annals of Statistics,* 23:1848–1861.

[114] Ferreira, P.E. 1975. A Bayesian analysis of a switching regression model: Known number of regimes. *J. Amer. Statist. Assoc.,* 70:370–374.

[115] Francq, C., and J.-M. Zakoyan. 2001. Stationarity of multivariate Markov-switching ARMA models. *Journal of Econometrics,* 102:339–364.

[116] Frees, E., and E. Valdez. 1998. Understanding relationships using copulas. *North Amerixcan Actuarian Journal,* 2:1–25.

[117] Fu, Y.-X., and R.N. Curnow. 1990. Maximum likelihood estimation of multiple change-points. *Biometrika,* 77:563–573.

[118] Fuh, C. D. 2003. SPRT and CUSUM in hidden Markov models. *Annals of Statistics,* 31:942–977.

[119] Fuh, C. D. 2004. Asymptotic operating characteristics of an optimal change-point detection in hidden Markov models. *Annals of Statistics,* 32:2305–2339.

[120] Fuh, C. D. 2006. Efficient likelihood estimation in state-space models. *Annals of Statistics,* 34:2026–2068.

[121] Gardner, L.A. 1969. On detecting changes in the mean of normal variables. *Ann. Math. Statist.,* 40:116–126.

[122] Genest, C., and A. Favre. 2007. Everything you always wanted to know abour copula modeling but were afraid to ask. *Journal of Hydrological Engineering,* 12:347–368.

[123] Genest, C., J. Quessy, and B. Remillard. 2006. Goodness-of-fit procedures for copula models based on the integral probability transformation. *Scandinavian J. of Statistics,* 33:337–366.

[124] Ghysels, E., A. Harvey, and E. Renault. 1996. Stochastic volatility. In: Maddala G., and Rao C. (eds.) *Handbook of Statistics* 14, Amsterdam: North Holland.

[125] Girshick, M.A., and H. Rubin. 1952. A Bayes approach to a quality control model. *Ann. Math. Statist.,* 23:114–125.

[126] Glidden, D. 1999. Checking the adequacy of the gamma family model for multivariate failure times. *Biometrica,* 86:381–393.

[127] Goldie, Ch. M., and P.E. Greenwood. 1986. Characterisations of set-indexed Brownian motion and associated conditions for finite-dimensional convergence. *Annals of Probability,* 14: 802–816.

[128] Goldie, Ch. M., and P.E. Greenwood. 1986. Variance of set-indexed sums of mixing random variables and weak convergence of set-indexed processes. *Annals of Probability,* 14:817–839.

[129] Goldfeld, S., and R. Quandt. 1973. A Markov model for switching regressions. *Journal of Econometrics,* 1:3–16.

[130] Gombay, E. 1999. Sequential testing of composite hypotheses. In: *Limit Theorems in Probability and Statistics,* Balatonlelle, II:107–125.

[131] Gombay, E., and L. Horvath. 1994. Limit theorems for changes in linear regression. *Journal of Multivariate Analysis,* 48:43–69.

[132] Gordon, L., and M. Pollak. 1994. An efficient sequential nonparametric scheme for detecting a change in distribution. *Annals of Statistics,* 22:763–804.

[133] Gordon, L., and M. Pollak. 1995. A robust surveillance scheme for stochastically ordered alternatives. *Annals of Statistics,* 23:1350–1375.

[134] Gordon, L., and M. Pollak. 1997. Average run length to false alarm for surveillance schemes designed with partially specified pre-change destribution. *Annals of Statistics,* 25:1284–1310.

[135] Gourieroux, C. 1997. *ARCH Models and Financial Applications,* NY:Springer.

[136] Haccou, P., E. Meelis, and S. van de Geer. 1987. The likelihood ratio test for the change-point problem for exponentially distributed r.v.'s. *Stoch. Proc. and Appl.,* 27:121–139.

[137] Hall, M., A. Oppenheim, and A. Willsky. 1983. Time varying parametric modelling of speech. *Signal Process,* 2:267–285.

[138] Hamilton, J. 1988. Rational Expectations: Econometric analysis of changes in regime: An investigation of the term structure of interest rates. *Journal of Economic Dynamics and Control,* 12:385–423.

[139] Hamilton, J. 2005. *Regime-Switching Models. Palgrave Dictionary of Economics.*

[140] Hansen, F.R., and H. Elliott. 1982. Image segmentation using simple Markov field models. *Computer Graphics and Image Processing,* 20:101–132.

[141] Harvey, A., E. Ruiz, and N. Shephard. 1984. Multivariate stochastic variance models. *Review of Economic Studies,* 61:247–264.

[142] Hawkins, D. 1977. Testing a sequence of observations for a shift in location. *J. Amer. Statist. Assoc.,* 72:180–186.

[143] Hines, W.G.S. 1976. Improving a simple monitor of a system with sudden parameter changes. *IEEE Trans. Inform. Theory,* 22:496–499.

[144] Hinkley, D.V. 1969. Inference about the intersection in two-phase regression. *Biometrika,* 56:495–504.

[145] Hinkley, D.V. 1970. Inference about the change-point in a sequence of random variables. *Biometrika,* 57:1–17.

[146] Hinkley, D.V. 1971. Inference about the change-point from cumulative sum tests. *Biometrika,* 58:509–523.

[147] Hinkley, D.V. 1971. Inference in two-phase regression. *J. Amer. Statist. Assoc.,* 66:736–743.

[148] Hoeffding, W. 1960. Lower bounds for the expected sample size and average risk of a sequential procedure. *Ann. Math. Statist.,* 31:352–368.

[149] Horn, R.A. 1986. *Matrix Analysis.* Cambridge: Cambridge Univ. Press.

[150] Horváth, L. 2012. *Inference for Functional Data.* NY:Springer.

[151] Horvath, L., M. Huskova, P. Kokoszka, and J. Steinebach. 2004. Monitoring changes in linear models. *Journal of Statistical Planning and Inference,* 126:225–251.

[152] Horvath, L., M. Kühn, and J. Steinebach. 2008. On the performance of the fluctuation test for structural change. *Sequential Analysis,* 27:126–140.

[153] Horvath, L., P. Kokoszka, and R. Zitikis. 2006. Sample and implied volatility in GARCH models. *J. of Financial Econometrics,* 136.

[154] Horvath, L., P. Kokoszka, and R. Zitikis. 2008. Distributional analysis of empirical volatility in GARCH processes. *J. of Statist. Inference and Planning,* 138:3578–3589.

[155] Horvath, L., M. Huskova, and P. Kokoszka. 2010. Testing the stability of the functional autoregressive process. *J. of Multivar. Analysis,* 101:352–367

[156] Horvath, L., M. Huskova, and M. Serbinowska. 1997. Estimators for the time of change in linear models. *Statistics,* 29:109–130.

[157] Horvath, L. 1989. The limit distributions of likelihood ratio and cumulative sum tests for a change in a binomial probability *J. Multivar. Analysis,* 31:149–159.

[158] Huber, P.J. 1981. *Robust Statistics* NY:Wiley.

[159] Hurst, H. 1951. Long-term storage capacity of reservoirs. *Transaction of American Society of Civil Engineers,* 116:770–808.

[160] Huskova, M. 1996. Estimation of a change in linear models. *Statist. Probab. Letters,* 26:13–24.

[161] Ibragimov, I.A., and Yu.V. Linnik. 1965. *Independent and Stationary Connected Variables.* Moscow: Nauka. *(Russian)*

[162] Ibragimov, I.A., and V. Yu. Linnik. 1971. *Independent and Stationary Sequences of Random Variables.* Groningen: Wolters-Noordhoff Publishing.

[163] Ibragimov, I.A., and R.Z. Khasminskii. 1979. *Asymptotic Estimation Theory* Moscow: Nauka (Russian).

[164] Ibragimov, I.A., and Yu. A. Rozanov. 1970. *Gaussian Random Processes* Moscow: Nauka *(Russian).*

[165] Inclan, C., and G.C. Tiao. 1994. Use of cumulative sums of squares for retrospective detection of changes of variances. *J. of Amer. Statist. Assoc.,* 89:913–923.

[166] James, B., K.J. James, and D. Siegmund. 1989. Asymptotic approximations for likelihood ratio tests and confidence regions for a change-point in the mean of a multivariate normal distribution. *Statist. Sinica,* 2:69–90.

[167] James, B., K.J. James, and D. Siegmund. 1988. Conditional boundary crossing probabilities with application to change-point problems. *Ann. Probab.,* 16:825–839.

[168] James, B., K.J. James, and D. Siegmund. 1987. Tests for a change-point. *Boimetrika,* 74:71–83.

[169] Jandhyala, V.K., and I.B. MacNeill. 1989. Residual partial sum limit process for regression models with applications to detecting parameter changes at unknown times. *Stoch. Proc. Appl.,* 33:309–323.

[170] Jeanne, O., and P. Masson. 2000. currency crises, sunspots, and Markov-switching regimes. *Journal of International Economics,* 50:327–350.

[171] Joanes, D.N. 1972. Sequential tests of composite hypotheses. *Biometrika,* 59:633–637.

[172] Joe, H. 1997. *Multivariate Models and Dependence Concepts.* L.:Chapmen and Hall.

[173] Johnson, R.A., and M.L. Bagshaw. 1974. The effect of serial correlation on the performance of CUSUM tests, I. *Technometrics,* 16:73–80.

[174] Johnson, R.A., and M.L. Bagshaw. 1975. The effect of serial correlation on the performance of CUSUM tests, II. *Technometrics,* 17:103–112.

[175] Johnson R.A., and M.L. Bagshaw. 1977. Sequential procedures for detecting parameter changes in a time-series model. *J. Amer. Statist. Assoc.,* 72:593–597.

[176] Juang, B., and L. Rabiner. 1985. Mixture autoregressive hidden Markov models for speech signals. *IEEE Transactions on Acoustics, Speech, and Signal Processing,* ASSP-30:1404–1413.

[177] Kander, A., and S. Zacks. 1966. Test procedure for possible changes in parameters of statistical distributions occuring at unknown time points *Ann. Math. Statist.,* 37:1196–1210.

[178] Kemp, K.W. 1967. Formal expressions which can be used for the determination of operating characteristics and average sample number of a simple sequential test. *J. Roy. Statist. Soc., B* 29:353–358.

[179] Kerestencioglu, F. 1993. *Change Detection and Input Design in Dynamic Systems* Taunton: Wiley.

[180] Khakhubia, Ts.G. 1986. The limit theorem for the maximum likelihood estimate of a change-point. *Teor. Veroyatn. Primen.,* 31:152–155 *(Russian).*

[181] Khan, R. A. 1995. Detecting changes in probabilities of a multicomponent process, *Sequential Analysis,* 14:375–388.

[182] Khmaladze, E.V., and A.M. Parjanadze. 1986. Functional limit theorems for linear statistics from sequential ranks *Probab. Th. Rel. Fields,* 73:585–595.

[183] Khmaladze, E.V. and A.M. Parjanadze. 1986. About asymptotic theory of statistics from sequential ranks. *Probability Theory and its Applications,* 31:758–772.

[184] Kiefer, J., and L. Weiss. 1957. Some properties of generalized sequential probability ratio tests. *Ann. Math. Statist.,* 28:57–74.

[185] Kim, H. J. and D. Siegmund. 1989. The likelihood ratio test for a change-point in simple linear regression. *Biometrika,* 76:409–423.

[186] Kim, H.J. 1994. Tests for a change-point in linear regression. In: IMS Lecture Notes — Monograph Series 23:170–176.

[187] Kim, H.J. 1993. Two-phase regression with nonhomogenous errors *Commun. Statist. – Theory and Methods,* 222:647–658.

[188] Kim, S., S. Cho, and S. Lee. 2000. On the cusum test for parameter changes in GARCH(1,1) models. *Commum Statist. Theory and Methods,* 29:445–462.

[189] Kim, D., and P. Perron. 2005. *Unit Root Tests with a Consistent Break Fraction Estimator.* Boston: Department of Economics Boston University.

[190] Kirch, J., and C. Aston. 2014. Change-points in high-dimensional settings. *arXiv 1409 177v1 [math ST] 5 sep.*

[191] Kligiene, N., and L. Telksnys. 1983. Methods of the change-point detection for random processes. *Avtom.Telemekh.,* 10:5–56. *(Russian)*

[192] Kokoszka, P., and R. Leipus. 2000 Change-point estimation in ARCH models. *Bernoulli,* 6(3):513–539.

[193] Kokoshka, P., and R. Leipus. 1999. Testing for parameter changes in ARCH models. *Lithuanian Mathematical Journal,* 39:231–247.

[194] Kolmogorov, A.N., Yu.V. Prokhorov, and A.N. Shiryaev. 1988. Probabilistic-statistical methods for detection of spontaneous effects. In *Math. Institute of the Academy of Sciences of the USSR, Collect. Articl.* 182:4–23. Moscow: Nauka*(Russian).*

[195] Koroliuk, V.S., N.I. Portenko, A.V. Skorokhod, and A.F. Turbin. 1985. *Reference Book on Probability Theory and Mathematical Statistics.* Moscow:Nauka. *(Russian)*

[196] Korostelev, A. 1997. Minimax large deviations risk in change-point problems. *Mathematical Methods of Statistics,* 6:365–374.

[197] Krämer, W., W. Ploberger, and R. Alt. 1988. Testing for structural change in dynamic models. *Econometrica,* 56:1355–1369.

[198] Kruglov, V.M., and B.Yu. Korolev. 1990. *Limit Theorems for Random sums.* Moscow:Nauka.*Russian*

[199] Kwiatkowski, D., P. Phillips, P. Schmidt, and Y. Shin. 1992. Testing the null hypothesis of stationarity against the alternative of a unit root: How sure we are that economic time series have a unit root. *J. of Econometrics,* 54:159–178.

[200] Lai, T.L. 1973. Gaussian processes, moving averages and quickiest detection problems. *Ann. Prob.,* 1:825–837.

[201] Lai, T.L. 1974. Control charts based on weighted sums. *Ann. Statist.,* 2:134–147.

[202] Lai, T.L. 1995. Sequential change-point detection in quality control and dynamical systems. *J. R. Statist. Soc. B,* 57:613–658.

[203] Lai, T.L. 1973. Optimal stopping and sequential tests which minimize the maximum expected sample size. *Ann. Statist.,* 1:659–673.

[204] Lai, T.L. 1988. Nearly optimal sequential tests of composite hypotheses. *Ann. Statist.,* 16:856–886.

[205] Lai, T.L. 2000. Sequential multiple hypothesis testing and efficient fault detection-isolation in stochastic systems. *IEEE Trans. on Information Theory,* 46:595–608.

[206] Lai, T.L. 1988. Information bounds and quick detection of parameter changes in stochastic systems. *IEEE Transactions on Information Theory,* 44:2917–2929.

[207] Lee, A.F.S., and S.M. Heghinian. 1977. A shift in the mean level in a sequence of independent normal random variables — a Bayesian approach. *Technometrics* 19:503–506.

[208] Lee, L., and J. Porter. 1984. Switching regression models with imperfect sample separation information. *Econometrica,* 52:391–418.

[209] Lee, S., J. Ha, and O. Na. 2003. The CUSUM test for parameter change in time series models. *Scandinavian J. of Statistics,* 30:781–796.

[210] Lee, S., and S. Park. 2001. The cusum of squares test for scale changes in infinite order moving average. *Scand. J. Statist.,* 28:625–644.

[211] Leisch, F., K. Hornik, and C.M. Kuan. 2000. Monitoring structural changes with the generalized fluctuation test. *Econometric Theory,* 16:835–854.

[212] Leipus, R. 1989. Functional limit theorems for rank statistics in the change-point problem. *Litov. Mat. Sborn.,* 29:733–744 *(Russian).*

[213] Leipus, R. 1990. Functional central limit theorem for nonparametric estimates of the spectre and the change-point problem for the spectral function. *Litov. Mat. Sborn.,* 30:674–697 *(Russian).*

[214] Lindgren, G. 1978. Markov regime models for mixed distributions and switching regressions. *Scandinavian Journal of Statistics,* 5:81–91.

[215] Lipster, R., and A. Shiryaev. 1974. *Statistics of Random Processes,* Moscow: Nauka. *(Russian)*

[216] Loader, C. 1996. Change-point estimation using nonparametric regression. *Annals of Statistics,* 24:4.

[217] Lombard, F. 1987. Rank tests for change-point problems. *Biometrika,* 74:615–624

[218] Lombard, F. 1988. Detecting change-points by Fourier analysis. *Technometrics,* 30:305–310.

[219] Lorden, G. 1971. Procedures for reacting to a change in distribution. *Ann. Math. Statist.,* 42:897–1008.

[220] Lorden, G., and I. Eisenberger. 1973. Detection of failure rate increases. *Technometrics,* 15:167–175.

[221] Lorden, G. 1976. 2-SPRT's and the modified Kiefer-Weiss problem of minimizing an expected sample size. *Ann. Statist.,* 4:281–291.

[222] Lorden, G. 1977. Nearly optimal sequentil tests for finitely many parameter values. *Ann. Statist.,* 5:1–21.

[223] Lorden, G., and M. Pollak. 2005. Non-anticipating estimation applied to sequential analysis and change-point detection. *Annals of Statistics,* 44:1422–1454.

[224] Lucas, J.M., and R.B. Crosier. 1992. Robust cusum: A robustness study for cusum quality control schemes. *Commun. Statist., Theory and Methods,* 11:2669–2687.

[225] Lütkepohl, H., T. Terasvirta, and J. Wolters. 1999. Investigating stability and linearity of a German money demand function. *Journal of Applied Econometrics,* 14:511–525

[226] Maddala, G., and I. Kim. 1998. *Unit Roots, Cointegration, and Structural Change.* Cambridge:Cambridge Univ. Press.

[227] Malevergne, Y., and D. Sornette. 2003. Testing the Gaussian copula hypothesis for financial assets dependencies. *Quantitative Finance,* 3:231–250.

[228] McNeill, A., R. Frey, and P. Embrechts. 2005. *Quantitative Risk Management.* Princeton: Princeton University Press.

[229] Mann, H., and A. Wald. 1943. On stochastic limit and order relationships. *Ann. Math. Statist.,* 14:217–226.

[230] Maronna, R., and V. Yohai. 1978. A bivariate test for the detection of a systematic change in mean. *Journal of American Statistical Association,* 73: 640–645.

[231] Mei, Y. 2006. Sequential change-point detection when unknown parameters are present in the pre-change distribution. *Annals of Statistics,* 34: 92–122.

[232] McDonald, D. 1988. *A CUSUM Procedure Based on Sequential Ranks.* Ottawa:Preprint Univ. of Ottawa.

[233] Miao, B.Q. 1988. Inference in a model with at most one slope change-point. *Journal of Multivariate Analysis,* 27:375–391.

[234] Mikosch T., and C. Starica. 2004. Non-stationarities in financial time series: The long-range dependence, and the IGARCH effects. *Review of Economics and Statistics,* 86:378–390.

[235] Montanes, A., and M. Reyes. 1999. The asymptotic behavior of the Dickey-Fuller test under the crash hypothesis. *Statistics and Probability Letters,* 42:81–89.

[236] Montanes, A., and M. Reyes. 2000. Structural breaks, unit roots, and methods for removing the autocorrelation pattern. *Statistics and Probability Letters,* 43:401–409.

[237] Moustakides, G.V. 1986. Optimal stopping times for detecting changes in distribution. *Ann. Stat.,* 14:1379–1387.

[238] Müller, H.G. 1992. Change-points in nonparametric regression analysis. *Annals of Statistics,* 20:737–761.

[239] Nadler, J., and N.B. Robbins. 1971. Some characteristics of Page's two-sided procedure for detecting a change in a location parameter. *Annals of Statistics,* 2:538–551.

[240] Nelson, C., and C. Plosser. 1982. Trends and random walks in macroeconomic time series: Some evidence and implications. *J. of Monetary Economics,* 10:130–162.

[241] Nelsen, R. 2006. *An Introduction to Copulas* NY:Springer.

[242] Newbold, P.M., and Y.Ch. Ho. 1968. Detection of changes in characteristics of a Gauss-Markov process. *IEEE Trans.Aerospace and Electronic Systems,* AES-4:707–718.

[243] Nikiforov, I.V. 1980. Modification and investigation of CUSUM test. *Avtom. Telemekh.,* 9:74–80 *(Russian).*

[244] Nikiforov I.V. 1983. *Sequential Detection of Changes in Characteristics of Time Series.* Moscow: Nauka *(Russian).*

[245] Nikiforov, I. 1995. A generalized change detection problem. *IEEE Transactions on Information Theory,* 41:171–187.

[246] Nikiforov, I. 1997. Two strategies in the problem of change detection and isolation. *IEEE Transactions on Information Theory,* 43:770–776.

[247] Novikov, A.A. 1980. The first exit time of the autoregressive process beyond a level and an application to the disorder problem. *Teor. Veroyatn. Primen.,* 35:282–292. *(Russian)*

[248] Novikov, A., and B. Ergashev. 1988. Analytical approach to the computation of characteristics for the exponential smoothing method of the change-point detection. In *Stat. Probl. Upr.,* 83:110–113. Vilnius: Inst. Math. Cybern., Lithuan. Acad. Scienc *(Russian).*

[249] Page, E.S. 1954. Continuous inspection schemes. *Biometrika,* 41:100–115.

[250] Page, E.S. 1955. A test for a change in a parameter occuring at an Unknown Point. *Biometrika,* 42:523–526.

[251] Page, E.S. 1967. On problem in which a change in a parameter occurs at an unknown point. *Biometrika,* 44:248–252.

[252] Pardzhanadze, A.M. 1987. Functional limit theorems in the problem of aposteriori disorder detection. *Theory Probab. Appl.,* 31:355–358.

[253] Paul, S.R., and K.Y. Fung. 1951. A generalized extreme studentized residual multiple-oulier-detection procedure in linear regression *Technometrics,* 33:339–348.

[254] Pavlov, I.V. 1988. A sequential procedure for testing many composite hypotheses. *Theory Prob. Appl.,* 33:138–142.

[255] Peligrad, M. 1982. Invariance principles for mixing sequences of random variables. *Ann. Probab.,* 10:968–981.

[256] Peligrad, M. 1982. On the central limit theorem for weakly dependent sequences with a decomposed strong mixing coefficient. *Stochastic Processes and their Applications,* 42:181–193.

[257] Pelkovitz, L. 1987. The general Markov chain disorder problem. *Stochastics,* 21:113–130.

[258] Pellegrini, S., and A. Rodriguez. 2007. Financial Econometrics and SV Models. Universidad Carlos III de Madrid: Madrid. halwebuc3m.es/esp/Curso Cordoba/TutorislGuide.pdf.

[259] Perron, P., and T. Vogelsang. 1992. Nonstationarity and level shifts with an application to purchasing power parity. *Journal of Business and Economic Statistics,* 10:301–320.

[260] Perron, P. 1989. The great crash, the oil price shock, and the unit root hypothesis. *Econometrica,* 57:1361–1401.

[261] Perron, P. 1994. Trend, unit root and structural change in macroeconomic time series. In *Cointegration for Applied Economist,* Rao, B.B. (eds.), 113-146. Basingstoke:MacMIllan Press.

[262] Perron, P. 2005. Dealing with Structural Breaks. In *Palgrave Handbook of Econometrics,* Patterson, K., and T. Mills (eds.). Economic theory. 1 NY:Palgrave.

[263] Perron, P. 1997. Further evidence from breaking trend functions in macroeconomic variables. *J. of Econometrics,* 80:355–385.

[264] Perron, P., and T. Yabu. 2009. Estimating deterministic trends with integrated or stationary noise component. *J. of Business and Economic Statistics,* 27.

[265] Petrov, V.V. 1972. *Sums of Independent Random Variables.* Moscow:Nauka *(Russian).*

[266] Petrov, V.V. 1975. *Sums of Independent Random Variables.* NY:Springer.

[267] Pettitt, A.N. 1979. A non-parametric approach to the change-point Problem. *Appl. Statist.,* 28:126–135.

[268] Pettitt, A.N. 1980. A simple cumulative sum type statistic for the change-point problem with zero-one observations. *Biometrika,* 67:79–84.

[269] Phillips, P. and P. Perron. 1986. Testing for a unit root in time series regression. *Biometrica,* 75:335–346.

[270] Picard, D. 1985. Testing and estimating change-points in time series. *Advances in Applied Probability,* 17:841–867.

[271] Ploberger, W., W. Kramer, and K. Kontrus. 1989. A new test for structural stability in the linear regression model. *Journal of Econometrics,* 40:307–318.

[272] Ploberger, W., and W. Krämer. 1992. The CUSUM test with OLS residuals. *Econometrica,* 60:271–285.

[273] Pollak, M., and D. Siegmund. 1975. Approximations to the expected sample size of certain sequential tests. *Ann. Statist.,* 3:1267–1282.

[274] Pollak, M. 1985. Optimal detection of a change in distribution. *Ann. Statist.,* 13:206–227.

[275] Pollak, M. 1987. Average run length of an optimal method of detecting a change in distribution. *Ann.Statist.,* 15:749–779.

[276] Pollak, M., and D. Siegmund. 1985. A diffusion process and its application to detecting a change in the drift of Brownian motion process. *Biometrika,* 72:267–280.

[277] Poritz, A. 1982. Linear Predictive Hidden Markov Models and Speech Signals. Acoustics, Speech and Signal Processing. *IEEE Conference on ICASSP* '82.7:1291–1294.

[278] Praagman, J. 1988. Bahadur efficiency of rank tests for the change-point problem. *Ann.Statist.,* 16:198–217.

[279] Prakasa Rao, B.L.S. 1987. *Asymptotic Theory of Statistical Inference.* NY:Wiley.

[280] Quandt, R.E. 1958. The estimation of parameters of a linear regression system obeying two separate regimes. *J. Amer. Statist. Assoc.,* 50:873–880.

[281] Quandt, R.E. 1972. A new approach to estimating switching regressions. *J. Amer. Statist. Assoc.,* 67:306–310.

[282] Quandt, R.E., 1960. Tests of the hypothesis that a linear regression system obeys two separate regimes. *J. Amer. Statist. Assoc.,* 55:324–330.

[283] Rabiner, L. 1989. A Tutorial on Hidden Markov Models and Selected Applications in Speech Recognition. *Proceedings of IEEE* 77:257–286.

[284] Rao, P.S. 1972. On two-phase regression estimator. *Sankhya,* 34:473–476.

[285] Rappoport, P., and L. Reichlin. 1989. Segmented trends and non-stationary time series. *Economic Journal,* 99:168–177.

[286] Ritov, Y. 1990. Decision theoretic optimality of the CUSUM procedure. *Annals of Statistics,* 18:1464–1469.

[287] Robbins, H., and D. Siegmund. 1970. Boundary crossing probabilities for the Wiener process and sample sums. *Ann. Math. Statist.,* 41:1410–1429.

[288] Roberts, S.W. 1966. A comparison of some control chart procedures. *Technometrics,* 8:411–430.

[289] Roberts, S.W. 1958. Control chart tests based on geometric moving average. *Technometrics,* 1:239–250.

[290] Robinson, P.B., and T.Y. Ho. 1978. Average run lengths of geometric moving average charts by numerical methods. *Technometrics,* 20:85–93.

[291] Ronzhin, A.F. 1987. Limit theorems for disorder times of the sequence of i.r.v.'s. *Teor. Veroyatn. Primen.*, 32:309–316 *(Russian).*

[292] Segen, J., and A. Sanderson. 1980. Detecting change in time series. *IEEE Trans. Inform. Theory,* IT-26:250–255.

[293] Sen, A., and M.S. Srivastava. 1975. On tests for detecting changes in mean. *Ann. Statist.*, 3:98–108.

[294] Sen, P.K. 1981. *Sequential Nonparametrics.* NY:Wiley.

[295] Shaban, S.A. 1980. Change-point problem and two-phase regression: An annotated bibliography. *Intern. Statist. Review,* 48:83–93.

[296] Shephard, N. 1996. Statistical aspects of ARCH and stochastic volatility. In: *Time Series Models in Econometrics, Finance, and Other Fields* Cox, D., D. Hinkley, and O. Barndorf-Nielsen. L. (eds.): Chapman and Hall.

[297] Shewhart, W.A. 1931. *Economic Control of Quality of Manufactured Product.* NY:D.van Nostrand.

[298] Shih, J. 1985. A goodness-of-fit test for association of bivariate survival model. *Biometrica,* 85:189–200.

[299] Shiryaev, A.N. 1961. Detection of spontaneous effects. *Dokl. Acad. Nauk SSSR,* 138:799–801. *(Russian)*

[300] Shiryaev, A.N. 1963. On optimal methods in the quickiest detection problems. *Teor. Veroyatn. Primen.*, 8:26–51 *(Russian).*

[301] Shiryaev, A.N. 1963. To disorder detection for a technological process, I. *Teor. Veroyatn. Primen.*, 8:264–281, 431-443 *(Russian).*

[302] Shiryaev, A.N. 1965. Some precise formulas in change-point problems. *Teor. Veroyatn. Primen.*, 10:380–385 *(Russian).*

[303] Shiryaev, A.N. 1976. *Statistical Sequential Analysis.* Moscow:Nauka *(Russian).*

[304] Shiryaev, A.N. 1996. Minimax optimality of the CUSUM method for continuous time. *Adv. Math. Sci.*, 310:173–174 *(Russian).*

[305] Shiryaev, A.N. 1980. *Probability.* Moscow:Nauka *(Russian).*

[306] Shiryaev, A.N., 1978. *Optimal Stopping Rules.*, NY:Springer-Verlag.

[307] Siegmund, D. 1988. Confidence sets in change-point problems. *Int. Stat. Review,* 56:31–48.

[308] Siegmund, D. 1986. Boundary crossing probabilities and statistical applications. *Ann. Stat.*, 14:361–404.

[309] Siegmund, D. 1985. *Sequential Analysis. Tests and Confidence Intervals.* NY:Wiley

[310] Siegmund, D. 1975. Error probability and average sample delay of the sequential probability ratio test. *J. Roy. Stat. Soc., B.,* 37:394–401.

[311] Sims, C., and T. Zha. 2004. Were There Switches in U.S. Monetary Policy? Working Paper, Princeton: Princeton University Press.

[312] Simons, G. 1967. Lower bounds for average sample number of sequential multihypothesis tests. *Ann. Math. Statist.,* 38:1343–1364.

[313] Smith, A.F.M. 1975. A Bayesian approach to inference about a change-point in a sequence of r.v.'s. *Biometrika,* 62:407–416.

[314] Srivastava, M., and Y. Wu. 1993. Comparison of EWMA, CUSUM, and Shiryaev-Roberts procedures for detecting a shift in the mean. *Annals of Statistics* 21:645–670.

[315] Srivastava, M.S., and Y. Wu. 1999. Quasi-stationary biases of change-point and change magnitude estimation after sequential CUSUM test. *Sequential Analysis,* 18:203–216.

[316] Tartakovsky, A.G. 1998. Asymptotic optimality of certain multihypothesis sequential tests: Non-i.i.d. case. *Statistical Inference for Stochastic Processes,* 1:265–295.

[317] Tartakovsky A.G., R. Li, and G. Yaralov. 2003. Sequential detection of targets in multichannel systems. *IEEE Trans. on Inform. Theory,* 49:425–445.

[318] Tartakovsky, A.G., I. Nikiforov, and M. Basseville. 2014 *Sequential Analysis: Hypothesis Testing and Change-Point Detection.* NY:CRC Press.

[319] Taylor, S. 1986. *Modelling Financial Time Series.* NY:John Wiley and Sons.

[320] Taylor, S. 1994. Modelling stochastic volatility: a review and comparative study. *Mathematical Finance,* 4:183–204.

[321] Telksnys, L.A. 1969. On application of optimal Bayes algorithm of teaching to determination of moments of changes in properties of random signals. *Avtom. Telemekh.,* 6:52–58 *(Russian).*

[322] Tietjen, G., and H. Moore. 1972. Some Grubb's type statistics for the detection of several outliers. *Technometrics,* 14:583–597.

[323] Timmerman, A. 2000. Moments of Markov switching models. *Jounal of Econometrics,* 96:75–111.

[324] Tjostheim, D. 1986. Some doubly stochastic time series models. *Journal of Time Series Analysis,* 7:51–72.

[325] Tsukahara, H. 2005. Semiparametric estimation in copula models. *Canadian J. of Statistics,* 3(33):357–375.

[326] Vaman, H.G. 1985. Optimal on-line detection of parameter changes in two linear models. *Stoch. Proc. Appl.,* 20:343–351.

[327] Vogelsang, T. 1998. Testing for a shift in mean without having to estimate serial-correlation parameters. *Journal of Business and Economic Statistics,* 16:73–80.

[328] Vostrikova, L.Yu. 1983. Functional limit theorems for the disorder problem. *Stochastics,* 9:103–124.

[329] Vostrikova, L.Yu. 1981. Disorder detection for the Wiener process. *Teor. Veroyatn. Primen.,* 26:362–368 *(Russian).*

[330] Wald, A. 1947. *Sequential analysis.* NY:Wiley.

[331] Wald, A., and J. Wolfowitz. 1948. Optimum character of the sequential probability ratio test. *Ann. Math. Statist.,* 19: 326–339.

[332] Wang Y. 1995. Jump and sharp cusp detection by wavelets. *Biometrika,* (1995) 82 385-397

[333] Wichern, D.W., R.B. Miller, and D.A. Hsu. 1998. Changes in variances in first-order autorehresive time series models — with an application. *Appl. Statist.,* 25:248–256.

[334] Wolf, D.A., and E. Schechtman. 1984. Nonparametric statistical procedures for the change-point problem. *J. Statist. Plan. Inference,* 9:389–396.

[335] Woodroofe M. 1982. *Nonlinear Renewal Theory in Sequential Analysis.* Philadelphia:SIAM.

[336] Worsley, K.J. 1986. Confidence regions and tests for a change-point in a sequence of exponential faily random variables. *Biometrika,* 73:91–104.

[337] Wu, J.S., and C.K. Chu. 1993. Kernel-type estimators of jump points and values of a regression function. *Annals of Statistics,* 21:1545–1566.

[338] Yakir, B. 1998. On the average run length to false alarm in surveillance problems which possess an invariance structure. *Annals of Statistics,* 26:1198–1214.

[339] Yakir, B. 1996. Dynamic sampling policy for detecting a change in distribution, with a probability bound on false alarm. 24:2199–2214.

[340] Yakir, B. 1996. A lower bound on the ARL to detection of a change with a probability constraint on false alarm. *Annals of Statistics,* 24:431–435.

[341] Yang, M. 2000. Some properties of vector autoregressive processes with Markov-switching coefficients. *Econometric Theory,* 16:23–42.

[342] Yao, Y.-C. 1988. Estimating the number of change-points via Schwarz' criterion. *Statist. Probab. Lett.,* 6:181–189.

[343] Yao, Y.-C. 1987. A note on testing against a change-point alternative. *Ann. Inst. Stat. Math.,* 39:377–383.

[344] Yin, Y.Q. 1988. Detection of the number, locations and magnitudes of jumps. *Commun. Stat., Stochastic Models,* 4:445–455.

[345] Yosida, K. 1980. *Functional Analysis.* Berlin:Springer.

[346] Yu, J., and R. Meyer. 2006. Multivariate stochastic volatility models; Bayesian estimation and model comparison. *Econometric Reviews,* 25:361–384.

[347] Zacks, S. 1983. Survey of classical and Bayesian approaches to the change-point problem. In *Recent Advances in Statistics. Papers in Honour of Herman Chernoff.* Rizvi, H., J. Rustagi, and D. Siegmund (eds.). 245-269. NY:Academic Press.

[348] Zeileis, A., F. Leisch, C. Kleiber, and F. Hornik. 2005. Monitoring structural change in dynamic econometric models. *Journal of Applied Econometrics,* 20:99–121.

[349] Zivot, E., and D. Andrews. 1992. Further evidence on the great crash, the oil price shock, and the unit root hypothesis. *J. of Business and Economic Statistics,* 10:251–270.

Index

Printed in the United States
by Baker & Taylor Publisher Services

Printed in the United States
by Baker & Taylor Publisher Services